AF601553

Petroleum Engineering

The Springer series in Petroleum Engineering promotes and expedites the dissemination of new research results and tutorial views in the field of exploration and production. The series contains monographs, lecture notes, and edited volumes. The subject focus is on upstream petroleum engineering, and coverage extends to all theoretical and applied aspects of the field. Material on traditional drilling and more modern methods such as fracking is of interest, as are topics including but not limited to:

- Exploration
- Formation evaluation (well logging)
- Drilling
- Economics
- Reservoir simulation
- Reservoir engineering
- Well engineering
- Artificial lift systems
- Facilities engineering

Contributions to the series can be made by submitting a proposal to the responsible Springer Editor, Charlotte Cross at charlotte.cross@springer.com or the Academic Series Editor, Dr. Gbenga Oluyemi g.f.oluyemi@rgu.ac.uk.

More information about this series at http://www.springer.com/series/15095

Tan Nguyen

Artificial Lift Methods

Design, Practices, and Applications

Tan Nguyen
Petroleum Department
New Mexico Tech
Socorro, NM, USA

ISSN 2366-2646 ISSN 2366-2654 (electronic)
Petroleum Engineering
ISBN 978-3-030-40719-3 ISBN 978-3-030-40720-9 (eBook)
https://doi.org/10.1007/978-3-030-40720-9

This Springer imprint is published by the registered company Springer Nature Switzerland AG
The registered company address is: Gewerbestrasse 11, 6330 Cham, Switzerland

Preface

This book titled *Artificial Lift Methods—Design, Practices, and Applications* consists of seven chapters: (1) review, (2) gas lift, (3) electrical submersible pump, (4) progressing cavity pump, (5) sucker rod pump, (6) plunger lift, and (7) artificial lift selection methodology for vertical and horizontal wells in conventional and unconventional reservoirs. Each chapter or each artificial lift method is unique and almost independent of other chapters except Chap. 1 and Chap. 7. In Chap. 1, the author reviews some basics of math, physics, fluid properties, flow inside reservoirs, flow inside a tubing, and multiphase flow in a tubing which are directly and/or indirectly related to other chapters in the book. Therefore, readers are strongly encouraged to review Chap. 1 before reading other chapters. In Chap. 7, the author begins with reviewing the advantages and disadvantages of each common lift method. Next, the artificial lift selection methodology for vertical wells in conventional reservoirs is presented. Finally, the author presents the most recent trends of artificial lift selection methodology for horizontal wells in unconventional plays. Readers do not need to have a deep understanding of all the artificial lift methods to read Chap. 7. In other words, it could be reasonable for readers to get started reading Chap. 7 to get an idea of what the artificial lift selection methodology looks like before reading other chapters.

Each artificial lift chapter (from Chap. 2 to Chap. 6) begins with the fundamentals of the lift method where the author reviews the heart of the lift system and how it works. The lift system is explained in the concept of the flow in the reservoir, the flow in the tubing, and how the external energy helps to lift the liquid. Next, the chapter focuses on the uniqueness of each lift method and then the detailed design. The author then closes each chapter with examples so readers know how to apply the presented concepts into practical applications.

All of the equations in this book are labeled with three digits "(X.X.X)". The first digit represents the number of the chapter. The second digit represents the number of the section. And the last digit is for the order of the equations in the chapter. For example, Eq. (3.2.5) can be interpreted as Chap. 3, section 2, and equation number 5.

The page number is labeled as the following format "X-XX". The first digit is for the chapter number following with a dash. The digits following the dash are the page number in each chapter. The page number in each chapter always begins with page number 1. For example, page 2.5 represents for Chap. 2 and page number 5; page 4.5 can be found in Chap. 4 page number 5.

Socorro, USA Tan Nguyen

Acknowledgements

I would like to first give thanks to my God whom I serve and worship with all of my heart. Next, I would like to thank New Mexico Tech and the Petroleum and Natural Gas Engineering Department for allowing me to take one semester sabbatical leave and for giving me the service time so I can initiate, continue, and complete writing this book. I would also like to thank Sebastian Pivnicka for reviewing the sucker rod pump and the plunger lift chapters. I also want to thank my students who contributed to this book while taking the artificial lift class with me in Spring 2019: Kien Nguyen, Benjamin Adu-Gyamfi, Carson Healy, and Tonya Ross. I would also like to thank my wife for her continued and tireless support of my family and to my life. My wife also helped me to debug the spellings, labels of equations and figures, and prepared the Table of Contents. Finally, I want to thank my two boys, Ryan and Stephen, for being such a good boys and for letting daddy focus on writing this book.

Contents

Chapter 1
Review

1.1 Review of Math and Physics

This section gives a quick review of definitions of common terms used throughout this book.

Scalar is a quantity described fully by a magnitude.

Vector is a quantity described by both a magnitude and a direction.

Dot product of two vectors $\vec{a}$ and $\vec{b}$ is a scalar and defined as $c = \|a\|\|b\|\cos\theta$ where θ is the angle formed by the two vectors $\vec{a}$ and $\vec{b}$.

Cross product between two vectors $\vec{a}$ and $\vec{b}$ is a vector and defined as: $\vec{c} = \{\|a\|\|b\|\sin\theta\}\vec{n}$.

where $\vec{n}$ is the normal vector of the plane formed by the two vectors $\vec{a}$ and $\vec{b}$.

Momentum, M, is a vector and defined as follows:

$$\vec{M} = m\vec{u} \tag{1.1.1}$$

Force, F, is defined as rate of change of momentum respect to time.

$$\vec{F} = \frac{d\vec{M}}{dt} = \frac{d(m\vec{u})}{dt} \tag{1.1.2}$$

For incompressible fluids (constant density), Eq. (1.1.2) becomes:

$$\vec{F} = \frac{d\vec{M}}{dt} = \frac{d(m\vec{u})}{dt} = m\frac{d\vec{u}}{dt} = m\vec{a} \tag{1.1.3}$$

Pressure, P, is a scalar and defined as the potential energy stored per unit volume

$$P = \frac{Energy(J)}{Volume(m^3)} = \frac{Work(Nm)}{Volume} = \frac{\vec{F}.\vec{S}}{\vec{A}.\vec{S}} \tag{1.1.4}$$

T. Nguyen, *Artificial Lift Methods*, Petroleum Engineering,
https://doi.org/10.1007/978-3-030-40720-9_1

Therefore

$$\vec{F} = P\vec{A} \tag{1.1.5}$$

Note that the area, A, is a vector and defined using the cross product as Shown in Fig. 1.1b.

The gradient of a scalar pressure, which is a function of three directions x, y, and z, is a vector and can be expressed mathematically as:

$$\nabla P = \frac{\partial P}{\partial x}\vec{i} + \frac{\partial P}{\partial y}\vec{j} + \frac{\partial P}{\partial z}\vec{k} \tag{1.1.6}$$

where $\vec{i}, \vec{j}, \vec{k}$ are the three unit vectors in the Cartesian coordinate system.

The divergence of a velocity vector, which is a function of three directions x, y, and z, is a vector operator that produces a scalar quantity of a vector field and given as:

$$\nabla.\vec{u} = \frac{\partial u_x}{\partial x} + \frac{\partial u_y}{\partial y} + \frac{\partial u_z}{\partial z} \tag{1.1.7}$$

Partial time and total time derivative

Let consider temperature of an object at a fixed location

Let Q be a property of a fluid such as temperature, velocity, density, etc. In general, Q is a function of time, t, and space in three directions: x, y, and z. Then Q can be written as Q = Q(t, x, y, z). The total differential change of Q is expressed as:

$$dQ = \frac{\partial Q}{\partial t}dt + \frac{\partial Q}{\partial x}dx + \frac{\partial Q}{\partial y}dy + \frac{\partial Q}{\partial z}dz \tag{1.1.8}$$

Divide both sides of Eq. (1.1.8) by dt yields the total time derivative of Q:

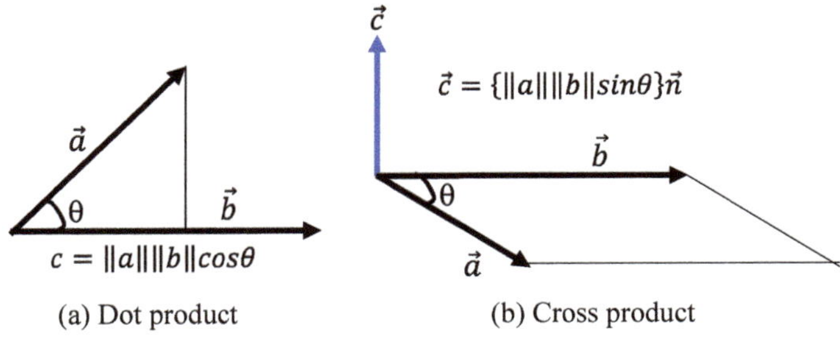

Fig. 1.1 Dot and cross product between two vectors

$$\frac{dQ}{dt} = \frac{\partial Q}{\partial t} + \frac{\partial Q}{\partial x}\frac{dx}{dt} + \frac{\partial Q}{\partial y}\frac{dy}{dt} + \frac{\partial Q}{\partial z}\frac{dz}{dt} \quad (1.1.9)$$

In Eq. (1.1.9), $\frac{\partial Q}{\partial t}$ and $\frac{dQ}{dt}$ are the partial and total time derivative, respectively. If Q is the temperature of an object at a fixed location, the temperature change of this object respect to time in the absence of any motion is defined as the partial time derivative (local time rate of temperature change). If this object moves into other regions of temperature, the temperature change of this object is defined as the total derivative or convective temperature change. Note that $dx/dt = u_x$, $dy/dt = u_y$, and $dz/dt = u_z$ are the three components of the local fluid velocity u. Equation (1.1.9) can be written as the substantial derivative:

$$\frac{DQ}{Dt} = \frac{\partial Q}{\partial t} + \frac{\partial Q}{\partial x}u_x + \frac{\partial Q}{\partial y}u_y + \frac{\partial Q}{\partial z}u_z \quad (1.1.10)$$

1.2 Review of Conservation of Mass

The conservation of mass states that the rate of mass accumulation in a control volume equals to the rate of mass out subtracted from the rate of mass. In other words, the rate of density increase in a control volume is equal to the net rate of mass flux per unit volume.

$$\frac{\partial \rho}{\partial t} = -\left(\frac{\partial \rho u_x}{\partial x} + \frac{\partial \rho u_y}{\partial y} + \frac{\partial \rho u_z}{\partial z}\right) = -(\nabla.\rho\vec{u}) \quad (1.2.1)$$

where $\rho\vec{u}$ is the mass flux in the unit of $\frac{kg/s}{m^2}$ that is the mass rate across an unit area of the control volume. Equation (1.2.1) can also be written as follows:

$$\frac{\partial \rho}{\partial t} + u_x\frac{\partial \rho}{\partial x} + u_y\frac{\partial \rho}{\partial y} + u_z\frac{\partial \rho}{\partial z} = -\rho\left(\frac{\partial u_x}{\partial x} + \frac{\partial u_y}{\partial y} + \frac{\partial u_z}{\partial z}\right) \quad (1.2.2)$$

In cylindrical coordinates (r, θ, z):

$$\frac{\partial \rho}{\partial t} + \frac{1}{r}\frac{\partial}{\partial r}(\rho r u_r) + \frac{1}{r}\frac{\partial}{\partial \theta}(\rho u_\theta) + \frac{\partial}{\partial z}(\rho u_z) = 0 \quad (1.2.2a)$$

Note that the left hand side of Eq. (1.2.2) is the substantial time derivative of density and hence this equation can be expressed as:

$$\frac{D\rho}{Dt} = -\rho(\nabla.\vec{u}) \tag{1.2.3}$$

If the fluid is incompressible then Eq. (1.2.3) is reduced to a much simpler form:

$$\nabla.u = \frac{\partial u_x}{\partial x} + \frac{\partial u_y}{\partial y} + \frac{\partial u_z}{\partial z} = 0 \tag{1.2.4}$$

1.3 Review of Conservation of Motion (Momentum Equation)

According to the second Newton's law, the rate of momentum change of a control volume equals to sum of all forces acting on that control volume.

$$\frac{D(m\vec{u})}{Dt} = \sum \vec{F} \tag{1.3.1}$$

$$\frac{D(\rho\vec{u})}{Dt} = \frac{1}{V}\sum \vec{F} \tag{1.3.2}$$

Neglecting the electrical and magnetic forces, there will be three main forces acting on the control volume including pressure force, viscous force, and gravitational force. Equation (1.3.2) can be re-written for the control volume per unit volume as follows:

$$\frac{D(\rho\vec{u})}{Dt} = -\nabla P - \nabla.\tau + \rho\vec{g} \tag{1.3.3}$$

where $\frac{D(\rho\vec{u})}{Dt}$ is the substantial time derivative of the mass flux defined in Eq. (1.1.10) and ∇P is the gradient of the scalar pressure defined in Eq. (1.1.6). Note that τ is a stress tensor which has nine components. For a Newtonian fluid (viscosity, μ, is a constant and independent to shear rate), a viscous stress tensor is defined as:

$$\tau_{ij} = 2\mu\varepsilon_{ij} = \mu\left(\frac{\partial u_i}{\partial x_j} + \frac{\partial u_j}{\partial x_i}\right) \tag{1.3.4}$$

where $\varepsilon_{ij} = \frac{1}{2}\left(\frac{\partial u_i}{\partial x_j} + \frac{\partial u_j}{\partial x_i}\right)$ is the rate of deformation. Therefore, the stress tensor is expressed as:

$$\tau_{ij} = \begin{pmatrix} \tau_{xx} & \tau_{xy} & \tau_{xz} \\ \tau_{yx} & \tau_{yy} & \tau_{yz} \\ \tau_{zx} & \tau_{zy} & \tau_{zz} \end{pmatrix} = \begin{pmatrix} 2\mu\frac{\partial u_x}{\partial x} & \mu\left(\frac{\partial u_y}{\partial x} + \frac{\partial u_x}{\partial y}\right) & \mu\left(\frac{\partial u_z}{\partial x} + \frac{\partial u_x}{\partial z}\right) \\ \mu\left(\frac{\partial u_x}{\partial y} + \frac{\partial u_y}{\partial x}\right) & 2\mu\frac{\partial u_y}{\partial y} & \mu\left(\frac{\partial u_z}{\partial y} + \frac{\partial u_y}{\partial z}\right) \\ \mu\left(\frac{\partial u_x}{\partial z} + \frac{\partial u_z}{\partial x}\right) & \mu\left(\frac{\partial u_y}{\partial z} + \frac{\partial u_z}{\partial y}\right) & 2\mu\frac{\partial u_z}{\partial z} \end{pmatrix} \tag{1.3.5}$$

In general, the momentum equation in rectangular coordinates in three directions (x, y, and z) can be given as follows:

$$\frac{D(\rho u_x)}{Dt} = -\frac{\partial P}{\partial x} + \mu\left(\frac{\partial^2 u_x}{\partial x^2} + \frac{\partial^2 u_x}{\partial y^2} + \frac{\partial^2 u_x}{\partial z^2}\right) + \rho g_x \tag{1.3.6}$$

$$\frac{D(\rho u_y)}{Dt} = -\frac{\partial P}{\partial y} + \mu\left(\frac{\partial^2 u_y}{\partial x^2} + \frac{\partial^2 u_y}{\partial y^2} + \frac{\partial^2 u_y}{\partial z^2}\right) + \rho g_y \tag{1.3.7}$$

$$\frac{D(\rho u_z)}{Dt} = -\frac{\partial P}{\partial z} + \mu\left(\frac{\partial^2 u_z}{\partial x^2} + \frac{\partial^2 u_z}{\partial y^2} + \frac{\partial^2 u_z}{\partial z^2}\right) + \rho g_z \tag{1.3.8}$$

If the fluid is incompressible (constant density) then Eqs. (1.3.6), (1.3.7), and (1.3.8) becomes:

$$\rho\left(\frac{\partial u_x}{\partial t} + u_x\frac{\partial u_x}{\partial x} + u_y\frac{\partial u_x}{\partial y} + u_z\frac{\partial u_x}{\partial z}\right) = -\frac{\partial P}{\partial x} + \mu\left(\frac{\partial^2 u_x}{\partial x^2} + \frac{\partial^2 u_x}{\partial y^2} + \frac{\partial^2 u_x}{\partial z^2}\right) + \rho g_x \tag{1.3.9}$$

$$\rho\left(\frac{\partial u_y}{\partial t} + u_x\frac{\partial u_y}{\partial x} + u_y\frac{\partial u_y}{\partial y} + u_z\frac{\partial u_y}{\partial z}\right) = -\frac{\partial P}{\partial y} + \mu\left(\frac{\partial^2 u_y}{\partial x^2} + \frac{\partial^2 u_y}{\partial y^2} + \frac{\partial^2 u_y}{\partial z^2}\right) + \rho g_y \tag{1.3.10}$$

$$\rho\left(\frac{\partial u_z}{\partial t} + u_x\frac{\partial u_z}{\partial x} + u_y\frac{\partial u_z}{\partial y} + u_z\frac{\partial u_z}{\partial z}\right) = -\frac{\partial P}{\partial z} + \mu\left(\frac{\partial^2 u_z}{\partial x^2} + \frac{\partial^2 u_z}{\partial y^2} + \frac{\partial^2 u_z}{\partial z^2}\right) + \rho g_z \tag{1.3.11}$$

In cylindrical coordinates (r, θ, z), the momentum equation for an incompressible and Newtonian fluid is given as [7]:

$$\begin{aligned} &\rho\left(\frac{\partial u_r}{\partial t} + u_r\frac{\partial u_r}{\partial r} + \frac{u_\theta}{r}\frac{\partial u_r}{\partial \theta} - \frac{u_\theta^2}{r} + u_z\frac{\partial u_r}{\partial z}\right) \\ &= -\frac{\partial P}{\partial r} + \mu\left[\frac{\partial}{\partial r}\left(\frac{1}{r}\frac{\partial}{\partial r}(r u_r)\right) + \frac{1}{r^2}\frac{\partial^2 u_r}{\partial \theta^2} - \frac{2}{r^2}\frac{\partial u_\theta}{\partial \theta} + \frac{\partial^2 u_r}{\partial z^2}\right] + \rho g_r \end{aligned} \tag{1.3.12}$$

$$\rho\left(\frac{\partial u_\theta}{\partial t}+u_r\frac{\partial u_\theta}{\partial r}+\frac{u_\theta}{r}\frac{\partial u_\theta}{\partial \theta}+\frac{u_r u_\theta}{r}+u_z\frac{\partial u_\theta}{\partial z}\right)$$
$$=-\frac{1}{r}\frac{\partial P}{\partial \theta}+\mu\left[\frac{\partial}{\partial r}\left(\frac{1}{r}\frac{\partial}{\partial r}(ru_\theta)\right)+\frac{1}{r^2}\frac{\partial^2 u_\theta}{\partial \theta^2}+\frac{2}{r^2}\frac{\partial u_r}{\partial \theta}+\frac{\partial^2 u_\theta}{\partial z^2}\right]+\rho g_\theta \quad (1.3.13)$$

$$\rho\left(\frac{\partial u_z}{\partial t}+u_r\frac{\partial u_z}{\partial r}+\frac{u_\theta}{r}\frac{\partial u_z}{\partial \theta}+u_z\frac{\partial u_z}{\partial z}\right)$$
$$=-\frac{\partial P}{\partial z}+\mu\left[\frac{1}{r}\frac{\partial}{\partial r}\left(r\frac{\partial u_z}{\partial r}\right)+\frac{1}{r^2}\frac{\partial^2 u_z}{\partial \theta^2}+\frac{\partial^2 u_z}{\partial z^2}\right]+\rho g_z \quad (1.3.14)$$

Example 1.1 Please use the momentum equation to develop equations for calculating the hydrostatic pressure of a homogeneous liquid and a homogeneous gas in a tube with a length of L and a true vertical depth of h as shown in Fig. 1.2.

Solution
Under static conditions, the left hand side of Eq. (1.3.3) is equal to zero. In addition, the shear force (viscous force) also equals to zero. Therefore, Eq. (1.3.3) can be reduces to:

$$\nabla P=\rho\vec{g}$$

This is a one dimensional problem. Assuming that the z-direction is the same as that of the gravitational force, $\vec{g}$. Equation (1.3.14) becomes:

$$\frac{\partial P}{\partial z}=\rho g_z\sin\theta \quad or \quad \partial P=\rho g_z\sin\theta\partial z \quad (1.3.15)$$

Integrating this Eq. gives:

$$\Delta=g_z\sin\theta\int_1^2\rho(P,T)\partial z \quad (1.3.16)$$

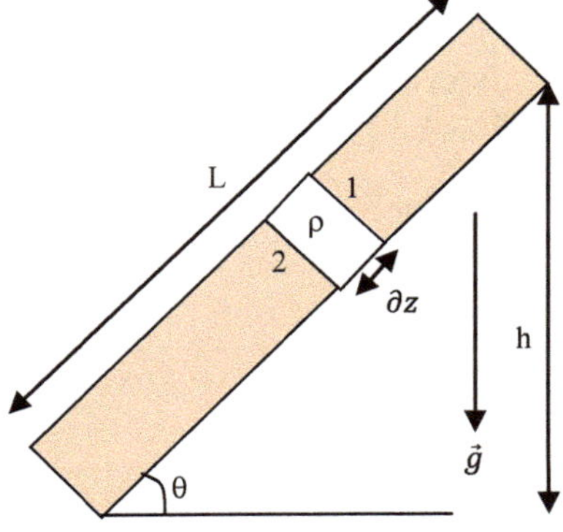

Fig. 1.2 Fluid in an inclined column

If the fluid is incompressible such as liquid with a density of ρ_L then Eq. (1.3.16) becomes:

$$\Delta P = \rho_L g_z L\sin\theta = \rho_L g_z h \quad (1.3.17)$$

If the unit of density is in kg/m^3, g = 9.81 m/s^2, and h is in m then the unit of ΔP in Eq. (1.3.17) is Pa or N/m^2.

In oil field unit, Eq. (1.3.17) becomes:

$$\Delta P = 0.052\rho_L h = 0.052\rho_L TVD \quad (1.3.18)$$

where TVD is true vertical depth in ft and density is in pounds per gallon (ppg).

If the fluid is compressible such as gas with a density of ρ_g, Eq. (1.3.15) becomes:

$$\int_1^2 \partial P = g_z \sin\theta \int_1^2 \frac{PM}{zRT} \partial z \quad (1.3.19)$$

where M is the molecular weight of gas; z is the compressibility factor, R is the gas constant, and T is the fluid temperature of the control volume. Assuming the fluid temperature is constant and equals to the average fluid temperature, T_{ave}, at the top and bottom of the fluid column. Rearrange Eq. (1.3.19) gives:

$$\int_1^2 \frac{\partial P}{P} = \frac{Mg_z \sin\theta}{zRT_{ave}} \int_1^2 \partial z \quad (1.3.20)$$

$$P_2 = P_1 e^{\frac{Mg_z L\sin\theta}{zRT_{ave}}} = P_1 e^{\frac{Mg_z h}{zRT_{ave}}} \quad (1.3.21)$$

In SI unit, the molecular weight, M, is in kg/mol, gravitational acceleration factor, g, is in m/s^2, height, h, is in meter, gas constant, R, is in $\frac{Pa \times m^3}{mol \times K}$, and temperature, T, is in K, then the unit of P_2 is the same as that of P_1.

In oil field unit, Eq. (1.3.21) becomes:

$$P_2 = P_1 e^{\left(\frac{0.01877\gamma_g h}{zT_{ave}}\right)} \quad (1.3.22)$$

where γ_g is the specific gravity of gas, h is depth or height of the liquid column in ft, and T is average temperature in Rankin, °R. The pressures P_1 and P_2 are in the units of psi.

Example 1.2 Using the momentum equation, please derive an equation to calculate the total pressure loss of an incompressible Newtonian fluid in pipe flow (Fig. 1.3).

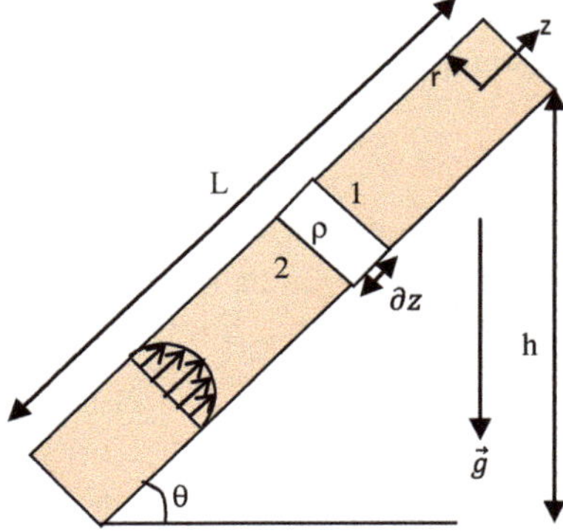

Fig. 1.3 Flow in an inclined pipe

Solution

This is again a one dimensional problem. Let z be the direction of the flow. Under steady state conditions and neglecting the acceleration, Eq. (1.3.14) becomes:

$$\begin{aligned}\left.\frac{\partial P}{\partial z}\right|_{Total} &= \left.\frac{\partial P}{\partial z}\right|_{friction} + \left.\frac{\partial P}{\partial z}\right|_{gravity} \\ \left.\frac{\partial P}{\partial z}\right|_{Total} &= \mu\left[\frac{1}{r}\frac{\partial}{\partial r}\left(r\frac{\partial u_z}{\partial r}\right)\right] + \rho g_z \sin\theta\end{aligned} \tag{1.3.23}$$

Equation (1.3.23) tells us that the total pressure loss of an incompressible Newtonian fluid flowing in an inclined pipe under laminar steady state conditions and neglecting acceleration consists of two components: pressure loss due to friction and pressure loss due to gravity.

The pressure loss due to friction per unit length (frictional pressure loss gradient):

$$\left.\frac{\partial P}{\partial z}\right|_f = \mu\left[\frac{1}{r}\frac{\partial}{\partial r}\left(r\frac{\partial u_z}{\partial r}\right)\right] \tag{1.3.24}$$

$$\partial\left(r\frac{\partial u_z}{\partial r}\right) = \frac{1}{\mu}\left.\frac{\partial P}{\partial z}\right|_f r\partial r \tag{1.3.25}$$

Integrating both side of Eq. (1.3.25) gives:

$$r\frac{\partial u_z}{\partial r} = \frac{1}{2\mu}\left.\frac{\partial P}{\partial z}\right|_f r^2 + C_1 \tag{1.3.26}$$

$$\partial u_z = \frac{1}{2\mu} \frac{\partial P}{\partial z}\bigg|_f r\partial r + \frac{C_1}{r} \partial r \tag{1.3.27}$$

At the center of the pipe: $@r = 0, \partial u_z = 0\, hence\, C_1 = 0$

$$u_z = \frac{1}{4\mu} \frac{\partial P}{\partial z}\bigg|_f r^2 + C_2 \tag{1.3.28}$$

No slip boundary condition at the pipe wall: $@r = R, u_z = 0\, hence\, C_2 = -\frac{1}{4\mu}\frac{\partial P}{\partial z}\big|_f R^2$. Therefore:

$$u_z = \frac{1}{4\mu} \frac{\partial P}{\partial z}\bigg|_f \left(r^2 - R^2\right) \tag{1.3.29}$$

Equation (1.3.29) reveals that the velocity profile of an incompressible Newtonian fluid in pipe flow under laminar flow conditions is a parabolic.

The liquid flow rate in pipe flow is defined as:

$$Q = \int_0^R u_z dA = \int_0^R u_z 2\pi r dr \tag{1.3.30}$$

Combining Eqs. (1.3.29) and (1.3.30) gives:

$$Q = \frac{\pi}{2\mu} \frac{\partial P}{\partial z}\bigg|_f \int_0^R \left(r^2 - R^2\right) r dr \tag{1.3.31}$$

$$Q = \frac{\pi}{8\mu} \frac{\partial P}{\partial z}\bigg|_f R^4 \tag{1.3.32}$$

Using the definition of average liquid velocity, $\bar{u}$, in pipe flow, the flow rate can be written as follows:

$$Q = \pi \bar{u} R^2 \tag{1.3.33}$$

Combining Eqs. (1.3.32) and (1.3.33) yields an equation to calculate for the frictional pressure loss in pipe flow under steady state and laminar flow conditions:

$$\frac{\partial P}{\partial z}\bigg|_f = \frac{8\mu\bar{u}}{R^2} = \frac{32\mu\bar{u}}{D^2} \tag{1.3.34}$$

If the unit of fluid viscosity, μ, is in Pas, average fluid velocity in pipe, $\bar{u}$, is in m/s, and pipe diameter is in m, the unit of $\frac{\partial P}{\partial z}\big|_f$ is in Pa/m.

In oil field unit, Eq. (1.3.34) becomes:

$$\left.\frac{\partial P}{\partial z}\right|_f = \frac{8\mu\bar{u}}{R^2} = \frac{\mu\bar{u}}{1500D^2} \tag{1.3.35}$$

where μ is in cp, $\bar{u}$ is in ft/s, D is the inner pipe diameter in inch, and $\left.\frac{\partial P}{\partial z}\right|_f$ is the frictional pressure drop gradient in psi/ft.

The pressure drop due to gravity is shown in Eq. (1.3.17).

The total pressure drop of an incompressible Newtonian fluid in an inclined pipe under steady state and laminar flow conditions without acceleration is given:

$$\left.\frac{\partial P}{\partial z}\right|_{Total} = \left.\frac{\partial P}{\partial z}\right|_{friction} + \left.\frac{\partial P}{\partial z}\right|_{gravity} = \frac{32\mu\bar{u}}{D^2} + \rho_L g \sin\theta \tag{1.3.36}$$

Equation (1.3.36) indicates that the frictional pressure drop is independent to fluid density and inclination angle. It depends on fluid viscosity, pipe geometry, and liquid flow rate. The pressure drop due to gravity is a function of fluid density and inclination angle.

Example 1.3 Using the momentum equation, please derive an equation for calculating the pressure drop of a fluid through an orifice. The pipe and orifice diameters are dp and do as shown in Fig. 1.4. Note that this example is a fundamental for showing how to predict the performance of an orifice gas lift valve.

Solution

Assumptions:

- Steady state flow, $\frac{\partial u_z}{\partial t} = 0$.
- One dimensional flow; z direction. $u_r = 0\, and\, u_\theta = 0$.
- Viscosity is small and hence the viscous force is neglected.
- Pressure drop due to gravity is much smaller than that due to convection.

With these assumptions, the pressure drop of the fluid through an orifice is mainly due to the change of the local velocity along the flow. Note that the fluid velocity not only changes its magnitude but also its direction when the fluid flows through the orifice. Equation (1.3.14) becomes:

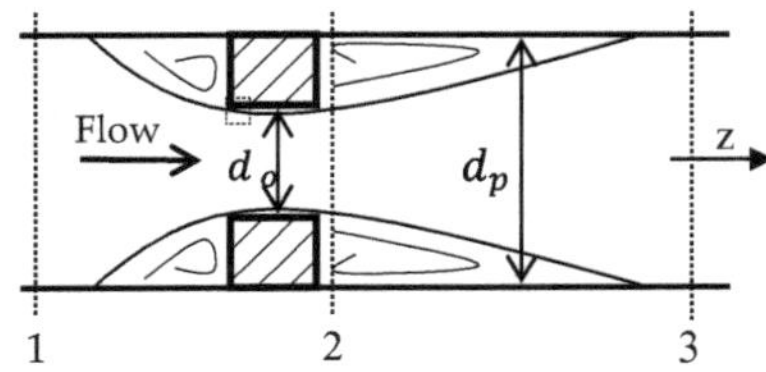

Fig. 1.4 Flow through an orifice

$$\rho\left(u_z\frac{\partial u_z}{\partial z}\right) = -\frac{\partial P}{\partial z} \tag{1.3.37}$$

At location 1 and 3:

$$\Delta P = P_1 - P_3 = \int_1^3 \rho u_z \partial u_z \tag{1.3.38}$$

If the fluid is incompressible, Eq. (1.3.38) can be expressed as:

$$\Delta P = P_1 - P_3 = \frac{\rho\left(u_3^2 - u_1^2\right)}{2} \tag{1.3.39}$$

Applying the continuity equation, Eq. (1.2.4), for this incompressible fluid yields:

$$\frac{\partial u_z}{\partial z} = 0 \tag{1.3.40}$$

Equation (1.3.40) tells us that if the pipe diameter is uniform, the velocity of an incompressible fluid along the pipe is the same. Therefore, if points 1 and 3 are far away from the orifice, then $u_1 = u_3$ and the total pressure drop through an orifice is, $\Delta P = 0$. This means the total pressure lost upstream of the orifice is the same as the total pressure gained downstream of the orifice.

The total pressure lost upstream of the orifice:

$$\Delta P_{lost} = P_1 - P_2 = \frac{\rho\left(u_2^2 - u_1^2\right)}{2} \tag{1.3.41}$$

The total pressure gained downstream of the orifice:

$$\Delta P_{gained} = P_2 - P_3 = \frac{\rho\left(u_3^2 - u_2^2\right)}{2} \tag{1.3.42}$$

In reality, when the fluid flows through an orifice, there are permanent pressure losses due to vortices, contraction and expansion before and after the orifice, and dead zones (no flow). Therefore, the pressure at point 1 is always greater than the pressure at point 3. In other words, the pressure gained downstream of the orifice is always less than the pressure lost upstream of the orifice.

If the fluid is compressible such as gas, Eq. (1.3.38) becomes:

$$\Delta P = P_1 - P_2 = \frac{1}{2}\rho_2 u_2^2 - \frac{1}{2}\rho_1 u_1^2 \tag{1.3.43}$$

Applying the continuity equation gives:

$$\frac{\partial}{\partial z}(\rho u_z) = 0 \tag{1.3.44}$$

Converting Eq. (1.3.44) to mass flux yields:

$$Q_m = \rho_1 u_1 A_1 = \rho_2 u_2 A_2 \tag{1.3.45}$$

Combining Eqs. (1.3.43) and (1.3.45) gives:

$$\Delta P = P_1 - P_2 = \frac{1}{2}\left[\frac{\rho_1}{\rho_2}\left(\frac{A_1}{A_2}\right)^2 - 1\right]\rho_1 u_1^2 \tag{1.3.46}$$

Equation (1.3.46) can be used to calculate for the total theoretical pressure drop of an incompressible fluid through an orifice.

In terms of volumetric flow rate upstream of the orifice, combining Eqs. (1.3.45) and (1.3.46) yields:

$$q_v = \frac{q_{mass}}{\rho_1} = \frac{\pi d_p^2}{4}\sqrt{2(P_1 - P_2)}\sqrt{\frac{\rho_2}{\rho_1^2\left[\frac{d_p^4}{d_o^4} - \frac{\rho_2}{\rho_1}\right]}} \tag{1.3.47}$$

Let $\beta = \frac{d_o}{d_p}$, Eq. (1.3.47) becomes:

$$q_v = C\frac{\pi d_p^2}{4}\sqrt{2(P_1 - P_2)}\sqrt{\frac{1}{\rho_1(1-\beta^4)}} \tag{1.3.48}$$

where C is the discharge coefficient factor, which takes care of the permanent losses. Equation (1.3.48) can be used to predict the performance of gas flowing through an orifice.

Example 1.4 Using the momentum equation, please derive an equation for calculating the pressure gain when an incompressible is rotated at a constant angular velocity in the impeller of a centrifugal pump (Fig. 1.5).

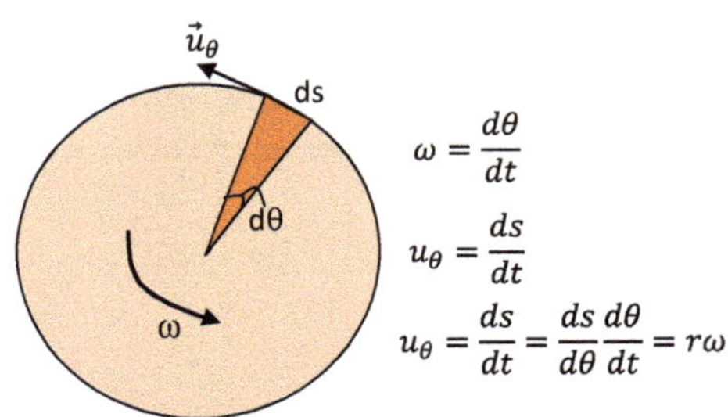

Fig. 1.5 Tangential and angular velocities

Solution

Assumptions:

- Because of the constant angular velocity of the object, the motion is under steady state conditions
- u_r and u_z equal to zero.
- $\frac{\partial u_\theta}{\partial \theta} = 0$, $\frac{\partial u_\theta}{\partial r} = 0$, and $\frac{\partial u_\theta}{\partial z} = 0$
- Neglecting the viscous and gravitational forces.

Equation (1.3.12) can now be reduced to:

$$\frac{u_\theta^2}{r} = \frac{1}{\rho}\frac{\partial P}{\partial r} \tag{1.3.49}$$

Equation (1.3.49) states that the centrifugal force $\left(\frac{u_\theta^2}{r}\right)$ acting radially outwards on the fluid element is balanced by the pressure gradient force directed intwards. Where u_θ is the tangential velocity component and the relationship between the tangential velocity and angular velocity is expressed as:

$$u_\theta = r\omega \tag{1.3.50}$$

Combining Eqs. (1.3.49) and (1.3.50) and integrating both sides of the equation gives:

$$\int_{P_i}^{P_d} \partial P = \int_{R_1}^{R_2} \rho\omega^2 r \partial r \tag{1.3.51}$$

$$\Delta P = P_d - P_i = \frac{\rho\omega^2}{2}\left(R_2^2 - R_1^2\right) = \frac{\rho}{2}\left(u_2^2 - u_1^2\right) \tag{1.3.52}$$

Pump head is defined as:

$$H = \frac{\Delta P}{\rho g} = \frac{\left(u_2^2 - u_1^2\right)}{2g} \tag{1.3.53}$$

where P_i and P_d are the impeller intake and impeller discharge pressures, respectively. u_1 and u_2 are the tangential fluid velocities at the intake and discharge of the pump impeller. Equation (1.3.53) tells us that the theoretical pump head or theoretical pressure gain across the impeller of a centrifugal pump is mainly due to the centrifugal force. Equation (1.3.53) can also be interpreted that the pressure gained across the impeller is due to the gain in kinetic energy of the fluid received from the rotation of the pump impeller.

1.4 Review of Basic Thermodynamic Properties of Liquid and Gas

1.4.1 Specific Gravity

The specific gravity of gas, γ_g, is defined as the ratio between gas density under actual conditions and air density at standard conditions.

Petroleum industry defines standard conditions at which the temperature is 15.5 °C (60 °F) and the pressure is 1 atm (14.7 psi). Density of air under standard conditions is 1.225 kg/m^3 = 0.001225 g/cm^3 = 0.0765 lbm/ft^3.

The specific gravity of liquid such as oil, is defined similar to that of gas but using water under standard conditions as a reference liquid instead of using air. The density of water under standard conditions is 1.0 g/mL = 1000 kg/m^3 = 62.4 lbm/ft^3 = 8.3 lbm/gal.

The petroleum industry uses API gravity with a unit of °API to measure of how heavy a petroleum fluid is compared to water. If the API gravity is smaller than 10, the petroleum fluid is heavier than water and it sinks. If the API gravity is greater than 10, the petroleum fluid is considered to be lighter than water and it floats. The relationship between the liquid specific gravity and API gravity is given as:

$$\gamma_l = \frac{141.5}{131.5 + {}^\circ API} \tag{1.4.1}$$

1.4.2 Bubble Point Pressure

Under original reservoir conditions, formation oils often contain some dissolved gas in solution. If the reservoir is produced for a period of time, the reservoir pressure will decline to a value that the dissolved gas begins to come out of solution and form bubbles. This pressure is known as the bubble point pressure, P_b. If the reservoir pressure is smaller than the bubble point pressure, the formation fluid will exhibit two phase: gas and liquid.

Bubble point pressure (P_b in psi) is a function of bubble point solution gas oil ratio (R_{sb} in scf/STBO), dissolved oil and gas specific gravity (γ_o, γ_g), and temperature (T in °F). Table 1.1 summarizes the common available correlations from 1947 to 1988 for predicting the bubble point pressure.

Table 1.1 Correlations for predicting bubble point pressure [2], [13], and [15]

Author	Origin	Corelation	Year	Eq.
Standing	California	$P_b = 18.2\left[\left(\frac{R_s}{\gamma_g}\right)^{0.83} \times 10^{0.00091T - 0.0125API} - 1.4\right]$	[17]	(1.4.2)
Glaso	North Sea	$P_b = 10^y$	[12]	(1.4.3)
		$y = 1.7669 + 1.7447logX - 0.3022(logX)^2$		(1.4.3a)
		$X = \left(\frac{R_s}{\gamma_g}\right)^{0.816}\left(\frac{T^{0.172}}{API^{0.989}}\right)$		(1.4.3b)
Al-Marhoun	Middle East	$P_b = \frac{5.38\times10^{-3} R_s^{0.715} \gamma_o^{3.144} (T+460)^{1.326}}{\gamma_g^{1.878}}$	1988	(1.4.4)

1.4.3 Solution Gas-Oil Ratio

Solution gas-oil ratio (R_s) is the amount of gas dissolved in one Stock Tank Barrel of Oil (STBO) at a specific pressure and temperature. Since oil volume is measured at atmospheric conditions STBO, the unit of R_s is scf/STB.

$$R_s\left(\frac{SCF}{STBO}\right) = \frac{V_{g,sc}(dissolved)}{V_{o,sc}(produced)} \tag{1.4.5}$$

As pressure increases, more gas is dissolved into oil and hence R_s increases approximately linearly with pressure. Oil under these conditions is said to be saturated. When the pressure reaches the bubble point pressure, no more gas is dissolved into oil (undersaturated oil) and hence Rs is a constant. This relationship is shown in Fig. 1.6. Below the bubble point pressure, R_s is a function of the reservoir pressure. Above the bubble point pressure, R_s is constant as R_{sb} called the solution gas-oil ratio at the bubble point pressure.

For practical applications, the solution gas oil ratio at the bubble point pressure is the value of interest. As mentioned, above the bubble pressure, the solution gas oil ratio remains the same. Solution gas-oil ratio (R_s in SCF/STBO) is a function of

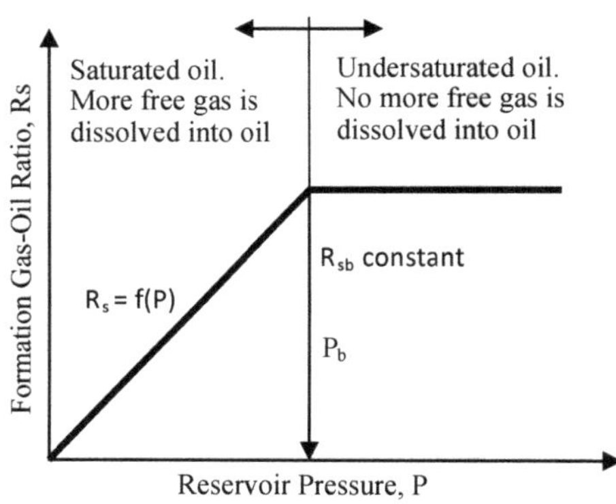

Fig. 1.6 Relationship between solution gas-oil ratio and pressure

Table 1.2 Correlations for predicting solution gas-oil ratio [1–3], [5], [13], [17–21]

Author	Origin	Corelation	Year	Eq.
Standing	California	$R_s = \gamma_g\left[\left(\frac{P}{18.2}+1.4\right)10^x\right]^{1.2048}$	[17]	(1.4.6)
		$x = 0.0125API - 0.00091T$		(1.4.6a)
Vasquez and Beggs	Worldwide	For API ≤ 30 : $R_s = 0.0362\gamma_{gc}P^{1.094}e^{\left(\frac{25.72API}{(T+460)}\right)}$	1977	(1.4.7)
		For API > 30 : $R_s = 0.0178\gamma_{gc}P^{1.187}e^{\left(\frac{23.93API}{(T+460)}\right)}$		(1.4.8)
		$\gamma_{gc} = \gamma_g\left[1+5.91\times10^{-5}API\times T_{sp}log\left(\frac{P_{sp}}{114.7}\right)\right]$ The subscript "sp" is for separator conditions		(1.4.8a)
Glasso	North Sea	$R_s = \gamma_g\left(10^x\times\frac{API^{0.989}}{T^{0.172}}\right)^{1.2255}$	[12]	(1.4.9)
		$x = 2.8869 - (14.1811 - 3.309logP)^{0.5}$		(1.4.9a)
Al-Marhoun	Middle East	$R_s = A^{1.3984}$	1985	(1.4.10)
		$A = 185.483P\gamma_g^{1.8778}\gamma_o^{-3.1437}(T+460)^{-1.3266}$		(1.4.10a)
Petrosky and Farshad	Gulf of Mexico	$R_s = \left[\left(\frac{P}{112.727}+12.34\right)\gamma_g^{0.8439}10^x\right]^{1.7318}$	[16]	(1.4.11)
		$x = 7.916\times10^{-4}API^{1.541} - 4.561\times10^{-5}T^{1.391}$		(1.4.11a)

dissolved gas and oil specific gravity (γ_o, γ_g), temperature (T in °F), and pressure (P in psi). It is determined by rearranging the bubble point pressure equations. Table 1.2 summarizes the common available correlations from 1947 to 1998 for predicting the solution gas-oil ratio. Note that the produced GOR, defined as the ratio between produced gas and produced oil, and the R_s will be the same if the reservoir pressure is higher than the bubble point pressure. If the reservoir pressure (or the flowing bottomhole pressure) is less than the bubble point pressure, the produced GOR will normally be higher than the R_s under this reservoir pressure ($P_r < P_b$). However, the produced GOR will be smaller than then initial R_s or the bubble point formation gas-oil ration R_{sb}.

Note that Vasquez and Beggs suggested that separator conditions affect the gas gravity and hence the new gas gravity, γ_{gc}, should be adjusted to a separator pressure of 114.7 psia. If the separator conditions is not available, temperature and pressure at the separator conditions are assumed to be 60 °F and 14.7 psia, respectively.

All the equations presented in Table 1.2 were derived from the bubble point pressure equations to obtain the bubble point formation gas-oil ratio.

1.4.4 Oil Formation Volume Factor

Oil formation volume factor, B_o, is defined as the ratio of the volume of oil at reservoir conditions to the volume of oil at surface conditions.

$$B_o = \frac{V_{o,reservoir}}{V_{o,surface}} \tag{1.4.12}$$

As oil travels up to surface, the fluid pressure reduces leading to more gas coming out of the solution. Therefore, the oil formation volume factor is normally greater than one. The oil formation volume factor increases as the reservoir pressure is increased until the reservoir pressure is the same as the bubble point pressure. This is because more gas is dissolved into the oil as the reservoir pressure is increased causing the volume of oil under reservoir conditions to swell. If the reservoir pressure is higher than the bubble point pressure, the oil formation volume factor decreases slightly because no more free gas is dissolved into the solution and the oil is compressed.

The formation volume factor for gas, B_g, is defined as the ratio of volume of one mole of gas at a given pressure and temperature to the volume of one mole gas at standard conditions.

$$B_g = \frac{V}{V_{sc}} = \frac{zp_{sc}T}{pT_{sc}} \tag{1.4.13}$$

For oil field unit, where B_g is in the unit of bbl/SCF, T is in °F, and p is in psia, B_g is given as:

$$B_g = 0.005\frac{z(T+460)}{p} \tag{1.4.13a}$$

The two-phase formation volume factor B_t is calculated as follows:

$$B_t = B_o + B_g(R_{si} - R_s) \tag{1.4.14}$$

The subscript "i" is for the initial reservoir conditions (Fig. 1.7).

Oil formation volume factor (B_o in bbl/STBO) is a function of dissolved gas and oil specific gravity (γ_o, γ_g), temperature (T in °F), and pressure (P in psi). Table 1.3 summarizes the common available correlations from 1947 to 1998 for predicting the oil formation volume factor.

1.4.5 Oil Viscosity

For Newtonian fluids, viscosity, μ, is a constant at a constant temperature and defined as the ratio between shear stress and shear rate. Viscosity alone is enough to characterize a Newtonian fluid. However, viscosity is not enough to characterize a non-Newtonian fluid because the viscosity of this fluid is a function of shear rate. Depending on the relationship between shear stress, τ, and shear rate, γ, a non-Newtonian fluid can be characterized as Bingham model, $\tau = \tau_y + \mu_p\gamma$, Power

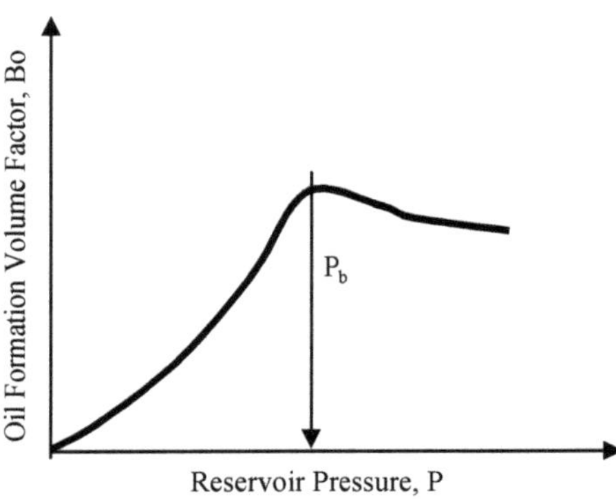

Fig. 1.7 Relationship between oil formation volume factor and pressure

Law model, $\tau = K\gamma^n$ or Yield Power Law model (Herschel Bulkley Model), $\tau = \tau_y + \mu_p\gamma^n$. Where μ_p, τ_y, K, and n are the Bingham plastic viscosity, yield point, consistency index and flow behavior index, respectively. Most of formation liquids are Newtonian fluids and hence this section will present correlations for predicting viscosity of Newtonian fluids only.

Before presenting viscosity correlations, let's review the terms live oil and dead oil. Imagine a situation where crude oil is placed in a container which is pressurized to the original reservoir pressure. The pressure in the container is gradually reduced until surface pressure is reached. At this time all the dissolved gas is released and the oil at this point is called 'dead' oil. Heavy oil and bitumen are good examples as dead oils because they have little or no gas that may be released at surface conditions. Live oil is the one that contains sufficient dissolved gas that may be released from the oil at surface conditions. Thus, live oil can be undersaturated (reservoir pressure is above bubble point pressure) and saturated (reservoir pressure is below bubble point pressure).

Common viscosity correlations of dead oil and live oil are presented in Table 1.4.

1.4.6 Density

Density, ρ, is defined as its mass, m, per unit volume, V.

$$\rho = \frac{dm}{dV} \tag{1.4.29}$$

For incompressible single-phase liquid:

Table 1.3 Correlations for predicting oil formation volume factor [2], [3], [11], [13], [14], [17–21]

Author	Origin	Corelation	Year	Eq.
Standing	California	Saturated: $B_o = 0.972 + 0.000147\left[R_s\left(\frac{\gamma_g}{\gamma_o}\right)^{0.5} + 1.25T\right]^{1.175}$	[17]	(1.4.15)
		Undersaturated: $B_o = B_{ob}e^{c_o(P_b - P)}$		(1.4.16)
		$c_o = \frac{-1433 + 5R_{sb} + 17.2T - 1180\gamma_g + 12.61\gamma_o}{10^5 P}$		(1.4.16a)
Vasquez and Beggs	Worldwide	API < 30 : $B_o = 1 + 4.677 \times 10^{-4}R_s + (T - 60)\left(\frac{API}{\gamma_{gc}}\right)(1.751 \times 10^{-5} - 1.81 \times 10^{-8}R_s)$	1977	(1.4.17)
		API > 30 : $B_o = 1 + 4.677 \times 10^{-4}R_s + (T - 60)\left(\frac{API}{\gamma_{gc}}\right)(1.1 \times 10^{-5} - 1.337 \times 10^{-9}R_s)$		(1.4.17a)
Glasso	North Sea	Saturated: $B_o = 1 + 10^a$	[12]	(1.4.18)
		$a = -6.585 + 2.913log(y) - 0.277(logy)^2$		(1.4.18a)
		$y = R_s\left(\frac{\gamma_g}{\gamma_o}\right)^{0.526} + 0.968T$ Undersaturated: Using Eqs. (1.4.16) and (1.4.16a)		(1.4.18b)

(continued)

Table 1.3 (continued)

Author	Origin	Corelation	Year	Eq.
Al-Marhoun	Middle East	Saturated: $B_o = 0.497 + 8.629 \times 10^{-4}(T + 460) + 1.8259 \times 10^{-3}y + 3.181 \times 10^{-6}y^2$	1985	(1.4.19)
		$y = R_s^{0.7424}\gamma_g^{0.3233}\gamma_o^{-1.2021}$ Undersaturated: Using Eqs. (1.4.16) and (1.4.16a)		(1.4.19a)
Petrosky and Farshad	Gulf of Mexico	Saturated $B_o = 1.011 + 7.205 \times 10^{-5} \times \left[R_s^{0.374}\left(\frac{\gamma_g^{0.291}}{\gamma_o^{0.626}}\right) + 0.246T^{0.537}\right]^{3.094}$	[16]	(1.4.20)

Table 1.4 Correlations for predicting formation fluid viscosity [5], [6], [13], [17], [19–21]

Author	Origin	Corelation	Year	Eq.
Beal	California	Dead oil (zero dissolved gas): $\mu_{o,d} = \left(0.32 + \frac{1.8\times10^7}{^\circ API^{4.53}}\right)\left(\frac{360}{T+200}\right)^a$	1946	(1.4.21)
		$a = 10^{\left(0.43+\frac{8.33}{^\circ API}\right)}$		(1.4.21a)
Beggs and Robinson	Worldwide	Dead oil: $\mu_{o,d} = 10^x - 1$	[6]	(1.4.22)
		$x = \frac{10^{(3.0324-0.0202\times^\circ API)}}{T^{1.163}}$		(1.4.22a)
		Saturated oil (below bubble point pressure) $\mu_{o,l} = \frac{10.715\mu_{o,d}^b}{(R_s+100)^{0.515}}$		(1.4.23)
		$b = 5.44(R_s+150)^{-0.338}$		(1.4.23a)
Vasquez and Beggs	Worldwide	Undersaturated oil (above bubble point pressure): $\mu_{o,d} = \mu_{o,b}\left(\frac{P}{P_b}\right)^m$	1977	(1.4.24)
		$m = 2.6 \times P^{1.187}e^{\left(-11.513-8.980\times10^{-5}P\right)}$		(1.4.24a)
Glasso	North Sea	Dead oil: $\mu_{o,d} = \frac{3.141\times10^{10}[\log(^\circ API)]^a}{T^{3.444}}$	[12]	(1.4.25)
		$a = 10.313 log(T) - 36.447\ where\ 50 < T < 300$		(1.4.25a)
Kartoatmokjo	Worldwide	Dead oil: $\mu_{o,d} = \frac{16\times10^8[log(^\circ API)]^{(5.753logT-26.972)}}{T^{2.818}}$	1991	(1.4.26)
		Undersaturated oil (above bubble point pressure): $\mu_o = \mu_{o,b} + 0.001127(P-P_b)\times \times\left(0.038\mu_{o,b}^{1.59} - 0.00652\mu_{o,b}^{1.815}\right)$		(1.4.27)
		Saturated oil (below bubble point pressure): $\mu_{o,l} = -0.06821 + 0.9824f + 0.0004034f^2$		(1.4.28)
		$f = \left(0.2 + 0.843 \times 10^{-0.000845R_s}\right)\mu_{o,d}^{(0.43+0.5165y)}$		(1.4.28a)
		$y = 10^{-0.00081R_s}$		(1.4.28b)

$$\rho = \frac{m}{V} \tag{1.4.30}$$

For gas, density is a function of temperature, T, and pressure, P.

$$\rho\frac{dm}{dV} = \frac{PM}{ZRT} \tag{1.4.31}$$

where Z is the compressibility factor, R is the gas constant. The value of the gas constant can be:

$$R = 10.7315 \frac{psia \times ft^3}{lbmol \times^\circ R} = 8.3144 \frac{kPa \times m^3}{kmol \times K}$$

For pressure in psia and temperature in °R, density will be in lbm/ft^3 as given:

$$\rho_g = 2.6988 \frac{P\gamma_g}{ZT} \tag{1.4.32}$$

For pressure in kPa, temperature in K, density in Kg/m^3 is given as:

$$\rho_g = 3.4834 \frac{P\gamma_g}{ZT} \tag{1.4.33}$$

The density of dead oil, $\rho_{o,d}$, that contains no dissolved gas is given as:

$$\rho_{o,d} = \rho_{w,sc}\gamma_o \tag{1.4.34}$$

where $\rho_{w,sc}$ is the density of water under standard conditions.

Oil that contains dissolved gas in solution is called live oil. The density of live oil is defined as:

$$\rho_{o,l} = \frac{m_{o,l}}{V_{o,l}} = \frac{m_g + m_o}{V_{o,l}} \tag{1.4.35}$$

For 1 Stock Tank Barrel of Oil (STBO) at surface, the total mass, volume, and density of live oil under reservoir conditions are calculated as:

$$m_{o,l} = m_g + m_o$$

$$m_{o,l} = 1STBO \times R_s \frac{SCF}{STBO} \times 0.0764\gamma_{g,d} \frac{lbm}{SCF} + 1STBO \times 5.615 \frac{SCF}{STBO} \times 62.4\gamma_{o,d} \frac{lbm}{SCF} \tag{1.4.36}$$

$$V_{o,l} = 1STBO \times B_o \frac{bbl}{STBO} \times 5.615 \frac{ft^3}{bbl} \tag{1.4.37}$$

$$\rho_{o,l} = \frac{\rho_{o,d} + \frac{0.0764 \times R_s \times \gamma_{g,d}}{5.615}}{B_o} \tag{1.4.38}$$

where R_s and B_o is the solution gas-oil ratio and oil formation volume factor. These two parameters will be discussed in the next section.

1.4.7 Examples

Example 1.5 Calculate the density of a live oil that has gravity of 35 °API, solution gas-oil ratio of 750 SCF/STBO, gas specific gravity of 0.8, temperature of 120 °F, and formation volume factor of 1.39 bbl/STBO.

Solution

From Eq. (1.4.1), the specific gravity of oil is:

$$\gamma_o = \frac{141.5}{131.5 + {}^\circ API} = 0.85$$

Dead oil density: $\rho_{o,d} = \rho_{w,sc}\gamma_o = 62.4 \times 0.85 = 53 \frac{lbm}{ft^3}$

The live oil density can be calculated using Eq. (1.4.37):

$$\rho_{o,l} = \frac{\rho_{o,d} + \frac{0.0764 \times R_s \times \gamma_{g,d}}{5.615}}{B_o} = \frac{53 + \frac{0.0764 \times 750 \times 0.8}{5.615}}{1.39} = 43.9 \frac{lbm}{ft^3}$$

Example 1.6 Use the PVT correlations to find the bubble point pressure, bubble point solution gas-oil ratio, formation volume factor, fluid viscosity of an oil of gravity 35 °API, formation gas-oil ratio of 750 SCF/STBO, gas specific gravity of 0.8, temperature of 120 °F, reservoir pressure of 4500 psia, temperature and pressure at the separator are 60 °F and 120 psia, respectively.

Solution

The specific gravity of oil is:

$$\gamma_o = \frac{141.5}{131.5 + {}^\circ API} = 0.85$$

- Bubble point solution gas-oil ratio using Standing correlation:

$$x = 0.0125API - 0.00091T = 0.0125 \times 35 - 0.00091 \times 120 = 0.3283$$

$$R_{sb} = \gamma_g \left[\left(\frac{P}{18.2} + 1.4 \right) 10^x \right]^{1.2048} = 0.8 \left[\left(\frac{4500}{18.2} + 1.4 \right) 10^{0.3283} \right]^{1.2048} = 1531 \frac{SCF}{STBO}$$

- Bubble point pressure using Standing correlation (Table 1.5):

Table 1.5 Summary of bubble point solution gas-oil ratio calculation using different correlations

1	Standing	x	0.328	
		R_{sb}	1531	SCF/STBO
2	Vasquez and Beggs	γ_{gc}	0.802	
		R_{sb}	1780	SCF/STBO
3	Glasso	x	1.440	
		R_{sb}	1263	SCF/STBO
4	Al-Marhoun	A	197.562	
		R_{sb}	1623	SCF/STBO
5	Petrosky & Farshad	x	0.154	
		R_{sb}	1261	SCF/STBO

$$P_b = 18.2\left[\left(\frac{R_s}{\gamma_g}\right)^{0.83} \times 10^{0.00091T-0.0125API} - 1.4\right]$$

$$= 18.2\left[\left(\frac{750}{0.8}\right)^{0.83} \times 10^{0.00091\times 120-0.0125\times 35} - 1.4\right] = 2478\,psia$$

Since the reservoir pressure of 4500 psia is greater than the bubble point pressure of 2478 psia, the formation fluid is undersaturated. However, the following calculations are carried out for both undersaturated and saturated fluid using different correlations given in Table 1.3. Readers can quickly recognize that there is not much a difference when using difference correlations as well as using undersaturated or saturated conditions (Table 1.6).

- Fluid viscosity using Beggs and Robinson correlation (Table 1.7):
  For dead oil (Undersaturated oil or above bubble point pressure):

$$x = \frac{10^{(3.0324-0.0202\times{}^\circ API)}}{T^{1.163}} = \frac{10^{(3.0324-0.0202\times 35)}}{120^{1.163}} = 0.808$$

$$\mu_{o,d} = 10^x - 1 = 10^{0.808} - 1 = 5.424\,cp$$

For live oil (Saturated oil or below bubble point pressure) (Table 1.8):

Table 1.6 Summary of bubble point pressure calculation using different correlations

1	Standing	P_b	2477.00	psia
2	Glasso	X	18.00	
		y	3.48	
		P_b	3028.17	psia
3	Al-Marhoun	P_b	2573.98	psia

Table 1.7 Summary of formation volume factor calculation using different correlations

1	Standing	Saturated	x	0.3283	
			R_{sb}	1530.511	SCF/STBO
			B_o	1.394	bbl/STBO
		Undersaturated	C_o	1.63339E-05	psi^{-1}
			B_o	1.349	bbl/STBO
2	Vasquez and Beggs	API > 30	γ_{gc}	0.802	
			B_o	1.377	bbl/STBO
3	Glasso	Saturated	y	842.688	
			a	–0.434	
			B_o	1.369	bbl/STBO
4	Al-Marhoun	Saturated	y	154.189	
			B_o	1.355	bbl/STBO
5	Petrosky & Farshad	Saturated	B_o	1.362	bbl/STBO

Table 1.8 Summary of viscosity calculation using different correlations

1	Beal	Dead oil	a	4.656	
			$\mu_{o,d}$	3.707	cp
2	Beggs and Robinson	Dead oil	x	0.808	
			$\mu_{o,d}$	5.424	cp
		Live oil	b	0.546	
			$\mu_{o,L}$	0.836	cp
3	Glasso	Dead oil	a	–15.004	
			$\mu_{o,d}$	3.203	cp

$$b = 5.44(R_s + 150)^{-0.338} = 5.44(750 + 150)^{-0.338} = 0.5458$$

$$\mu_{o,l} = \frac{10.715\mu_{o,d}^{b}}{(R_s + 100)^{0.515}} = \frac{10.715 \times 5.424^{0.5458}}{(750 + 100)^{0.515}} = 0.8359\, cp$$

Example 1.7 The oil rate flowing to a well was predicted as 1000 BOPD at downhole conditions. The downhole pressure and temperature are 1200 psia and 135 °F, respectively. What is the crude oil volume that will be measured at the surface conditions? Given: oil gravity of 35 °API and gas specific gravity of 0.7.

Solution

The specific gravity of oil under downhole conditions:

$$\gamma_o = \frac{141.5}{131.5 + {}^\circ API} = \frac{141.5}{131.5 + 35} = 0.85$$

Using Standing correlation to obtain the solution gas-oil ratio as follows:

$$x = 0.0125API - 0.00091T = 0.0125 \times 35 - 0.00091 \times 135 = 0.315$$

$$R_s = \gamma_g \left[\left(\frac{P}{18.2} + 1.4 \right) 10^x \right]^{1.2048} = 0.7 \left[\left(\frac{1200}{18.2} + 1.4 \right) 10^{0.315} \right]^{1.2048} = 267 \frac{SCF}{STBO}$$

Formation volume factor using Standing correlation for saturated oils:

$$B_o = 0.972 + 0.000147 \left[R_s \left(\frac{\gamma_g}{\gamma_o} \right)^{0.5} + 1.25T \right]^{1.175}$$

$$= 0.972 + 0.000147 \left[267 \left(\frac{0.7}{0.85} \right)^{0.5} + 1.25 \times 135 \right]^{1.175} = 1.145 \frac{bbl}{STBO}$$

The oil rate at surface conditions:

$$q_o@surface = \frac{1000}{1.145} = 873 \frac{STBO}{D}$$

In general, the relationship between the reservoir and surface volumetric flow rate of oil can be expressed as follows:

$$q_{o,sur} = B_o q_{o,res} \tag{1.4.39}$$

The gas flow rate under reservoir conditions is given as:

$$q_{g,res} = B_g (R_p - R_s) q_o \tag{1.4.40}$$

where R_p is the produced gas-oil ratio in SCF/STBO.

1.5 Review of Inflow Performance Relationship (IPR)

1.5.1 IPR for Undersaturated Oil Reservoirs ($P_r > P_b$)

If the reservoir pressure, P_r, is greater than the bubble point pressure, P_b, the formation fluid inside the reservoir can be considered as a single phase. Most of formation fluid is Newtonian fluid and characterized by viscosity. According to Darcy's law, the radial flow in a porous medium of an incompressible Newtonian fluid under steady state and laminar flow conditions can be described as follows (Fig. 1.8):

$$q = -\frac{kA}{\mu}\left(\frac{\partial P}{\partial r} - \rho g\right) \tag{1.5.1}$$

If the reservoir is thin and the gravity is neglected then Darcy's law can be expressed as:

$$q = \frac{2\pi kh}{\mu} r\left(\frac{\partial P}{\partial r}\right) \tag{1.5.2}$$

In oil field unit:

$$qB_o = 7.08 \times 10^{-3} \frac{kh}{\mu} r\left(\frac{\partial P}{\partial r}\right) \tag{1.5.3}$$

$$7.08 \times 10^{-3} \frac{kh}{qB_o\mu} \int_{p(r)}^{p_e} dp = \int_{r}^{r_e} \frac{\partial r}{r}$$

The reservoir pressure distribution can be expressed as:

$$p_e - p(r) = \frac{141.2qB_o\mu}{kh} ln\left(\frac{r_e}{r}\right) \tag{1.5.4}$$

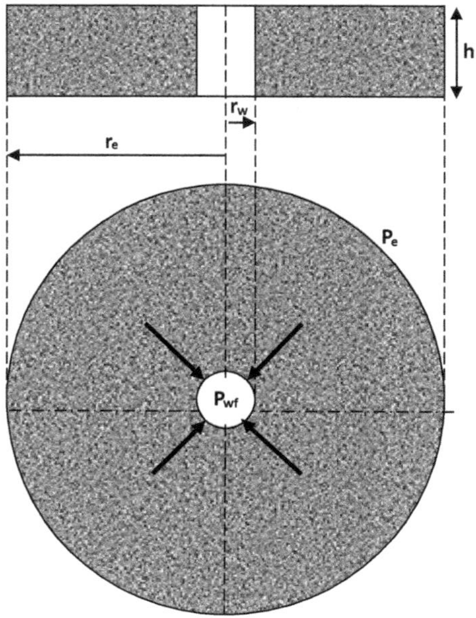

Fig. 1.8 Radial flow in porous media

For a damaged reservoir:

$$p_e - p_{wf} = \frac{141.2qB_o\mu}{kh}\left[ln\frac{r_e}{r_w} + S\right] \tag{1.5.5}$$

$S = \left(\frac{k}{k_s} - 1\right)ln\frac{r_s}{r_w}$ is defined as a skin factor.

where k is the formation permeability in mD; μ is the formation fluid viscosity in cp; B_o is the oil formation volume factor in bbl/STBO; h is the reservoir thickness in ft; r_w and r_e are the wellbore and reservoir radius in ft; P_{wf} and P_e are the flowing bottomhole and reservoir pressure in psi. Equations (1.5.4) or (1.5.5) describes the flow in porous media of an undersaturated reservoir and is called the Inflow Performance Relationship (IPR). IPR reveals that the production rate in the reservoir is proportional to the formation permeability, k, the thickness of the reservoir, h, the pressure drawdown, ΔP, and inversely proportional to the fluid viscosity, μ, and the skin factor, S.

The Productivity Index, PI (STBO/D/psi), is defined as the ratio of the liquid rate to the pressure drawdown. From Eq. (1.5.5) gives:

$$PI = \frac{q}{p_e - p_{wf}} = \frac{kh}{141.2B_o\mu\left[ln\frac{r_e}{r_w} + S\right]} \tag{1.5.6}$$

Using the volumetric weighted pressure definition gives the definition of average reservoir pressure:

$$\bar{p} = \frac{\int_{r_w}^{r_e} pdV}{V} = \frac{\int_{r_w}^{r_e} 2\pi rh\phi pdr}{\int_{r_w}^{r_e} 2\pi rh\phi dr} = \frac{2\int_{r_w}^{r_e} rpdr}{r_e^2 - r_w^2} \tag{1.5.7}$$

Expressing the IPR in terms of average reservoir pressure for steady state flow:

$$\bar{p} - p_{wf} = \frac{141.2qB_o\mu}{kh}\left(ln\frac{r_e}{r_w} - 0.5 + S\right) \tag{1.5.8}$$

Expressing the IPR in terms of average reservoir pressure for pseudo-steady state flow:

$$\bar{p} - p_{wf} = \frac{141.2qB_o\mu}{kh}\left(ln\frac{r_e}{r_w} - 0.75 + S\right) \tag{1.5.9}$$

1.5.2 IPR for Saturated Oil Reservoirs ($P_r < P_b$)

If the reservoir pressure is initially smaller than the bubble point pressure or the flowing bottomhole pressure is set smaller than the bubble pressure, there is an existence of free gas in the reservoir. The flow in the reservoir can now be considered as a two-phase flow. According to Vogel [22], the oil IPR for steady state flow of a two-phase reservoir is expressed as:

$$\frac{q_o}{q_{o,\max}} = 1 - 0.2\frac{P_{wf}}{\overline{P}} - 0.8\left(\frac{P_{wf}}{\overline{P}}\right)^2 \tag{1.5.10}$$

where $q_{o,\max}$ is the absolute open flow potential.

For pseudo-steady state flow:

$$q_o = \frac{k_o h\bar{p}\left[1 - 0.2\frac{P_{wf}}{\overline{P}} - 0.8\left(\frac{P_{wf}}{\overline{P}}\right)^2\right]}{254.2B_o(\bar{p})\mu_o(\bar{p})\left(ln\frac{r_e}{r_w} - 0.75 + S\right)} \tag{1.5.11}$$

If the reservoir pressure is higher than the bubble point pressure but the flowing bottomhole pressure is smaller than the bubble point pressure then the IPR will have two sections: a straight line following Eq. (1.5.6) for $P_{wf} > P_b$ and a curve following Vogel's equation for $P_{wf} < P_b$.

1.5.3 IPR for Gas Reservoirs

Under steady state flow in a gas reservoir and assuming Darcy flow is applicable; the relationship between gas flow rate under standard conditions, q_{sc}, and the flowing bottomhole pressure is expressed as follows:

$$q_{sc} = \frac{k_g h\left(p_e^2 - p_{wf}^2\right)}{1424\bar{\mu}\bar{z}T\left(ln\frac{r_e}{r_w} + S\right)} \tag{1.5.12}$$

Note that Eq. (1.5.12) is achieved by converting Eq. (1.5.6) from STB/D to MSCF/D and using average value of the gas formation volume factor as follows:

$$\overline{B}_g = \frac{0.0283\bar{z}T}{\frac{\left(p_e + p_{wf}\right)}{2}} \tag{1.5.13}$$

Under pseudo-steady state flow, the gas IPR is presented as:

$$q_{sc} = \frac{k_g h\left(\bar{p}^2 - p_{wf}^2\right)}{1424\bar{\mu}\bar{z}T\left(ln\frac{r_e}{r_w} - 0.75 + S\right)} \tag{1.5.14}$$

where q_{sc} is the gas rate in MSCF/D, k_g is the gas permeability in md, and $\bar{p}$ is the average reservoir pressure in psi.

If Darcy flow is not applicable in the reservoir, the Aronofsky and Jenkins model [4] can be used as an IPR for a stabilized gas pseudo-steady state flow in the reservoir.

$$\bar{p}^2 - p_{wf}^2 = \frac{1424\bar{\mu}_g\bar{z}T}{k_g h}\left(ln\frac{r_e}{r_w} - 0.75 + S\right)q_{sc} + \frac{1424\bar{\mu}_g\bar{z}TD}{k_g h}q_{sc}^2 \tag{1.5.15}$$

For steady state flow:

$$p_e^2 - p_{wf}^2 = \frac{1424\bar{\mu}_g\bar{z}T}{k_g h}\left(ln\frac{r_e}{r_w} + S\right)q_{sc} + \frac{1424\bar{\mu}_g\bar{z}TD}{k_g h}q_{sc}^2 \tag{1.5.16}$$

The first term on the right hand side of Eq. (1.5.15) is the IPR when Darcy law is applicable as shown in Eq. (1.5.14). The second term in Eq. (1.5.15) takes care of the non-Darcy effects. Equation (1.5.15) can be simplified as:

$$\bar{p}^2 - p_{wf}^2 = aq_{sc} + bq_{sc}^2 \tag{1.5.17}$$

For steady state flow:

$$p_e^2 - p_{wf}^2 = aq_{sc} + bq_{sc}^2 \tag{1.5.18}$$

where D is the non-Darcy coefficient in the unit of D/MSCF and given as:

$$D = \frac{6 \times 10^{-5}\gamma_g k_s^{-0.1} h}{\mu_g r_w h_{perf}^2} \tag{1.5.19}$$

where k_s is the near-wellbore permeability in mD, h and h_{perf} are the net and perforated thickness in ft.

1.6 Review of Outflow Performance Relationship (OPR)

1.6.1 Single Phase—Incompressible Fluid OPR

Outflow Performance Relationship (OPR) describes the relationship between following bottomhole pressure, P_{wf}, and flowrate, q, in a production tubing. Neglecting the pressure losses due to acceleration, the flowing bottomhole pressure can be calculated using the below equation:

$$p_{wf} = p_{wh} + \rho_o gh + \Delta p_f \tag{1.6.1}$$

where p_{wh} is the wellhead pressure, Δp_f is the frictional pressure drop in the production tubing, and h is the true vertical well depth. Under steady state conditions, the frictional pressure losses in the unit of Pa in a production tubing can be expressed in SI unit as follows:

$$\Delta p_f = \frac{2 f_F \rho_o u^2}{D} L \tag{1.6.2}$$

Where ΔP_f is the frictional pressure drop in the production tubing in Pa, f_F is the Fanning friction factor, u is the liquid velocity in the tubing in m/s, D is the inner diameter of the production tubing in meter, and L is the length of the tubing in meter.

For oil field unit:

$$\Delta p_f = \frac{f_F \rho_o u^2}{25.8D} L \tag{1.6.2a}$$

where ΔP_f is the frictional pressure drop in the production tubing in psi, f_F is the Fanning friction factor, u is the liquid velocity in the tubing in ft/s, D is the inner diameter of the production tubing in inch, and L is the length of the tubing in ft.

Under laminar flow regime, the Fanning friction factor is calculated as:

$$f_F = \frac{16}{Re} \tag{1.6.3}$$

For Newtonian fluids, the Reynolds number in SI unit is given as:

$$Re = \frac{\rho_o u D}{\mu_o} \tag{1.6.4}$$

where ρ_o and μ_o are the density and viscosity of oil in kg/m^3 and Pas, respectively.

In oil field unit:

$$Re = \frac{928\rho_o uD}{\mu_o} \tag{1.6.4a}$$

where ρ_o and μ_o are the density and viscosity of oil in ppg and cp, respectively. The inner tubing diameter, D, is in inch and the liquid velocity in the tubing, u, is in ft/s.

Under turbulent flow regime, the Fanning friction factor can be calculated using colebrook correlation [10]:

$$\frac{1}{\sqrt{f_F}} = -2\log_{10}\left(\frac{\varepsilon/D}{3.7} + \frac{2.51}{Re\sqrt{f_F}}\right) \tag{1.6.5}$$

where ε is the absolute pipe roughness which has the unit the same as that of the pipe diameter D. For smooth pipe and Reynolds number is smaller than 100000, the Fanning friction factor can be calculated using Blasius's equation [8]:

$$f_F = \frac{0.0791}{Re^{0.25}} \tag{1.6.6}$$

1.6.2 Single Phase—Compressible Fluid OPR

The flowing bottomhole pressure of a compressible fluid flow in a production tubing is a summation of the gas hydrostatic pressure and the frictional pressure losses. The hydrostatic pressure of the gas column can be calculated using Eq. (1.3.22) in Field Unit:

$$P_2 = P_{wh}e^{\left(\frac{0.01877\gamma_g h}{zT_{ave}}\right)} \tag{1.6.7}$$

where P_{wh} is the well head pressure and γ_g is the gas specific gravity.

The frictional pressure loss gradient in SI unit system (Pa/m) for gas in pipe flow is expressed as:

$$\Delta p_f = \frac{2f_F\rho_g u_g^2}{D}\Delta L \tag{1.6.8}$$

where gas density, ρ_g in kg/m^3, is a function of pressure and temperature and can be calculated using Eq. (1.4.33). u_g is the gas velocity in m/s and D is the inner diameter of the tubing in m. f_F is the Fanning friction factor obtained by using Eq. (1.6.5). Note that Eq. (1.6.2) is applicable for one fluid segment inside the tubing at which the temperature and pressure are assumed to be constant. Iteration process is needed to achieve the total pressure drop due to gravity and friction in the tubing. The flowing bottomhole pressure is the summation of pressures calculated using Eqs. (1.6.7) and (1.6.8).

1.6.3 Two-Phase Mixture OPR

Since the pressure, temperature, and hence velocity of the two-phase gas-liquid mixture change along the tubing, one should calculate the pressure drop gradient for a particular location in the tubing. Therefore, the total pressure drop in the tubing is normally presented as pressure traverse curves for specific operating conditions. Generally speaking, the total pressure drop gradient of an incompressible fluid flowing in pipes is calculated as follows:

$$\left.\frac{dp}{dl}\right|_T = \left.\frac{dp}{dl}\right|_g + \left.\frac{dp}{dl}\right|_f + \left.\frac{dp}{dl}\right|_a \tag{1.6.9}$$

where the subscripts T, g, f, and a are for total, gravity, friction, and acceleration.

The simplest model to predict the pressure drop of a two-phase flow in a pipe is the homogeneous no-slip model. This model neglects the flow patterns in conjunction with the following assumptions: the flow is under steady state and one direction; two phases are well mixed, and there is no slippage occurs between the phases.

The pressure drop gradient due to gravity is given as:

$$\left.\frac{dp}{dl}\right|_g = \rho_m g sin\theta \tag{1.6.10}$$

$$\rho_m = H_L\rho_L + (1 - H_L)\rho_g \tag{1.6.11}$$

where H_L is the liquid hold up and ρ_m is the mixture density.

The frictional pressure drop gradient is calculated as follows:

$$\left.\frac{dp}{dl}\right|_f = \frac{2f_F\rho_m u_m^2}{D} \tag{1.6.12}$$

Where u_m is the mixture velocity calculated as:

$$u_m = \frac{q_m}{A_p} = \frac{q_L}{A_p} + \frac{q_g}{A_p} = u_{SL} + u_{SG} \tag{1.6.13}$$

where A_p is the cross-sectional area of the tubing; q_L, q_g, and q_m are the flow rate of liquid, gas and mixture, respectively; u_{SL} and u_{SG} are the superficial velocities of liquid and gas, respectively.

The Fanning friction factor is calculated depending on the flow pattern inside the tubing. If the flow is laminar then Fanning friction factor is calculated using Eq. (1.6.3). If the flow is turbulent then the Fanning friction factor can be calculated using Eq. (1.6.5). The two-phase mixture Reynolds number is given as:

$$Re = \frac{\rho_m u_m D}{\mu_m} \tag{1.6.14}$$

The mixture viscosity is obtained as follows:

$$\mu_m = H_L \mu_L + (1 - H_L)\mu_g \tag{1.6.15}$$

In most cases, the pressure drop due to the acceleration is smaller in comparison to that due to the gravity and the friction. Therefore, the pressure drop gradient due to the acceleration can be neglected when calculating the total pressure drop gradient.

When considering two-phase flow regimes such as stratified flow, slug flow, annular flow, and dispersed bubble flow, there are many different two-phase gas-liquid correlations can be used to predict the total pressure drop gradient. The common two-phase correlations are Hagedorn and Brown, Duns and Ros, Beggs and Brill, Orkiszewski, Ansari. For detail of each model, readers are referred to the original works of each correlation. Figure 1.9 shows an example of the pressure traverse curves obtained using Beggs and Brill correlation [9].

1.6.4 Examples

Example 1.8 A well with a radius of 6 in. is producing at a bottomhole pressure of 3000 psi and the solution gas-oil ratio of 450 SCF/STBO. The average reservoir pressure is 5500 psi; the reservoir permeability is 20 mD; the oil volume factor is 1.2 bbl/STBO; the reservoir thickness is 30 ft; and the reservoir drainage radius is 1000 ft. Under reservoir conditions, the oil gravity is 35 °API; oil temperature is 165 °F, gas specific gravity is 0.75, and the bubble point pressure is 2500 psi.

a. What is the current oil production rate if the well is producing under steady state and pseudo-steady state conditions?
b. What is the current oil production rate under pseudo-steady state condition if the skin factor is 10?
c. What is the oil production rate if the bottomhole pressure is reduced to 2000 psi?

Solution

a. The well is producing at the flowing bottom hole pressure of 3000 psi which is greater than the bubble point pressure of 2500 psi. Therefore, single-phase flow will exist in the reservoir. Using Beggs and Robinson correlation for calculating the oil viscosity gives a value of 2.6 cp (please review the example 1.6 for the detail calculation).

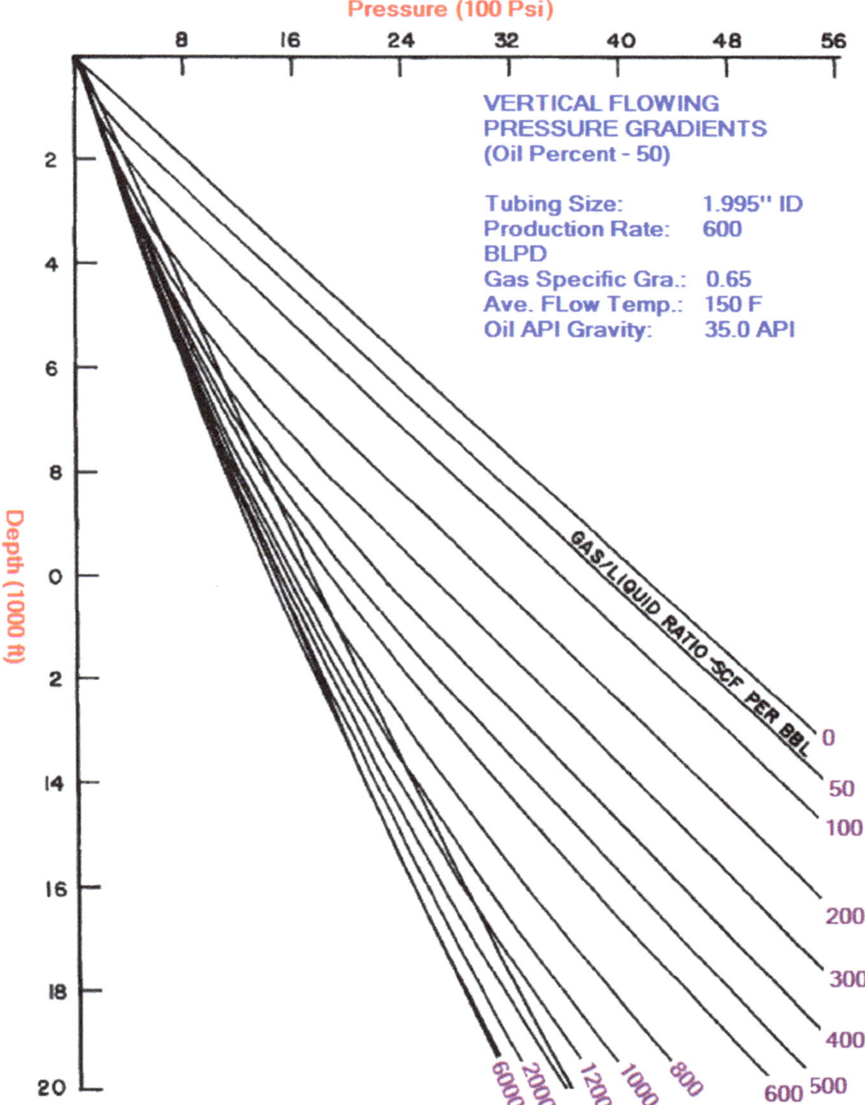

Fig. 1.9 Pressure traverse curves at different GLR using Beggs and Brill correlation

Expressing Eq. (1.5.8) in terms of average reservoir pressure for steady state flow and no formation damage gives the oil production rate as follows:

$$q_o = \frac{kh(\bar{p} - p_{wf})}{141.2\mu B_o\left(\ln\frac{r_e}{r_w} - 0.5\right)}$$

$$q_o = \frac{15 \times 50(5500 - 3000)}{141.2 \times 4 \times 1.2(\ln\frac{1000}{0.5} - 0.5)} = 596.5\frac{STBO}{D}$$

Under pseudo-steady state flow:

$$q_o = \frac{kh(\bar{p} - p_{wf})}{141.2\mu B_o\left(\ln\frac{r_e}{r_w} - 0.75\right)} = 618.3\frac{STBO}{D}$$

b. If the skin factor S = 10, the oil production rate is:

$$q_o = \frac{kh(\bar{p} - p_{wf})}{141.2\mu B_o\left(\ln\frac{r_e}{r_w} - 0.75 + S\right)} = 215.4\frac{STBO}{D}$$

c. If the bottomhole pressure is reduced to 2000 psi, two-phase flow in the reservoir exists. Using the Standing correlation for calculating the formation volume factor gives:

$$B_o = 0.972 + 0.000147\left[R_s\left(\frac{\gamma_g}{\gamma_o}\right)^{0.5} + 1.25T\right]^{1.175}$$

$$B_o = 0.972 + 0.000147\left[450\left(\frac{0.75}{0.85}\right)^{0.5} + 1.25 \times 165\right]^{1.175} = 1.257\frac{bbl}{STBO}$$

Applying Beggs and Robinson correlation for calculating the live oil viscosity gives a value of 0.758 cp.

For pseudo-steady state flow, the IPR for saturated oil is given in Eq. (1.5.11):

$$q_o = \frac{k_o h\bar{p}\left[1 - 0.2\frac{P_{wf}}{\bar{P}} - 0.8\left(\frac{P_{wf}}{\bar{P}}\right)^2\right]}{254.2B_o(\bar{p})\mu_o(\bar{p})\left(\ln\frac{r_e}{r_w} - 0.75 + S\right)}$$

$$q_o = \frac{15 \times 50 \times 5500\left[1 - 0.2\frac{2000}{5500} - 0.8\left(\frac{2000}{5500}\right)^2\right]}{254.2 \times 1.257 \times 0.758 \times (\ln\frac{1000}{0.5} - 0.75 + 10)} = 829.7\frac{STBO}{D}$$

Example 1.9 Given data for a reservoir:

Drainage area, A	120 acres	Reservoir temperature, T	150 °F = 640 °R
Reservoir pressure, P_e	4500 psi	Gas viscosity, μ_g	0.02 cp
Wellbore radius, r_w	0.328 ft	Compressibility factor, Z	0.95
Reservoir thickness, h	75 ft	Skin factor, S	1
Permeability, k	0.15 mD	Non-Darcy coefficient	Varied

Develop the gas reservoir IPR for two cases: Darcy Flow and non-Darcy flow.

Solution

a. The IPR for gas reservoir under steady state and Darcy flow is given:
The reservoir drainage radius:

$$r_e = \sqrt{\frac{A}{\pi}} = \sqrt{\frac{5227200ft^2}{\pi}} = 1290\,ft$$

$$q_{sc} = \frac{k_g h\left(p_e^2 - p_{wf}^2\right)}{1424\bar{\mu}\bar{z}T\left(\ln\frac{r_e}{r_w} + S\right)}$$

$$q_{sc} = \frac{0.15 \times 75\left(p_e^2 - p_{wf}^2\right)}{1424 \times 0.02 \times 0.95 \times 640\left(\ln\frac{1290}{0.328} + 1\right)} = 7.3476 \times 10^{-5}\left(p_e^2 - p_{wf}^2\right)$$

b. The IPR for gas reservoir under steady state and non-Darcy flow:
For steady state flow:

$$p_e^2 - p_{wf}^2 = \frac{1424\bar{\mu}_g\bar{z}T}{k_g h}\left(\ln\frac{r_e}{r_w} + S\right)q_{sc} + \frac{1424\bar{\mu}_g\bar{z}TD}{k_g h}q_{sc}^2$$

Or

$$p_e^2 - p_{wf}^2 = aq_{sc} + bq_{sc}^2$$

$$a = \frac{1424\bar{\mu}_g\bar{z}T}{k_g h}\left(\ln\frac{r_e}{r_w} + S\right) = 13609.8$$

$$b = \frac{1424\bar{\mu}_g\bar{z}TD}{k_g h} = 1467D$$

$$p_e^2 - p_{wf}^2 = 13609.8q_{sc} + 1467Dq_{sc}^2$$

The results of this example are shown in Fig. 1.10 at different non-Darcy coefficient values of 0.0005, 0.005, and 0.05 D/MSCF. At the flowing bottomhole pressure, Pwf = 0 psi, the maximum gas rate is about 1500 MSCF/D. As the non-Darcy coefficient increases, the production rate decreases because of higher flow restriction. Aronofsky and Jenkins model also predicts that depending on the value of the non-Darcy coefficient, there is a minimum value

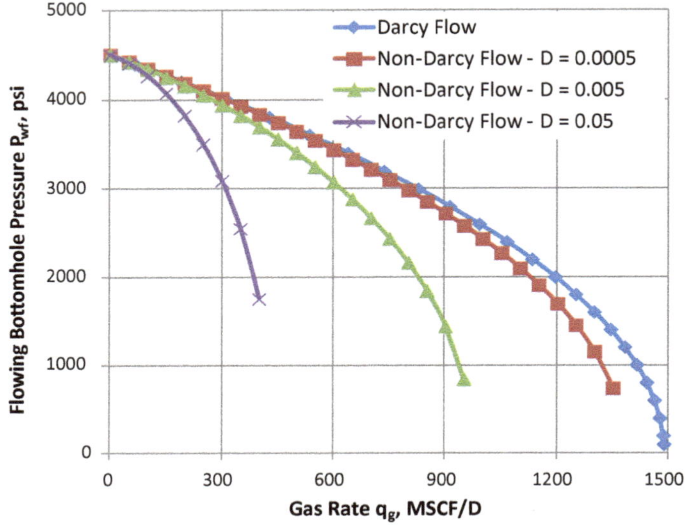

Fig. 1.10 Effect of non-Darcy coefficient on the gas reservoir IPR

of the flowing bottomhole pressure at which the maximum production rate will be achieved. Lower than this minimum value, the reservoir is not able to produce. For example, if D = 0.005 D/MSCF, the minimum value of P_{wf} is 837 psi and the maximum gas rate is 950 MSCF/D.

References

1. Al-Marhoun MA (1992) New correlations for formation volume factors of oil and gas mixtures. J Can Pet Technol 31(3):22. PETSOC-92-03-02
2. Al-Shammasi AA (2001) A review of bubblepoint pressure and oil formation volume factor correlations. SPE Res Eval Eng 4(2):146–160. SPE-71302-PA
3. API Manual of Petroleum Measurement Standards (2004) Chapter 11—Physical properties data and other volume correction factor standards for hydrocarbons
4. Aronofsky J, Jenkins R (1954) A simplified analysis of unsteady radial gas flow. J Pet Technol 6. SPE 271 G
5. Beal C (1970) The viscosity of air, water, natural gas, crude oil and its associated gases at oil field temperatures and pressures, vol 3. Reprint series (Oil and Gas Property Evaluation and Reserve Estimates), SPE, Richardson, Texas, pp 114–127
6. Beggs HD, Robinson JR (1975) Estimating the viscosity of crude oil systems. J Pet Technol 27(9):1140–1141. SPE-5434-PA
7. Bird R, Stewart W, Lightfoot E (2007) Transport pheonomena–Second Edition. John Wiley & Son Publisher. ISBN 0-470-11539-4
8. Blasius PRH (1913) Das Aehnlichkeitsgesetz bei Reibungsvorgangen in Flüssigkeiten. Forschungsheft 131:1–41

9. Brown K (1980) The technology of artificial lift methods, vol 2a. PennWell Publishing Company, Tulsa
10. Colebrook CF, White CM. (1937) Experiments with fluid friction in roughened pipes. Proc R Soc London Ser A Math Phys Sci 161(906):367–381
11. Frashad F, LeBlanc JL, Garber JD et al (1996). Empirical PVT correlations for colombian crude oils. In: Presented at the SPE Latin American and Caribbean Petroleum Engineering Conference, Port of Spain, Trinidad and Tobago, 23–26 April. SPE-36105-MS
12. Glasø Ø (1980) Generalized pressure-volume-temperature correlations. J Pet Technol 32 (5):785–795. SPE-8016-PA
13. Kanu A, Ikiensikimama S (2014) Globalization of black oil PVT correlations. In: Presented at the SPE Nigeria annual international conference and exhibition held in Lagos, Nigerio, August 2014. SPE 172494-MS
14. Kartoatmodjo TRS, Schmidt Z (1991) New correlations for crude oil physical properties. Society of Petroleum Engineers, unsolicited paper 23556-MS
15. Lasater JA (1958) Bubble point pressure correlations. J Pet Technol 10(5):65–67. SPE-957-G
16. Petrosky GE, Farshad F (1998) Pressure-volume-temperature correlations for Gulf of Mexico crude oils. SPE Reservoir Evaluation & Engineering—51395-PA
17. Standing MB (1947) A pressure-volume-temperature correlation for mixtures of California oils and gases. API Drill Prod Pract 1947:275–287
18. Standing MB (1981) Volumetric and phase behavior of oil field hydrocarbon systems, 9th edn. Society of Petroleum Engineers of AIME, Richardson
19. Sutton RP, Farshad F (1990) Evaluation of empirically derived PVT properties for Gulf of Mexico crude oils. SPE Res Eng 5(1):79–86. SPE-13172-PA
20. Vazquez ME (1976) Correlations for fluid physical property prediction. MS thesis, University of Tulsa, Tulsa
21. Vazquez M, Beggs HD (1980) Correlations for fluid physical property prediction. J Pet Technol 32(6):968–970. SPE-6719-PA
22. Vogel JV (1968) Inflow performance relationships for solution-gas drive wells. JPT 1968:83–92

Chapter 2
Gas Lift

2.1 Fundamentals of Gas Lift System

2.1.1 Continuous Gas Lift

Continuous gas lift is a method at which gas is injected continuously into the lower part of the production tubing to improve the well flow potential. As gas mixes with the formation fluids, the fluid hydrostatic pressure and the frictional pressure drop inside the production tubing decrease leading to a reduction in the flowing bottomhole pressure and hence increasing the production.

Figure 2.1 shows a simple continuous gas lift system setup. Gas-oil mixture production is collected at the production manifold. This gas-oil mixture is separated at the separator: the liquid oil is transported and stored in the storage tank; the free gas is transported to the gas compressor at which the free gas is compressed to achieve higher pressure. This high pressure gas is injected into the well annulus and forced through the gas lift valve. The expansion of this high pressure gas provides an addition work for lifting the liquid column and aerates the liquid phase in the production tubing to reduce the liquid density (reduce the back pressure). The combination of these two actions causes a reduction in the flowing bottomhole pressure and hence improving production.

Gas lift can be used effectively to accomplish the following tasks:

- Produce wells that do not have natural flow;
- Increase production for producing wells;
- Unload wells that later flows naturally;
- Liquid loading for gas wells.

Figure 2.2 illustrates the main principle of how a continuous gas lift system works. In this illustration, the upper part of the plot shows the relationship between pressure and depth; the lower part shows the relationship between flowrate and bottomhole pressure. The curve in the lower part of Fig. 2.2 is the IPR. The well is

T. Nguyen, *Artificial Lift Methods*, Petroleum Engineering,
https://doi.org/10.1007/978-3-030-40720-9_2

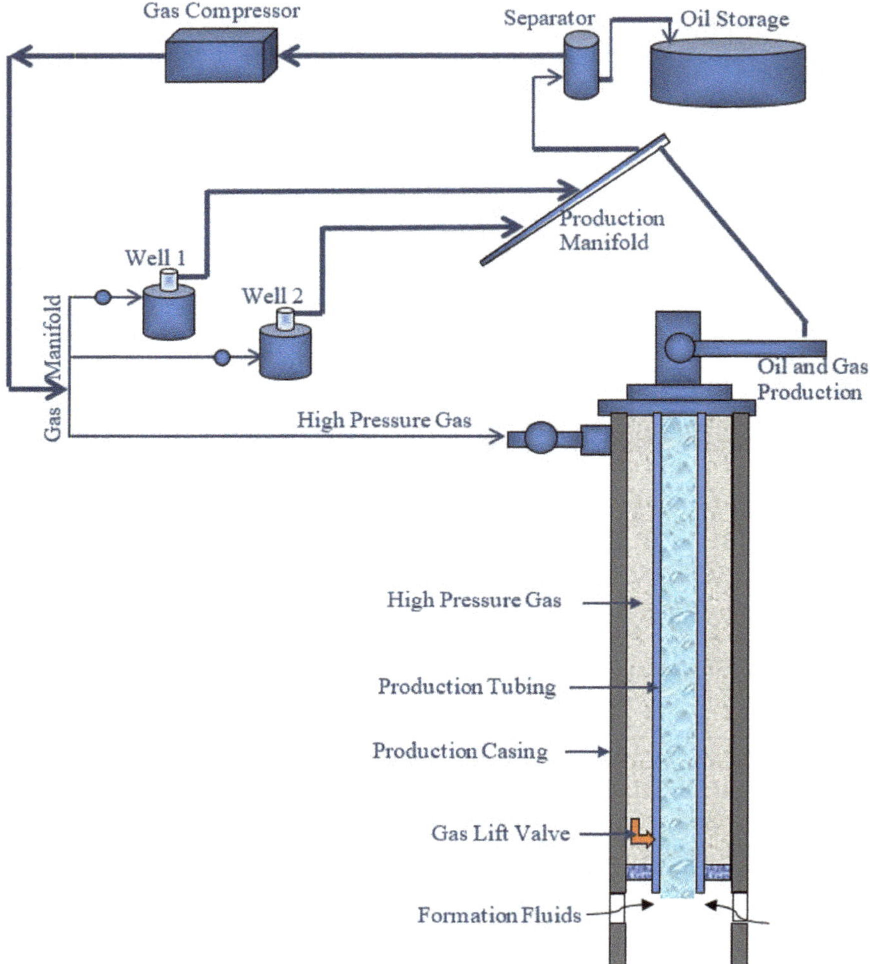

Fig. 2.1 Continuous gas lift system

being produced at a flow rate of Q_1 at the flowing bottomhole pressure P_{wf1}. The initial formation fluid gradient is represented as the solid dark blue line. As gas is injected into the production tubing at the valve depth, there are two different liquid columns flowing inside the production tubing: formation fluid from the perforation to the valve depth and injected gas and liquid mixture from the valve depth to surface. The pressure gradient of the formation fluid is the same as that of the initial reservoir fluids as shown in Fig. 2.2; these two lines are parallel. When high-pressure gas aerates the formation liquid, the hydrostatic pressure and frictional pressure drop inside the tubing reduce causing a big reduction in the flowing

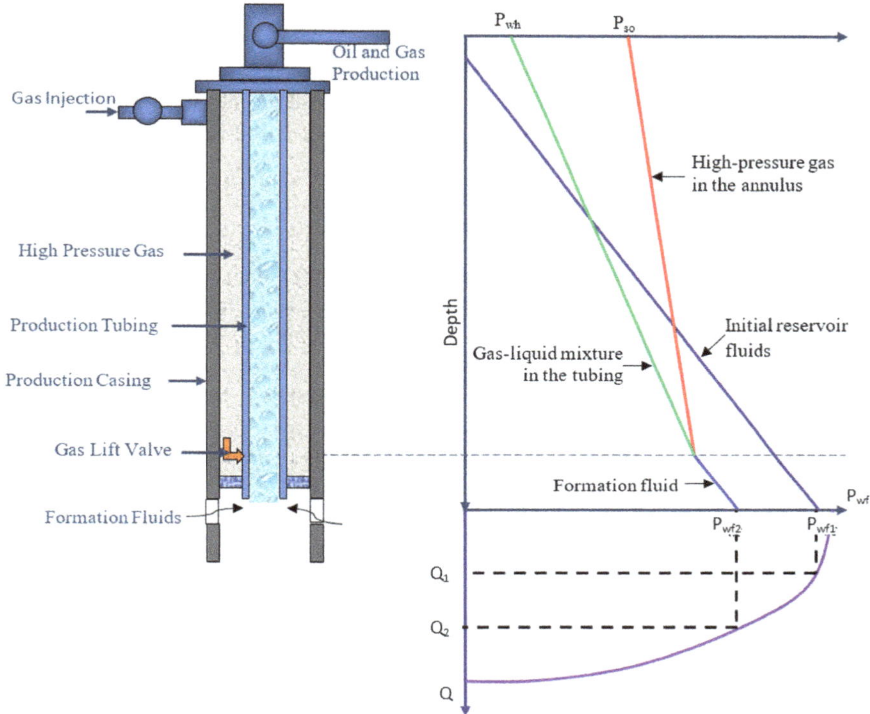

Fig. 2.2 Principle of continuous gas lift system

bottomhole pressure from P_{wf1} to P_{wf2}. This lower flowing bottomhole pressure allows more formation fluid to enter the well. The well is now being produced at a higher rate of Q_2 and lower bottomhole pressure of P_{wf2} when gas is injected.

Advantages of Continuous Gas Lift

The advantages of continuous gas lift can be summarized as follows:

- Gas lift design and installation are considered to be one of the most forgiving forms of artificial lift methods. As long as gas is injected into the production tubing, the well will normally produce some liquid. The only concern is the efficiency of the gas energy used to lift a unit volume of liquid.
- Gas lift is an excellent method for wells with high sand production and high formation gas-liquid ratio. Sand production may cause severe damages for downhole pumps. In addition, high formation gas-liquid ratio may significantly reduce the efficiency of downhole pumps due to gas bubbles, gas pockets, and gas lock.
- Gas lift system is relatively simple with few moving parts. The only downhole moving part is the valve stem, which can be made and controlled very reliably from the surface.

- Using wireline, gas lift valves can be replaced without the need of killing the well or pulling the tubing.
- Subsurface gas lift components are relatively inexpensive and the surface gas injection control equipment is simple and required very minimum space for installation. Therefore, gas lift is a good candidate for offshore wells, which have a very limited surface footprint.

Disadvantages of Continuous Gas Lift

- Gas lift system strongly depends on the gas source which is from the formation gas and the outside supplied source. If formation gas is limited then continuous gas lift method relies mainly on the outside supplied source, which can be expensive and not reliable.
- Space for installing gas compressors can be a problem for offshore platforms.
- If the spacing between wells is wide and the number of wells is low, gas lift application can be limited. In other words, gas lift system should be applied for small spaced wells as well as high number of wells.
- Gas lift is not recommended for heavy oil wells.
- Gas lift is basically a low energy efficiency method when compared to other lift methods.

2.1.2 Intermittent Gas Lift

If gas is injected into the production tubing periodically to displace a liquid slug, this method is called intermittent gas lift. Intermittent gas lift does not work based on lowering the fluid density to have production. It uses high-pressure gas expansion to displace liquid slugs to the surface. As each liquid slug is produced, gas injection is interrupted to allow the fluid volume to build up again.

Intermittent gas lift is generally used for wells which have a low productivity index or low reservoir pressure [8]. In other words, because of the low reservoir pressure or low productivity index, the flow potential in the reservoir is very low and hence continuous gas lift is not available (Fig. 2.3).

There are three main stages during a complete gas lift cycle including: liquid accumulation, lift, and after flow as shown in Figs. 2.4 and 2.5.

- **Liquid accumulation**: as shown in Fig. 2.4a, formation fluid flows through the standing valve into the production tubing and accumulate above the gas lift valve. The wellhead valve and the flowline remain open to maximize the flow into the well. The tubing pressure may be higher than the atmospheric pressure because of the separator pressure. The surface controller and the operating gas lift valve are in close positions. The surface pressure in the annulus (casing pressure) increases slightly because of the gas compression in the annulus as liquid is being invading into the wellbore. When the more liquid accumulates

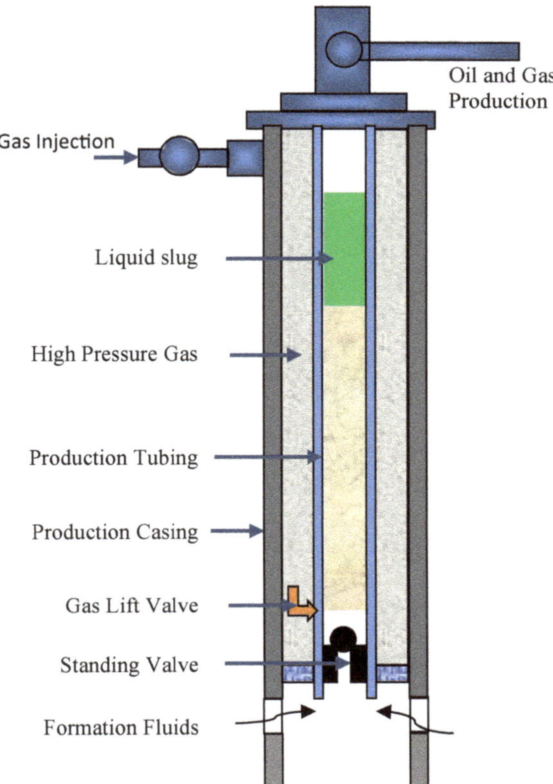

Fig. 2.3 Intermittent gas lift system

inside the production tubing, the hydrostatic pressure of the accumulated fluid is high enough to cause the standing valve to close resulting a constant casing pressure. At this point, the surface controller is open to allow high-pressure gas entering the annulus. The casing pressure increases rapidly until the gas lift operating valve opens.

- **Lift**: Fig. 2.4b shows the second stage of an intermittent gas lift cycle. As gas enters the production tubing and lifts the liquid slug upward, the casing pressure decreases and tubing pressure increases slightly. As soon as enough gas has been injected to remove the formation fluids, the surface controller closes leading to a sudden decrease in the casing pressure. This reduction in the casing pressure will cause the gas lift operating valve to close. The casing pressure will remain fairly constant after this point.
- **After flow**: as the liquid exits the well and flows into the flowline, the pressure of the gas slug dramatically dissipates causing a rapid reduction in the tubing pressure. At this time, the operating valve still remains closed and the standing valve is open allowing another liquid accumulation period.

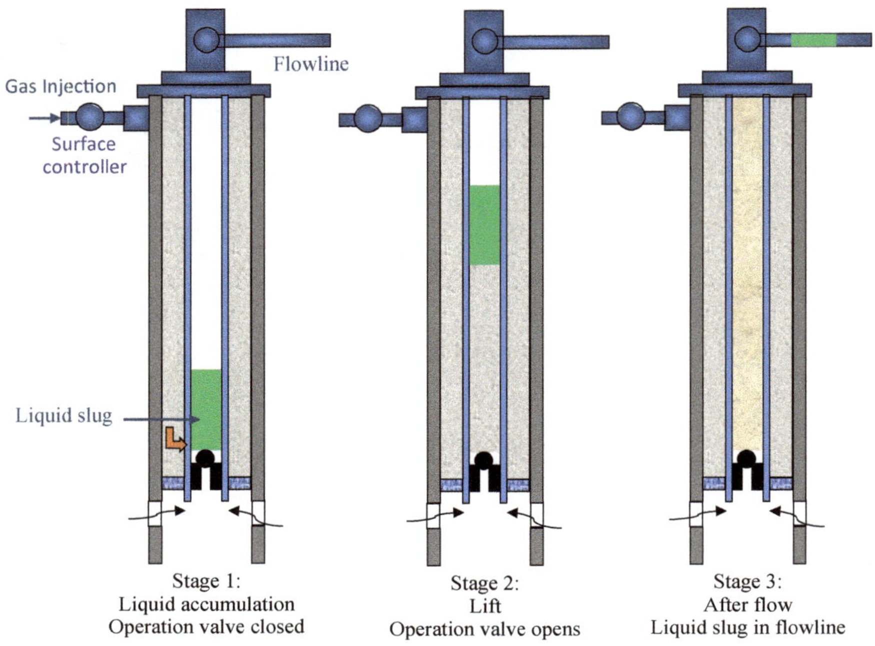

Fig. 2.4 Main stages during a complete cycle of an intermittent gas lift

Fig. 2.5 Pressure response during a cycle of an intermittent gas

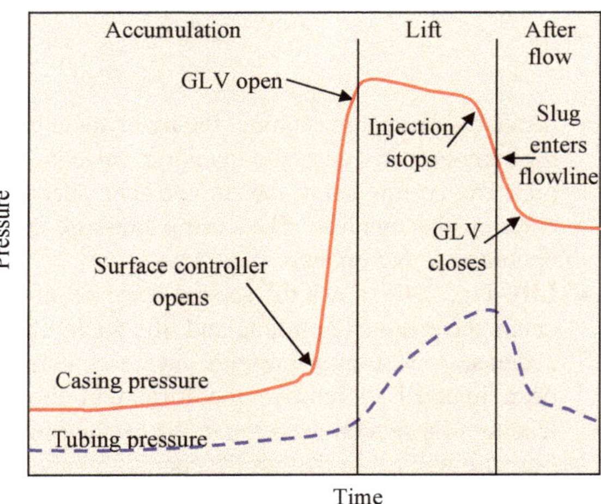

Advantages of Intermittent Gas Lift

- Intermittent gas lift can be applicable for wells that have very low bottomhole pressure.
- Prolong life of wells that are practically abandoned if continuous gas lift is used.

Disadvantages of Intermittent Gas Lift

- This method is only applicable for very low production wells.
- The efficiency of this method is quite low when compared with the continuous gas lift for lifting one unit volume of liquid.
- This method causes more sand production in comparison with the continuous gas lift because of the fluctuation of the production.
- Optimization for an intermittent gas lift system is not a simple task.

2.2 Gas Lift Equipment

Gas lift equipment can be classified as surface and subsurface equipment.

2.2.1 Surface Equipment

The main surface equipment of a gas lift system is the gas injection controller. The surface gas injection controller is designed to control the amount of gas needed to inject into the casing under various conditions. For continuous gas lift system, the gas injection controller is normally a pneumatic control valve, which uses air to control the opening of the valve and hence to control the gas injection rate. For intermittent gas lift system, the gas injection controller is normally a valve used to regulate the on and off injection gas cycle. This device is sometimes called the "intermitted". There is a clock-driven mechanism in this valve that causes a motor valve to open at regular intervals.

2.2.2 Subsurface Equipment

The subsurface equipment of a gas lift system includes mandrels, gas lift valves, valve latches, and kick-over tools.

2.2.2.1 Side Pocket Mandrels

Side pocket mandrel is subsurface equipment in which gas lift valves are installed and operated. The mandrels are normally 1 or 1 ½ in. inside diameter pocket profile. Each mandrel can be manufactured to be tubing flow, annular flow, bypass, and chemical-treating injection systems. Mandrel can be classified as tubing retrievable and wireline retrievable mandrel. In a tubing retrievable mandrel configuration (Fig. 2.6a), gas lift valves are run together with tubing string and tubing string must be pulled to repair or replace a gas lift valve. This configuration is costly and

Fig. 2.6 Gas lift mandrel

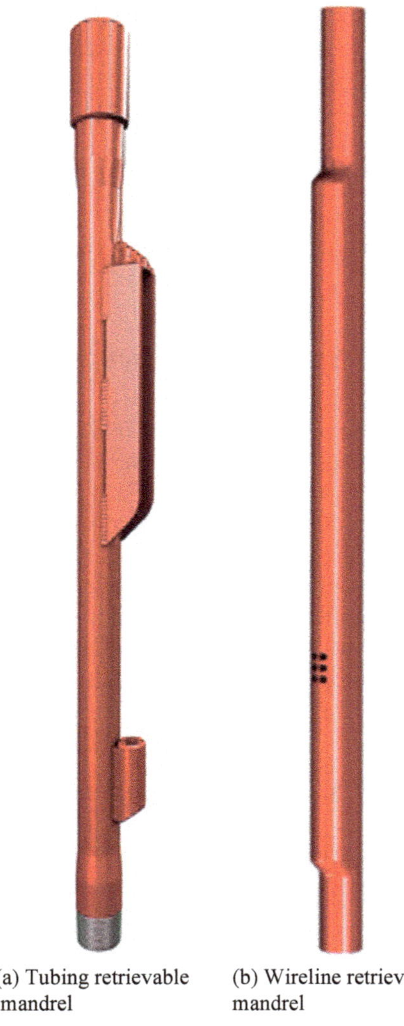

(a) Tubing retrievable mandrel (b) Wireline retrievable mandrel

time-consuming when a wellbore intervention to change a gas lift valve is needed. Another type of mandrel is the wireline retrievable one. This is sometimes called the side pocket mandrel because the side pocket is offset the centerline of the tubing as shown in Fig. 2.6b. This mandrel allows installing or replacing a gas lift valve using normal wireline operations without the need of pulling the tubing for valve repairs or adjustments.

2.2.2.2 Gas Lift Valve

Gas lift valve is a device located inside the gas lift mandrel and provide communication between production tubing and tubing annulus. The main function of a gas lift valve is to control the flow of injected gas from surface into the production tubing. Gas lift valve can be considered as the heart of a gas lift system because the optimization of a gas lift system mainly depends on the number and position of gas lift valves as well as the initial settings of this valve.

A conventional configuration of a gas lift valve installed inside a side-pocket mandrel (wireline retrievable mandrel) is shown in Fig. 2.7. High pressure injected gas is forced into the annulus between the production tubing and the production casing. If the annular gas pressure at the valve depth is high enough, the gas lift valve will open allowing gas to enter the production tubing through the gas lift valve. This injected gas will mix with the fluids inside the production tubing to reduce the hydrostatic pressure and hence reduce the flowing bottomhole pressure. The classification of gas lift valve will be discussed in the Gas Lift Valve section.

2.2.2.3 Valve Latch

Gas lift valve latches are designed to lock or retrieve a gas lift valve in or from the appropriate side pocket mandrel profile. To retrieve the latch and attached valve, the operator jars the toolstring upward to release the shear pin by shearing and disengaging the locking mechanism from the latch plug pocket profile.

Latches normally have o-ring to prevent debris from entering the pocket of the mandrel.

Common latches are available for 1 in. and 1 ½ in OD gas lift valves (Fig. 2.8).

2.2.2.4 Gas Lift Kick-Over Tools

Gas lift kick-over tools are wireline devices used to selectively install and retrieve gas lift valves from side pocket mandrels with an orienting sleeve. Standard wireline techniques can most of the time be used to run kick-over tools into the well. The orienting sleeve guides the kickover tool into perfect alignment above the

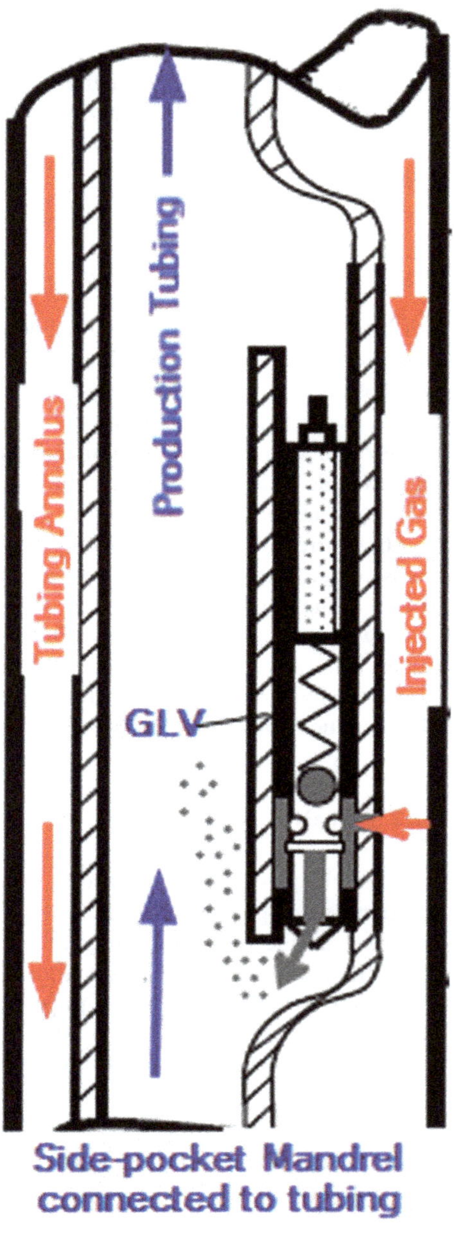

Fig. 2.7 GLV inside a pocket mandrel

pocket of the mandrel and hence enables precise installation or retrieval of gas lift or dummy valves. Gas lift kick-over tools are also used to install or retrieve control devices for chemical injection and water-flooding applications.

Fig. 2.8 Gas lift valve latch

The following kick-over tool running procedure is adopted from Weatherford [9]:

1. Install the pulling tool onto the kickover tool. Make up the kickover tool onto the bottom of the wireline tool string, and install the assembly into the lubricator.
2. Lower the unit into the tubing until the kickover tool is below the selected mandrel, the depth of which is known from well records.
3. Raise the tools slowly through the tubing until they stop, which indicates that the locating finger in the kickover tool has contacted the top of the orienting sleeve of the mandrel.
4. Pull tension on the wireline until the weight indicator of the wireline unit indicates enough weight to actuate the kickover tool to its kicked over position above the pocket in the mandrel.
5. Slowly lower the tools until a weight loss is registered on the weight indicator. Weight loss indicates that the kickover tool has kicked over and located the pocket of the mandrel and the flow control device in it. No weight loss indicates that the kickover tool did not release to the kicked over position; in this case, repeat Steps 3, 4, and 5.
6. Jar downward to secure the pulling tool to the running head of the latch.
7. Jar upward to pull the flow control device from the pocket of the mandrel. If it is not possible to pull the valve, jar downward to release the pulling tool from the running head of the latch. The tool string can now be removed from the well. As

the kickover tool is pulled upward through the mandrel, the locating finger in the kickover tool will stop in the orienting sleeve of the mandrel.

8. Jar or pull upward to release the locating finger and permit the kickover tool to pass through the mandrel (Fig. 2.9).

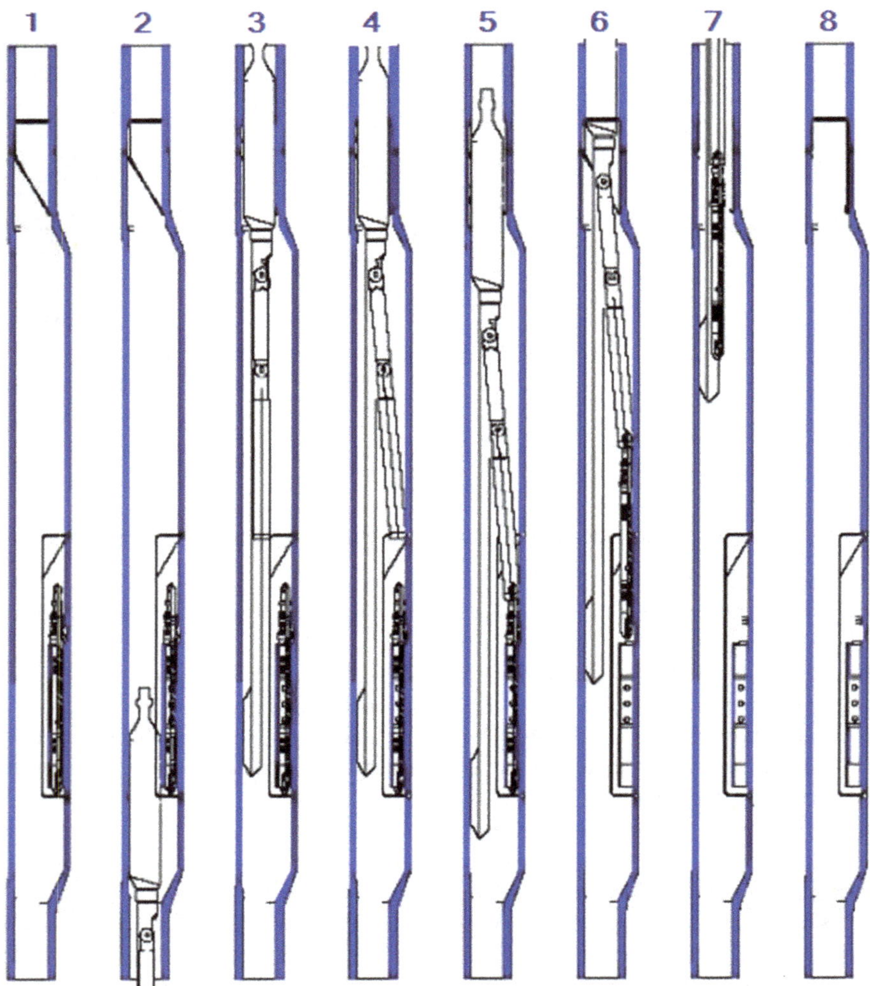

Fig. 2.9 Kick-over tools pulling procedure

2.3 Gas Lift Installation

Generally speaking, a gas lift system should be designed with enough flexibility to minimize the number of workovers required over a well's producing life. Types of downhole gas lift installations for continuous gas lift are usually based on the following factors:

- Types of seal: open installation, semi-closed installation, closed installation, and chamber installation. The last two installations are applicable mainly for intermittent gas lift system.
- Production flow: tubing flow installation and annular flow installation.
- Based on casing diameter and wellbore deviation: dual installation, coiled tubing installation, and macaroni installation.

2.3.1 Open Installation

An open gas lift installation is one in which the tubing string is suspended in the well without a packer. The casing and tubing are freely in communication. This is the oldest type of gas lift installation. It has several major disadvantages:

- If the injected gas pressure is high enough, the high-pressure gas may flow into the formation leading to a very low gas lift efficiency system.
- An additional backpressure will apply on the formation causing a reduction in production rate.
- When production resumes, the formation liquid may flow back through the gas lift valves. This will cause the valves to wear out much faster.
- There is no zonal isolation in this installation system (Fig. 2.10).

2.3.2 Semi-closed Installation

A semi-closed installation has a packer installed in the tubing to seal off the tubing-casing annulus.

This is the most common type of installation for continuous gas lift wells. It eliminates most of the characteristic disadvantages of the open installation.

The packer keeps produced fluids from entering the annulus, and prevents the casing pressure from directly communicating with the formation (Fig. 2.11).

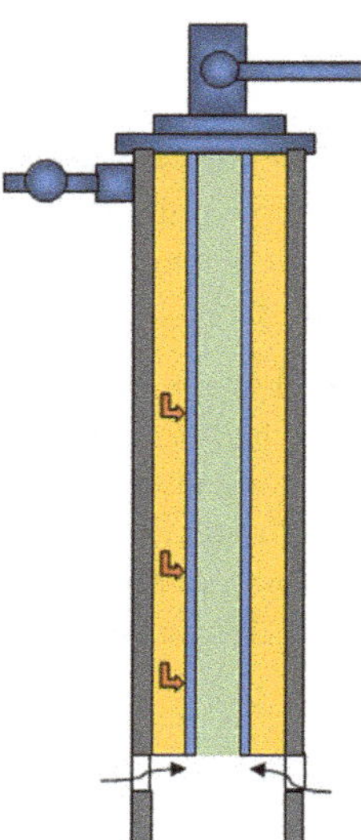

Fig. 2.10 Open installation

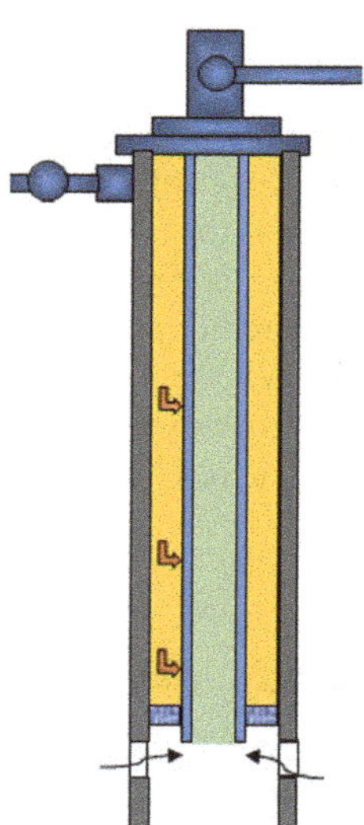

Fig. 2.11 Semi-closed installation

2.3.3 Closed Installation

A closed installation is similar to a semi-closed installation, except that a standing valve is placed in the tubing string below the bottom gas lift valve to prevent fluids from moving downward.

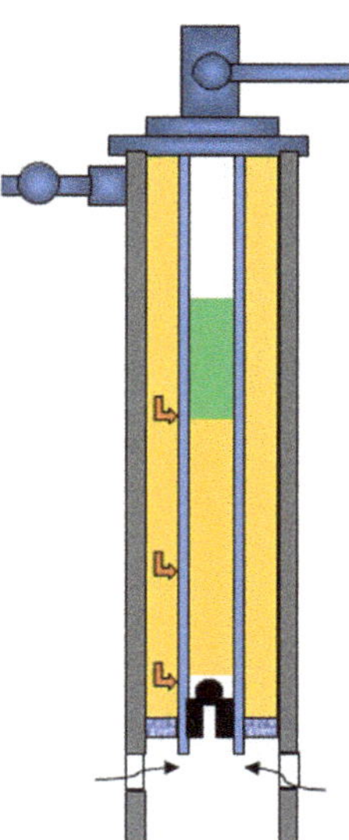

Fig. 2.12 Closed installation

Thus, high-pressure gas injected into the tubing from the annulus cannot increase backpressure on the formation, and any produced fluids standing in the tubing will not flow back into the formation.

These features make the closed installation the option of choice for intermittent gas lift (Fig. 2.12).

2.3.4 Chamber Installation

Chamber installation is mainly used for intermittent gas lift operations at which the reservoir has a low bottomhole pressure and high productivity index. This well may have high fluid volumes if high drawndown is created at the bottom.

The chamber uses the casing volume to store fluids. An inserted chamber can also be run to allow more fluid volume to be produced than the same length of tubing.

The chamber installation consists of lower and upper packers, a perforated nipple, a bleed valve, and an operating chamber valve. The operation of this installation is similar to that of intermittent gas lift and works as follows:

- *Liquid accumulation*: As the chamber fills with fluid, gas in the chamber passes through a bleed valve into the tubing.
- *Lift*: When the chamber is filled, a slug of gas is injected down the annulus to open the operating chamber valve. The high-pressure gas in the chamber forces the fluid to enter the tubing through a perforated nipple above the bottom

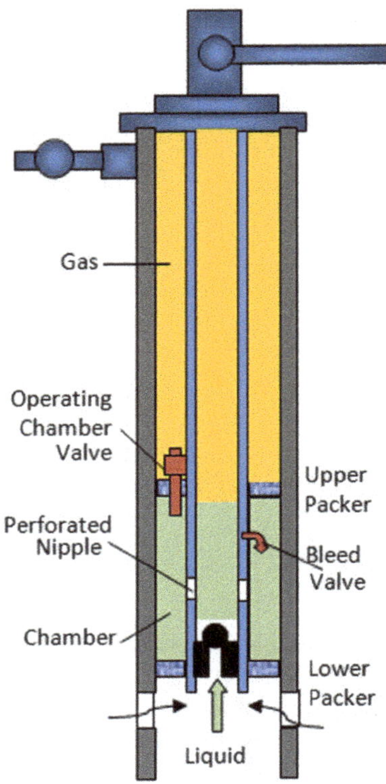

Fig. 2.13 Chamber installation

packer. When all the fluid in the chamber above the nipple is forced into the tubing, gas follows behind the slug and forces it to the surface.

- *After flow*: The liquid exits the well and flows into the flowline. At this time, the operating chamber valve should close and the standing valve is open allowing another liquid accumulation period. In other words, the filling cycle begins again (Fig. 2.13).

2.3.5 Tubing Flow Installation

Tubing flow is the most common type of gas lift installation. High-pressure gas is injected into the annulus and the formation liquid is produced in the production tubing. In most cases, the cross-sectional area of the production tubing is smaller than that of the annulus. Therefore, the multiphase flow in the tubing requires smaller gas liquid ratio than that of in the annulus case.

When corrosion is a problem, producing through the production tubing is much safer in comparison to producing through the annulus. This is because producing

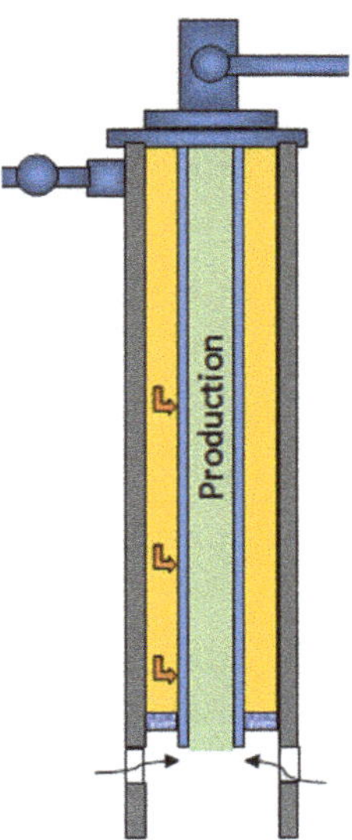

Fig. 2.14 Tubing flow installation

through the tubing can protect the internal casing wall as well as external tubing surface. In addition, workover can be done with production tubing but it is almost impossible with casings (Fig. 2.14).

2.3.6 Annular Flow Installation

If the liquid production rate is very high, the annular flow installation can be considered. In this installation, the high-pressure gas is injected into the production tubing and the formation liquid is produced through the annular space.

It is obvious that the external surface of the tubing and internal wall of the production casing are exposed directly to the formation fluids. When completing these wells, high quality casings must be considered to prevent corrosion. It may be better to spend more money in the beginning than to abandon the well after a short production well's life because of casing's damage.

In addition, problems related to paraffin, hydrate, sand production cannot be easily controlled or removed in case of annular flow installation (Fig. 2.15).

2.3.7 Dual Installation

Dual installation is used to produce two different zones. Gas is injected through the annulus and into the two separate tubing strings. This installation is only applicable

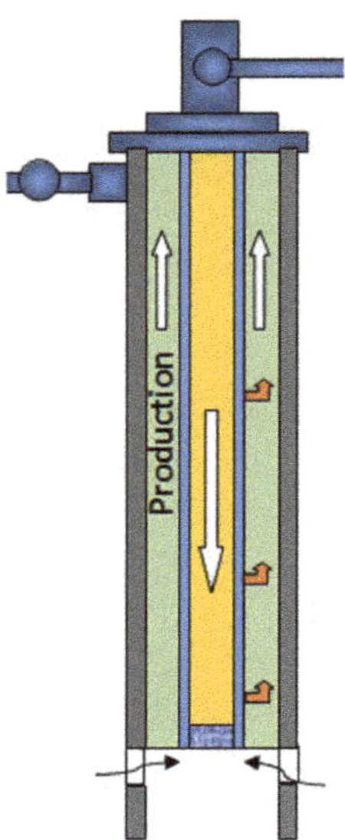

Fig. 2.15 Annular flow installation

when the inner diameter of the production casing must be high enough to install two tubing strings. The ideal installation would be to have gas lift valves on each tubing string that allows production independently on each other. This is usually accomplished by using at least one string with valves that operate independently of the casing injection pressure (PPO valve).

The dual installation can be designed with parallel as shown in Fig. 2.16 or concentric dual tubing strings. The two most common configurations are: (1) parallel strings of 2 3/8 in. OD tubing inside 7 in. casing and (2) parallel strings of 3 1/ 2 in. OD tubing inside 9 5/8 in. casing.

For the dual installation to work, the valves must be spaced to prevent interference between the two zones and selected so that the desired amounts of gas are injected for each zone. Therefore, dual installation is very difficult to design and operate.

2.3.8 Coiled Tubing Installation

Coiled tubing installation is used to convert a flowing well to gas lift system without pulling the main production tubing string. This installation can be done by simply installing conventional gas lift mandrels at the appropriate depths along the coiled tubing string. Then the coiled tubing technique can be applied to run this string inside the production tubing. This method strongly depends on the inner diameter of the production tubing. If the coiled tubing diameter is too small, the production rate will be limited.

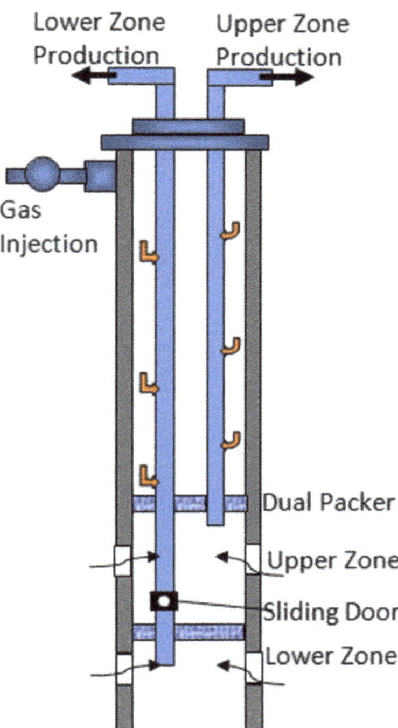

Fig. 2.16 Dual installation

2.3.9 Macaroni Installation

Macaroni Installation is used for ultra-slim holes. Usually this is done by using a 2 3/8″, 2 7/8″ or 3 ½″ tubing as the casing. A smaller tubing (1″–1 1/2″) with gas lift valves installed may be run to produce the well. This smaller tubing is usually known as macaroni tubing. The reduced drilling and completion costs can sometimes make this installation competitive.

Macaroni installation can be used to have two small tubing strings run inside the production casing to produce two different producing zones. These two small tubing strings can be configured as concentric or parallel (Fig. 2.17).

2.4 Gas Lift Valve

Gas lift valve is a subsurface device in a gas lift system installed inside a mandrel to control the high-pressure gas into the production tubing. A gas lift valve consists of five main components:

1. Nitrogen-charged dome: acts on the bellows to hold the valve (stem or ball seat) in the closed position.

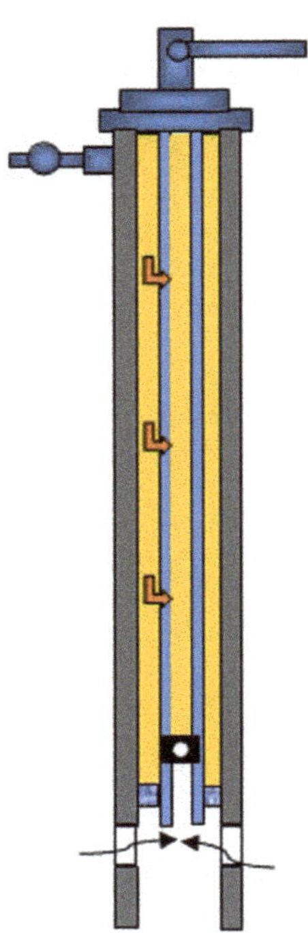

Fig. 2.17 Macaroni installation

2. Bellows: serve as a flexible or responsive element. Movement of the bellows causes the valve stem to move up or down causing the ball to open or close over the port.
3. Valve stem and stem tip: moves up or down to admit gas through the port. In the close position, the stem tip (or ball) seats on the port.
4. Check valve: prevents liquid flow back into the production casing annulus.
5. Port: allow or prevent gas to flow through depending on the position of the valve stem (Fig. 2.18).

Depending on the position of the valve installed in the tubing string, a gas lift valve may be used as an operating valve or an unloading valve. The operating valve is normally the deepest one in the tubing. The main function of the operating valve is to control the amount of gas injected into the tubing to maximize the efficiency the gas lift system. The unloading valves are the ones above the operating valve. They are set at predetermined depths to progressively reduce the hydrostatic pressure of the fluid inside the tubing during the unloading process. In other words, the unloading valves help to reduce the surface gas injection pressure.

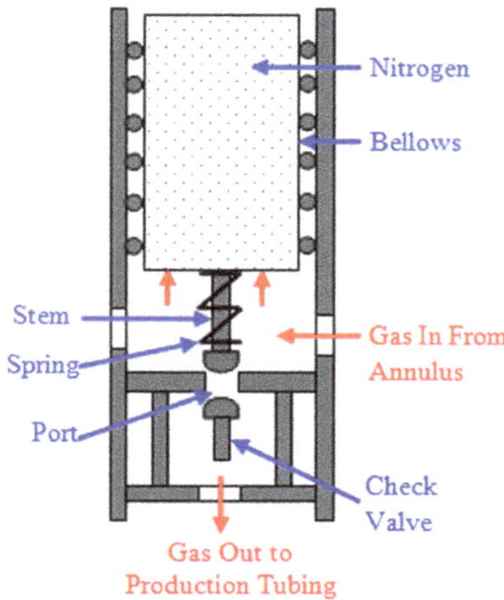

Fig. 2.18 Main component of a GLV

2.4.1 *Gas Lift Valve Classification*

There are two main types of gas lift valves: Injection Pressure Operated Valve (IPO) and Production Pressure Operated Valve (PPO). Other types of valves available in the market are throttling and pilot valves.

2.4.1.1 Injection Pressure Operated Valve (IPO)

Injection Pressure Operated Valve (IPO) by far is the most common valve used in gas lift systems. The schematic of the valve is shown in Fig. 2.19. Gas is injected into the casing annulus and forced into the gas lift valve through the gas entrance. Because the area of the bellows is much larger than that of the port, and since the bellows is exposed to the high-pressure gas in the casing, the valve is much more sensitive to the casing pressure than the production fluid in the tubing. In other words, casing pressure controls the opening of the gas lift valve. If the casing pressure is higher than the valve opening pressure, the valve stem moves up allowing gas to pass through the port and enter the production tubing. If the casing pressure at valve depth is smaller than the valve closing pressure, the valve stem moves down preventing gas passing through the port. The valve closing and opening pressure will be discussed in the next section.

2.4.1.2 Production Pressure Operated Valve (PPO)

Production pressure operated valves (PPO valves) are operated primarily by changing pressure of the production fluid. For a PPO valve, the port is exposed to the high-pressure gas injection and the bellows is exposed to the production

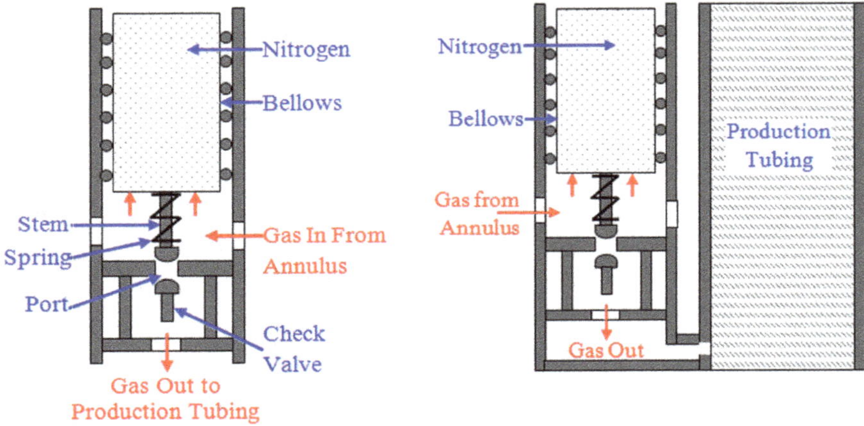

Fig. 2.19 Injection pressure operated valve (IPO valve)

pressure from the production tubing as shown in Fig. 2.20. Because the area of the bellows is much larger than that of the port, this valve is much more sensitive to the production pressure than the high-pressure gas in the casing annulus. Therefore, it is the production pressure that controls the opening or closing of the valve. If the production pressure at valve depth is higher than the PPO-valve opening pressure as shown in Fig. 2.20b, the valve stem moves up allowing gas to pass through the port and enter the production tubing as illustrated in red arrows. If the production pressure at valve depth is smaller than the valve closing pressure, the valve stem moves down preventing gas passing through the port as presented in Fig. 2.20a.

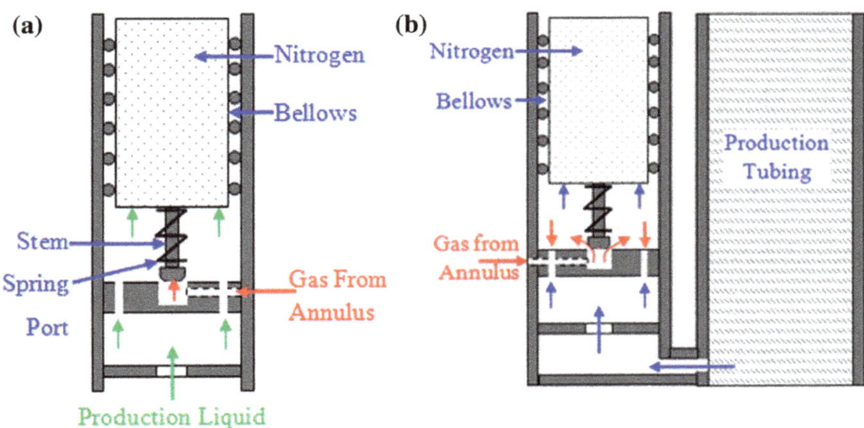

Fig. 2.20 Production pressure operated valve (IPO valve)

PPO valves are used primarily: (1) when source of pressure fluctuates; (2) in dual completions where two production tubing strings are installed in the same well to produce two different producing zones; (3) for intermittent gas lift system where the operating valve is designed to remain shut until a sufficient fluid load is present in the tubing. After this point, the operating valve opens allowing gas to be injected into the tubing to lift up the liquid slug.

2.4.1.3 Throttling Valve

Throttling valve is often called a continuous flow valve. The throttling valve's design is similar to that of the IPO valve. The only difference is the modification of the stem tip (the ball) to make it more tubing sensitive when in the opening position as shown in Fig. 2.21.

For a flow through an orifice, at a constant injection pressure (upstream pressure or casing pressure), the gas flow rate increases as the downstream pressure (tubing pressure) decreases during subcritical flow. If the flow reaches critical conditions, the gas flow rate remains constant despite further decreases in downstream pressure as shown in Fig. 2.22. In other words, the valve is not sensitive to the tubing pressure (downstream pressure). When the valve stem tip is modified to make it more sensitive to the downstream pressure under opening position, the valve may exhibit throttling flow region. Under the throttling flow region, at a constant injection pressure, the gas flow rate increases with decreasing downstream pressure until it reaches a maximum and then decrease with decreasing downstream pressure as shown in Fig. 2.22.

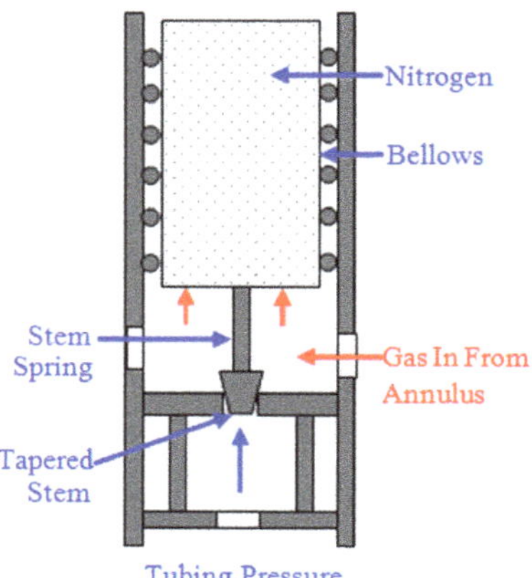

Fig. 2.21 Throttling valve

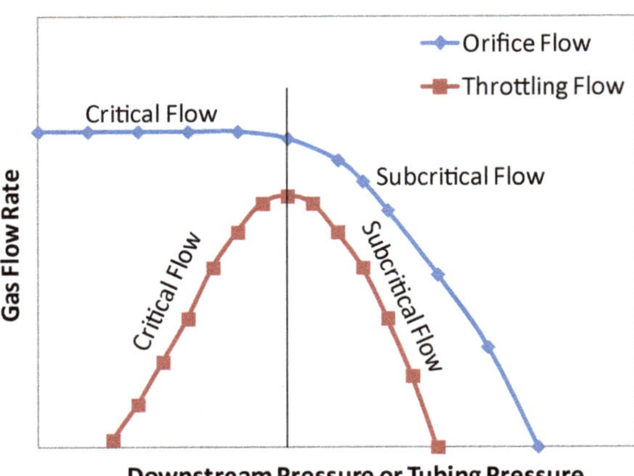

Fig. 2.22 Orifice and throttling flow regime under critical and subcritical flow conditions

2.4.1.4 Pilot Valve

Pilot valve is operated using high pressure gas in the casing annulus similar to an IPO valve. In other words, pilot valve is more sensitive to gas pressure in the casing annulus than liquid pressure in the tubing. It is used when a large gas injection rates are required. This valve is often used in intermittent gas lift system.

A pilot valve has two main sections: pilot section and power section as shown in Fig. 2.23. When the gas pressure in the casing annulus at valve depth is higher than the dome pressure, the valve stem in the pilot section moves up and allow high pressure gas from the casing annulus to enter the power section. This high pressure gas pushes the piston in the power section downward. A large volume of gas from the larger ports in the power section can now flow into the well. The power section is closed when the casing pressure at valve depth is smaller than the valve closing pressure.

2.4.2 Gas Lift Valve Performance

The relationship between gas flow rate and pressure drop across the valve (or upstream and downstream pressure ratio) is defined as the performance of a gas lift valve. Under subcritical conditions, as more gas passes the valve, the pressure drop across the valve is higher. If the critical flow condition is reached, the gas rate passing an orifice gas lift valve is no longer a function of the downstream pressure. In this section, the performance of an orifice is first presented. Then this theory will be applied to predict the performance of an actual gas lift valve performance.

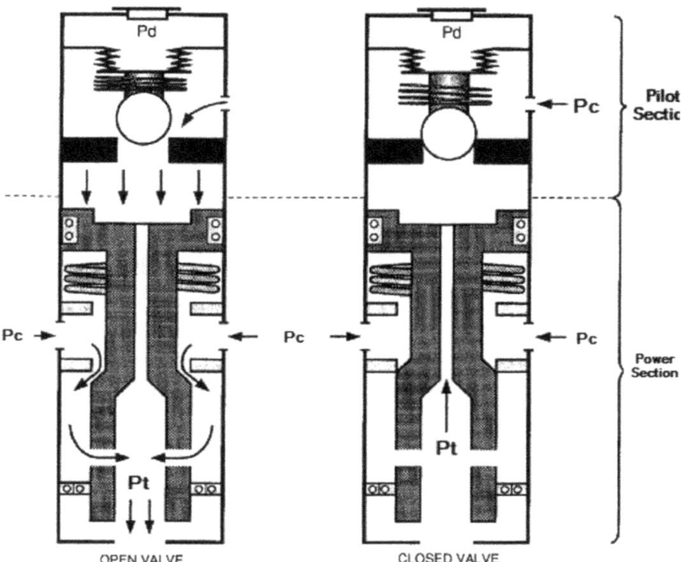

Fig. 2.23 Pilot valve

2.4.2.1 Fundamentals of Gas Passage Through an Orifice

The gas flow rate through an orifice depends mainly on the upstream pressure, downstream pressure, geometry of the orifice, and the gas properties (Fig. 2.24).

According to Example 1.3, under steady state flow conditions of an invicid fluid flowing in one dimensional direction, the pressure drop across the orifice is presented as Eq. (1.3.43):

$$\Delta P = P_1 - P_2 = \frac{1}{2}\rho_2 u_2^2 - \frac{1}{2}\rho_1 u_1^2 \tag{1.3.43}$$

Combining with the continuity equation gives an equation to predict the volumetric gas rate through an orifice as shown in Eq. (1.3.48):

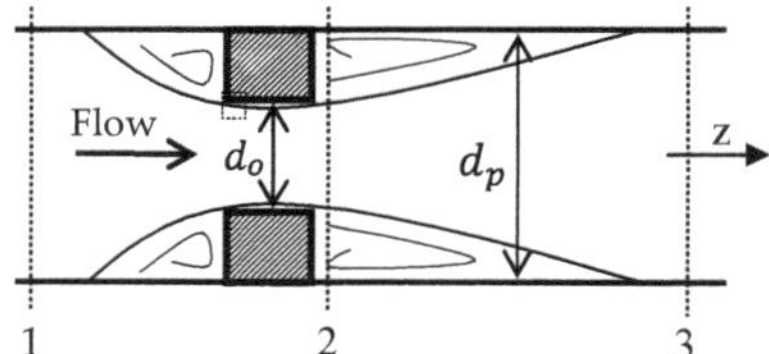

Fig. 2.24 Flow through an orifice

$$q_v = C\frac{\pi d_p^2}{4}\sqrt{2(P_1 - P_2)}\sqrt{\frac{1}{\rho_1(1-\beta^4)}} \tag{1.3.48}$$

where $\beta = \frac{d_o}{d_p}$ is the ratio between the orifice and upstream pipe diameters. For more detail of the derivation of these equations, readers are recommended to review Example 1.3 in Chap. 1.

Note that Eq. (1.3.48) is applicable only when the flow is under subcritical conditions. At a certain gas rate, the gas velocity equals to the sound velocity in the gas. Under this condition, the pressure waves are not able to be propagated upstream. Therefore, the gas rate through the orifice is now only a function of the upstream pressure and the flow is said critical. In other words, for subcritical flow, the gas rate through an orifice depends on both upstream and downstream pressures. For critical flow, the gas rate through an orifice depends only on the upstream pressure. Potter and Wiggert [5] presented an equation to determine the critical gas flow conditions as follows:

$$\left(\frac{P_d}{P_u}\right)_{critical} = \left(\frac{2}{k+1}\right)^{\frac{k}{k-1}} \tag{2.4.1}$$

where k is the specific heat capacity ratio, $k = \frac{C_p}{C_v}$. If the ratio P_d/P_u is greater than the value calculated using Eq. (2.4.1), the flow is under critical condition. For natural gas, $\kappa = 1.32$, then Eq. (2.4.1) becomes:

$$\left(\frac{P_d}{P_u}\right)_{critical} = 0.546 \tag{2.4.2}$$

The specific heat capacity of different gases is presented in Table 2.1.

2.4.2.2 Orifice Gas Lift Valve Performance

Craft et al. [2] applied Eq. (1.3.48) to derive an equation for calculating the gas rate through an orifice under standard conditions in the unit of thousand standard cubic feet per day (MSCF/D) as follows:

$$Q_g^{sc}\left(\frac{MSCF}{D}\right) = 0.239C_d d_o^2 P_u\sqrt{\frac{1}{\gamma_g(T_u+460)}}\sqrt{\frac{k}{k-1}}\sqrt{\left(\frac{P_d}{P_u}\right)^{\frac{2}{k}} - \left(\frac{P_d}{P_u}\right)^{\frac{k+1}{k}}} \tag{2.4.3}$$

Table 2.1 Specific heat capacity values of different gases

Gas	$\gamma = c_p/c_v$	Gas	$\gamma = c_p/c_v$	Gas	$\gamma = c_p/c_v$
Acetylene	1.30	Ethane	1.18	Methane	1.32
Air, standard	**1.40**	Ethyl alcohol	1.13	Methyl alcohol	1.20
Ammonia	1.32	Ethyl chloride	1.19	Methyl butane	1.08
Argon	1.66	Ethylene	1.24	Methyl chloride	1.20
Benzene	1.12	Helium	1.66	**Natural gas**	**1.32**
N-butane	1.18	N-heptane	1.05	Nitric oxide	1.40
Iso-butane	1.19	Hexane	1.06	Nitrogen	1.40
Carbon dioxide	1.28	Hydrochloric acid	1.41	Nitrous oxide	1.31
Carbon disulphide	1.21	Hydrogen	1.41	N-octane	1.05
Carbon monoxide	1.40	Hydrogen chloride	1.41	Oxygen	1.40
Chlorine	1.33	Hydrogen sulfide	1.32	N-pentane	1.08

Bold values refers to the common gases used during the unloading process

where C_d is the discharged coefficient factor, d_o is the orifice diameter in 64th of an inch, P_u and P_d are the upstream and downstream pressures in psia, T_u is the upstream fluid temperature in °F, and k is the specific heat capacity as given in Table 2.1.

Thornhill-Craver Company (1946) originally reported an equation for calculating the gas rate under standard conditions through a 6 in. bean choke with rounded entrance. The value of the discharged coefficient C_d will be lower if the entrance changes to sharp-edged. For the square-edged orifice and assuming the full area of the orifice is open to flow, this model suggested a value of $C_d = 0.865$. Combining with Eq. (2.4.2) with a value of $k = 1.32$, gas specific gravity $\gamma_g = 0.65$, upstream temperature $T_u = 60$ °F, Thornhill-Craver gives the final equations for predicting the performance of a sharp-edged orifice under subcritical and critical flow conditions as follows:

$$\begin{cases} \text{Subcritical: } \dfrac{P_d}{P_u} > 0.546 : Q_g^{sc}\left(\dfrac{MSCF}{D}\right) = 0.43 d_o^2 P_u \sqrt{\dfrac{1}{\gamma_g (T_u + 460)}} \sqrt{\left(\dfrac{P_d}{P_u}\right)^{1.52} - \left(\dfrac{P_d}{P_u}\right)^{1.76}} \\ \text{Critical: } \mathit{If}\ \dfrac{P_d}{P_u} \le 0.546 : Q_g^{sc}\left(\dfrac{MSCF}{D}\right) = 0.43 d_o^2 P_u \sqrt{\dfrac{1}{\gamma_g (T_u + 460)}} \sqrt{0.546^{1.52} - 0.546^{1.76}} \end{cases} \tag{2.4.4, 2.4.5}$$

where the orifice diameter, d_o, is in the unit of 64th of an inch. For example, $d_o = 12/64 = 0.1875$ in.

Under critical flow conditions, the gas rate reaches the maximum value, Q_{max}, and can be calculated as Eq. (2.4.5). Normalizing Eqs. (2.4.4 and 2.4.5) by dividing them to Q_{max} gives:

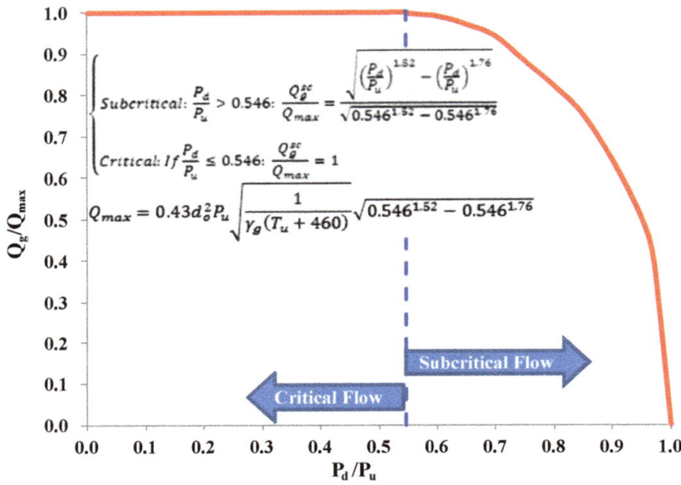

Fig. 2.25 Normalization of a square-edged orifice performance using Thornhill-Craver equation

$$\begin{cases} \text{Subcritical: } \dfrac{P_d}{P_u} > 0.546: \dfrac{Q_g^{sc}}{Q_{\max}} = \dfrac{\sqrt{\left(\frac{P_d}{P_u}\right)^{1.52} - \left(\frac{P_d}{P_u}\right)^{1.76}}}{\sqrt{0.546^{1.52} - 0.546^{1.76}}} \\ \text{Critical: } \mathit{If}\ \dfrac{P_d}{P_u} \le 0.546: \dfrac{Q_g^{sc}}{Q_{\max}} = 1 \end{cases} \quad (2.4.6, 2.4.7)$$

where

$$Q_{\max} = 0.43 d_o^2 P_u \sqrt{\frac{1}{\gamma_g (T_u + 460)}} \sqrt{0.546^{1.52} - 0.546^{1.76}} \quad (2.4.8)$$

The normalized results are illustrated in Fig. 2.25.

Thornhill-Craver equation described in Eqs. (2.4.4) and (2.4.5) can be plotted and presented in Fig. 2.26. Note that Thornhill-Craver equation was not intended to use for non-orifice type gas-lift valves. To use the plot in Fig. 2.26, readers choose an upstream pressure value such as, P_u = 1000 psig; draw a vertical line starting from the P_u to reach the downstream pressure value curve (the red line gives an example of P_d = 900 psig); then draw a horizontal line to reach the orifice-diameter line such as 16/64″; from this intersection, draw a vertical line to obtain the gas rate in MSCF/D.

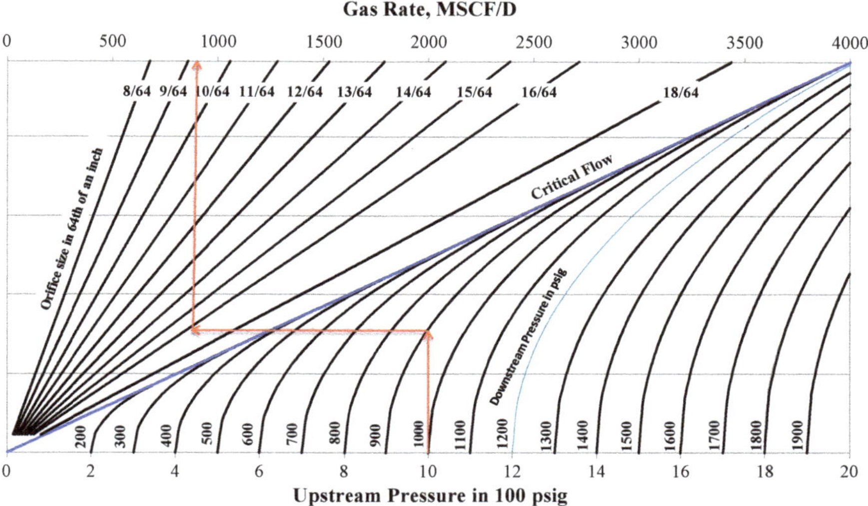

Fig. 2.26 Thornhill-craver Solution for square-edged orifice performance

2.4.2.3 Throttling Gas Lift Valve Performance

Most of IPO valves do not snap shut. Gas lift valves are very rarely fully open when flowing gas. Increasing the compressed gas pressure in the casing at valve depth until it reaches the valve opening pressure, the ball (stem tip) very slowly moves up and allows gas to pass through the port. Gas rate keeps increasing as the ball moves up higher. In addition, the downstream pressure will decline because of the reduction in hydrostatic pressure due to more gas is being injected (higher GLR) into the tubing string. At a certain condition, the gas flow through the gas lift valve will reach the critical condition at which the gas rate theoretically remains constant. However, if throttling flow occurs, the gas rate will reduce when the flow is under critical conditions.

In the throttling flow region of the gas lift valve performance, the downward force on the stem and bellows assembly, resulting from the nitrogen pressure in the dome, becomes sufficient to depress the stem, thus restricting gas passage through the valve. Hepguler et al. [4] conducted an experimental work and realized that as the production pressure decreases, there is sufficient flow area initially to allow an increase in flow rate. After the critical condition is reached, the flow rate declines approximately linearly as the production pressure decreases (upstream pressure remains constant) as shown in Fig. 2.27. In other words, users have to be very careful if orifice-flow is applied to an actual gas lift valve. The actual performance of a gas lift valve may be very different from the predicted one which assumes gas lift valve behaves as an orifice.

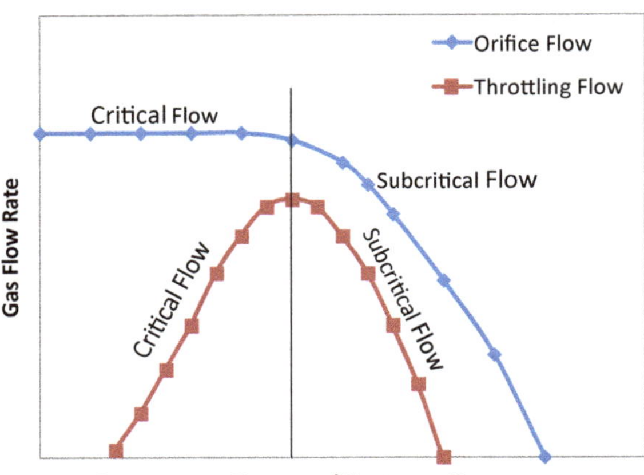

Fig. 2.27 Orifice and throttling flow

One way to obtain reliable data in orifice and throttling flow regions is to perform experiments on the gas lift valves being used. This approach assumes that the valve is treated as a black box and the volumetric flow rate is reported as a function of valve-setting parameters and the differential pressure across the valve. This experimental approach is very time-consuming because of the combination of parameters affecting gas-passage performance of a valve. Gas lift valve testing is recommended to follow:

- API Recommended Practice 11V2 1st Edition January 1, 1995
- API Recommended Practice 11V2 2nd Edition March 2001
- ISO 17078.2.

The API 11V2 RP recommends conducting three tests for a gas lift valve including:

- Probe Test: to determine the loadrate of the valve and the maximum amount of stem travel available for a set of pressure. The loadrate of a valve is a measure of the amount of opening that can be obtained for an incremental increase in pressure acting on the bellows.
- Flow Coefficient Test: to determine the valve's flow capacity as a function of the stem position. If the stem position of the valve is known, the flow coefficient, Cv, can be used to calculate the flow rate for both liquid and gas.
- Dynamic Test: to reveal a valve's performance characteristics including flow capacity. The shape of the performance curve reveals the valve's sensitivity to pressure (Fig. 2.28).

Another approach is based on modeling the valve on physics principles. This method allows a significant reduction in the number of tests needed to characterize valve performance.

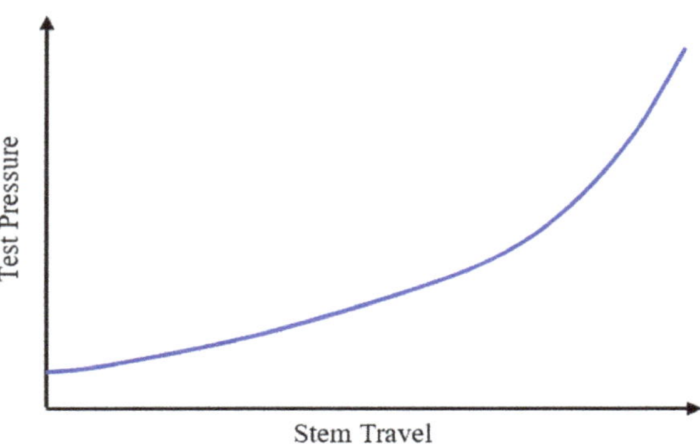

Fig. 2.28 Probe test according to API 11V2RP

2.4.3 *Operation of a Gas Lift Valve*

In this section, the opening and closing pressure of an IPO and PPO valve will be discussed following [1]. In addition, the effect of the tubing and casing pressure on the opening or closing of IPO and PPO valves will also be investigated.

Before going into details discussion of the valve's operation, let's review the relationship between pressure, force, and area. Pressure is a scalar and defined as energy per unit volume and presented in Eq. (1.15). Pressure has the unit of Pascal or force per area. As the area increases, the force acting on this area increases at the same amount of pressure. For a gas lift valve, the bellows and the port are two main areas that need to be considered when doing the force-balance analysis. In most cases, the bellow area is much higher than the port area; the difference can be up to ten times. Therefore, the bellows area is the control parameter on the sensitivity of the gas lift valve. If the high-compressed gas in the casing is in contact with the bellows, the valve will be sensitive to the gas-injection casing pressure and it is called Injection Pressure Operated valve (IPO). If the production fluid in the tubing string is in contact with the bellows, the valve will be sensitive to the tubing pressure and it is called Production Pressure Operated valve (PPO).

2.4.3.1 Opening Pressure of IPO Valves

When the IPO valve is on the close position, there are two forces trying to close the valve as shown in Fig. 2.29: one force is due to the dome pressure acting on the bellows area; the other force is due to gas from the casing annulus acting on the port area. Therefore, the resultant force trying to hold the valve closed is:

$$F_c = P_d A_b + P_c A_p \tag{2.4.9}$$

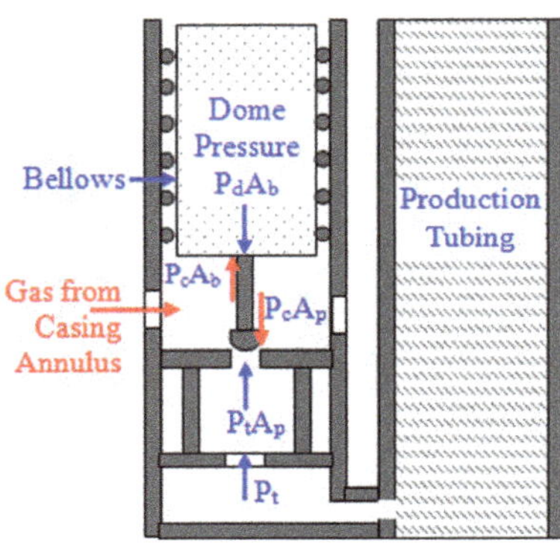

Fig. 2.29 Force analysis on IPO valve: close position

At the same time, there are two forces trying to open the valve. One force is due to gas from the casing annulus acting on the bellows; the other force is due to the production fluid acting on the port area. The resultant force trying to open the valve is:

$$F_o = P_c A_b + P_t A_p \tag{2.4.10}$$

For the valve to open, the F_o must be at least equal or greater than the F_c ($F_o \geq F_c$). Therefore, the gas pressure from the casing annulus at the valve depth needs to open an IPO valve can be obtained by equalizing Eq. (2.4.9) and Eq. (2.4.10) yields:

$$P_c = P_{vo} = \frac{P_d A_b - P_t A_p}{A_b - A_p} \tag{2.4.11}$$

$$\text{Let R} = \frac{\text{A}_\text{p}}{\text{A}_\text{b}} \text{ be the dimensionless area} \tag{2.4.12}$$

$$P_c = P_{vo} = \frac{P_d - RP_t}{1 - R} = \frac{P_d}{1 - R} - \frac{R}{1 - R} P_t \tag{2.4.13}$$

The second term in Eq. (2.4.13) is defined as the Tubing Effect (TE).

$$TE = \frac{R}{1 - R} P_t \tag{2.4.14}$$

The Tubing Effect Factor is defined as:

$$TEF = \frac{R}{1 - R} \tag{2.4.15}$$

Equation (2.4.13) tells us that the gas pressure in the annulus at valve depth used to open an IPO valve (P_{vo}) depends on the dome pressure, tubing pressure, and the dimensionless area, R. If the tubing pressure is high (high TE), the pressure required

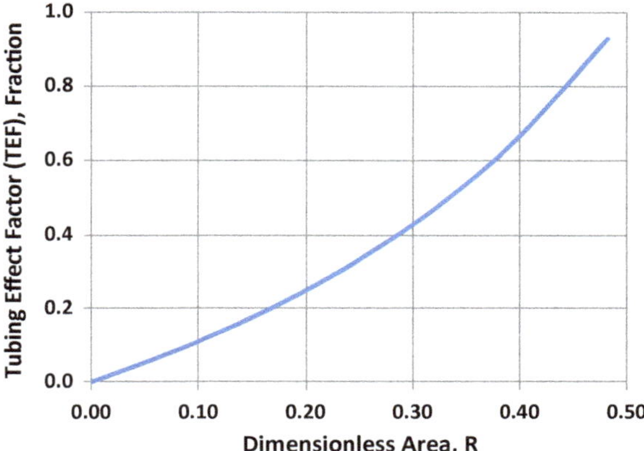

Fig. 2.30 Effect of R on the TEF

to open the valve is less. Figure 2.30 shows how dimensionless area, R, affects the TEF. The TEF increases exponentially as the dimensionless area is higher than 0.2.

2.4.3.2 Closing Pressure of IPO Valves

When the IPO valve is in the open position as shown in Fig. 2.31, gas is flowing through the port. There are two forces trying to close the valve: one force is due to the dome pressure acting on the bellows area; the other force is due to gas from the casing annulus acting on the port area. Therefore, the resultant force trying to hold the valve closed can be expressed as:

$$F_c = P_d A_b + P_c A_p \tag{2.4.16}$$

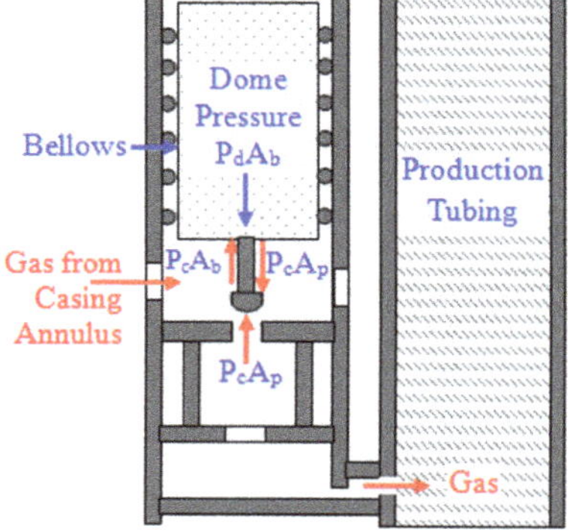

Fig. 2.31 Force analysis on IPO valve: open position

At the same time, there are two forces trying to open the valve: One force is due to gas from the casing annulus acting on the bellows; the other force is also due to gas from the casing annulus acting on the port area. The resultant force trying to open the valve is:

$$F_o = P_c A_b + P_t A_p \tag{2.4.17}$$

For the valve to close, the F_o must be at least equal or smaller than the F_c ($F_o \leq F_c$). Therefore, the gas pressure from the casing annulus at the valve depth needs to open an IPO valve can be obtained by equalizing Eqs. (2.4.16) and (2.4.17) yields:

$$P_c = P_{vc} = P_d \tag{2.4.18}$$

In other words, for a IPO valve to close, the gas pressure inside the casing annulus at valve depth must be reduced back to the dome pressure.

2.4.3.3 Opening Pressure of PPO Valves

When a PPO valve is on the close position as shown in Fig. 2.32, there are three forces trying to close the valve: one force is due to the dome pressure acting on the bellows area; the second force is due to the production fluid acting downward on the port area; and the third force is due to the tension of the spring, S_t, acting downward. Therefore, the resultant force trying to hold the valve close is:

$$F_c = P_d A_b + P_t A_p + S_t (A_b - A_p) \tag{2.4.19}$$

At the same time, there are two forces trying to open the valve. One force is due to gas from the casing annulus acting on the port area; the other force is due to the production fluid acting on the bellows area. The resultant force trying to open the valve is:

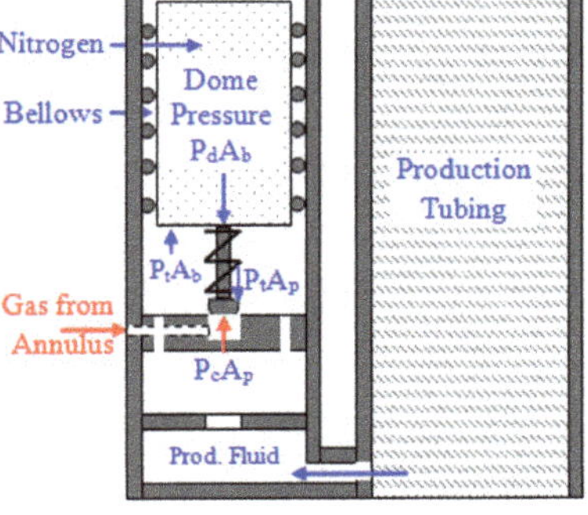

Fig. 2.32 Force analysis on PPO valve: close position

$$F_o = P_c A_p + P_t A_b \tag{2.4.20}$$

For the valve to open, the F_o must be at least equal or greater than the F_c ($F_o \geq F_c$). Therefore, the pressure of the production fluid inside the tubing at valve depth needs to open an PPO valve can be obtained by equalizing Eqs. (2.4.19) and (2.4.20) yields:

$$P_t = P_{vo} = \frac{P_d - RP_c + S_t(1-R)}{1-R} = \frac{P_d}{1-R} - \frac{R}{1-R} P_c + S_t \tag{2.4.21}$$

The second term in Eq. (2.4.21) is defined as the Casing Effect (CE). Equation (2.4.21) tells us that the pressure of the production fluid inside the production tubing depends on the dome pressure, gas pressure inside the casing annulus, and the spring tension.

2.4.3.4 Closing Pressure of PPO Valves

A similar force analysis for a PPO valve when the valve is on the opening position yields:

The resultant force trying to hold the valve close is:

$$F_c = P_d A_b + P_t A_p + S_t (A_b - A_p) \tag{2.4.22}$$

The resultant force trying to open the valve is:

$$F_o = P_t A_p + P_t A_b \tag{2.4.23}$$

Combining Eqs. (2.4.22) and (2.4.23) give an equation to calculate for the tubing pressure needed to close a PPO valve as follows:

$$P_t = P_{vc} = P_d + S_t(1-R) \tag{2.4.24}$$

2.4.3.5 Spread of Gas Lift

Spread in gas lift is defined as the difference between the opening and closing pressure of a gas lift valve.

For IPO valves:

$$\begin{aligned} Spread &= \Delta P_{IPO} = P_{vo}^{IPO} - P_{vc}^{IPO} = \frac{P_d - RP_t}{1-R} - P_d \\ Spread &= \Delta P_{IPO} = \frac{R}{1-R}(P_d - P_t) = TEF(P_d - P_t) \end{aligned} \tag{2.4.25}$$

For PPO valves:

$$\begin{aligned} Spread &= \Delta P_{PPO} = P_{vo}^{PPO} - P_{vc}^{PPO} = \frac{P_d}{1-R} - \frac{R}{1-R}P_c + S_t - P_d \\ Spread &= \Delta P_{PPO} = \frac{R}{1-R}(P_d - P_c) + RS_t = TEF(P_d - P_c) + RS_t \end{aligned} \quad (2.4.26)$$

For IPO valves, if the spread is known then the valve opening pressure equals to the summation of the dome pressure and the spread. Spread can be important in continuous flow but it is particularly important in intermittent gas lift installations where underbalanced pressure valves are used. PPO valves are much more common than IPO valves because the amount of injected gas can be controlled easier for each cycle. Most of the PPO valves are under balanced pressure valves, which the spread is negative. From Eq. (2.4.26), one can recognize that if the spring tension is zero then the spread is negative when the gas pressure inside the casing annulus is higher than the dome pressure. In addition, the magnitude of the spread is strongly dependent on the TEF or the dimensionless area, R. As the dimensionless area ($R = A_p/A_b$) increases (bigger port size), the TEF increases exponentially as shown in Fig. 2.30. Therefore, reducing the spread (reducing the port size) can limit the amount of gas passage per cycle but it will lead to an increase in horsepower and decrease in lift efficiency per lift cycle. For this reason, pilot valves are made in such a way that it can provide a small R value to have a small value of spread yet do not restrict the gas flowing through the port.

2.4.3.6 Dome Charge Pressure Corrections

Before sending gas lift valves to clients, the manufacturers charge the dome with nitrogen to a specific pressure under a controlled surface temperature; normally it is 60 °F. These valves are then used under downhole conditions where the valves experience a much higher temperature. Winkler and Eads [10] introduced correlations to correct the dome pressure from surface conditions to downhole conditions.

For the surface dome pressure, P_d', is smaller than 1,238 psi, the dome pressure under downhole conditions, P_d, can be predicted as follows:

$$P_d = P_d' + \left[-0.00226 + 0.001934P_d' + 3.054 \times 10^{-7}\left(P_d'\right)^2\right](T_{vd} - 60) \quad (2.4.27)$$

For $P_d' \geq$ 1,238 psi:

$$P_d = P_d' + \left[-0.267 + 0.002298P_d' + 1.84 \times 10^{-7}\left(P_d'\right)^2\right](T_{vd} - 60) \quad (2.4.28)$$

where P_d and $P_d^{'}$ are the dome charged pressure under downhole conditions and at surface 60 °F. T_{vd} is the temperature at valve depth in Fahrenheit. Ganzarolli and Altemani [3] also introduced temperature correction factors to correct the dome pressure from surface conditions to downhole conditions.

2.4.4 *Test Rack Opening Pressure*

The test rack opening pressure, P_{TRO}, is the gas pressure to open a gas lift valve under surface conditions at which the dome temperature is maintained at 60 °F and the downstream pressure is zero psig. This is not an actual gas pressure at valve depth in the casing annulus or an actual production fluid pressure at valve depth inside the production tubing to open IPO or PPO valves, respectively. The actual pressure to open an IPO valve or a PPO valve can be predicted using Eqs. (2.4.12) or (2.4.21), respectively.

For IPO valves, the downstream pressure is the tubing pressure; for PPO valves, the downstream pressure is the casing pressure. Recall Eqs. (2.4.12) and (2.4.21) for calculating the opening pressure for IPO valves and PPO valves:

$$\text{For IPO}: P_c = P_{vo} = \frac{P_d}{1-R} - \frac{R}{1-R}P_t$$
$$\text{For PPO}: P_t = P_{vo} = \frac{P_d}{1-R} - \frac{R}{1-R}P_c + S_t$$

When the downstream pressure is set equal to zero then the above two equations become:

$$\text{For IPO}: P_c = P_{vo} = \frac{P_d}{1-R} \tag{2.4.29}$$

$$\text{For PPO}: P_t = P_{vo} = \frac{P_d}{1-R} + S_t \tag{2.4.30}$$

The only difference in the above two equations is the tension of the spring. In other words, the test rack opening pressure does not depend on the type of valve or the tubing or casing pressure; instead it depends solely on the dome pressure and the tension of the spring. Therefore, one can write the general test rack opening pressure equation as follows:

$$P_{TRO} = \frac{P_d^{'}}{1-R} + S_t \tag{2.4.31}$$

where $P_d^{'}$ in psi is the nitrogen dome charged pressure at 60 °F; S_t is the tension of the spring in psi.

Generally speaking, when designing a gas lift system, the opening pressure at valve depth is known. Applying Eqs. (2.4.29) or (2.4.30) to calculate the dome charged pressure, P_d, at valve depth temperature. Equations (2.4.27) or (2.4.28) is used next to obtain the dome pressure ($P_d^{'}$) at 60 °F. This is the prediction of the test rack opening pressure. In the lab, charge the dome at the pressure $P_d^{'}$, conduct the test to confirm the test rack opening pressure prediction.

To correct the dome pressure at valve depth conditions to surface conditions at 60 °F, the non-ideal gas law can also be applied beside the Winkler and Eads model. The dome pressure at 60 °F can be estimated as follows:

$$P_d^{'} = \frac{520 P_d Z_{60F}}{Z_{vd} T_{vd}} \quad (2.4.32)$$

where the constant 520 = (60 °F + 460) is the surface temperature in Rankin; Z is the gas compressibility factor; the subscripts d, v_d, and 60 °F are for dome, valve depth, and surface condition at 60 °F, respectively. For gas lift practical applications, the ratio between Z_{60F}/Z_{vd} is around 0.985.

2.5 Multiple Valve Gas Lift Unloading

2.5.1 *Fundamentals of Unloading Process*

After production tubing is installed and packer is set in place, the well is normally full of completion fluid inside the production tubing and inside the annulus formed by the production tubing and the production casing. This completion fluid in the annulus, usually treated free-solid weighted brine salt water, must be removed before the well is put on production and gas lift system can be function. This process is called unloading for a gas lift well.

Unloading process may cause damages to gas lift valves than any time during the life of a gas lift well. This is because of the liquid flow through the valves during the unloading process. Keeping mind that gas lift valves are not designed for liquid flow; it is for gas flow only. Therefore, when the unloading occurs, the high velocity-liquid may cut the seat assemblies and damage the check valve causing a leak during the normal operations of the well. When there is a long term or short

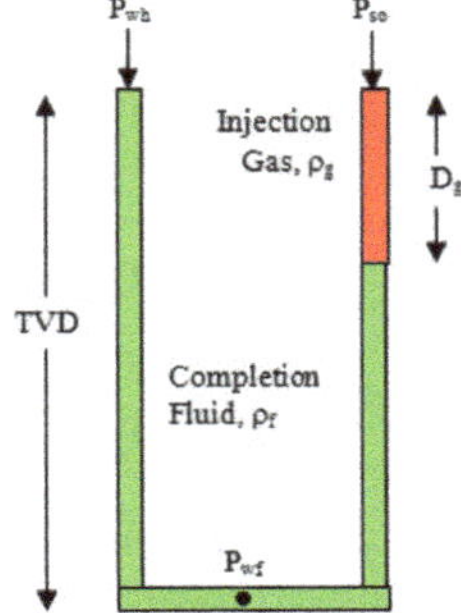

Fig. 2.33 U-tubing effect

term shutdown, the fluid from the production tubing may enter the casing annulus because of this leak. If this happens, the well must be partially or fully unloaded.

If the gas injection pressure is not stable at surface, it is highly recommended to install a packer for the well to prevent the multiple unloading processes during the lift of the well. In other words, installing a packer will eliminate the need for unloading the well after a shutdown due to the interruption of the gas supply, power supply, hurricane, etc.

For better understanding the basic concept of unloading an oil or gas well, let consider a u-tube which initially is full of completion fluid with density of ρ_f. Gas with density of ρ_g is injected into one side of the u-tube at a pressure of P_{so} as shown in Fig. 2.33. Under static conditions, the pressures at the bottom on two legs of the u-tube, P_{wf}, are the same and can be written as follows:

$$P_{wh} + 0.052\rho_f TVD = P_{so} + 0.052\rho_g D_g + 0.052\rho_f (TVD - D_g)$$

Rearrange this equation gives:

$$P_{so} = P_{wh} + 0.052(\rho_f - \rho_g) D_g \tag{2.5.1}$$

Equation (2.5.1) reveals that the surface gas operating pressure depends on the well head pressure, the difference in densities between liquid and gas, and the interfacial depth between liquid and gas, D_g.

For a particular oil or gas well, we can assume the well head pressure and the differential density are more or less constant. The gas surface operating pressure is mainly dependent on the position of the interface between gas and liquid. Assuming the operating valve depth is 10,000 ft (D_g = 10,000 ft); the well head pressure is maintained at 400 psi; the completion fluid and gas densities are 10 ppg and 0.006 ppg, respectively. If only one gas lift valve at the depth of 10,000 ft is used to unload the well, the required gas surface operating pressure calculated using Eq. (2.5.1) would be 5,600 psi. This surface pressure would be too high for conventional gas lift system at which the gas surface operating pressure is available around 1,500 psi. Therefore, unloading process for gas lift well is normally carried out with multiple valves called unloading valves to reduce the gas surface operating pressure. This technique is called multiple valve gas lift unloading and used to differentiate with the single point gas lift unloading. The maximum gas surface operating pressure happens when gas does the U-turn at the first unloading valve (the smallest valve depth). This pressure is called the kick-off pressure. Reconsider this example, if the depth of the first unloading valve is 1500 ft, then the maximum gas surface operating pressure (kick-off pressure) is 1,180 psi, which is very reasonable for a conventional gas lift system.

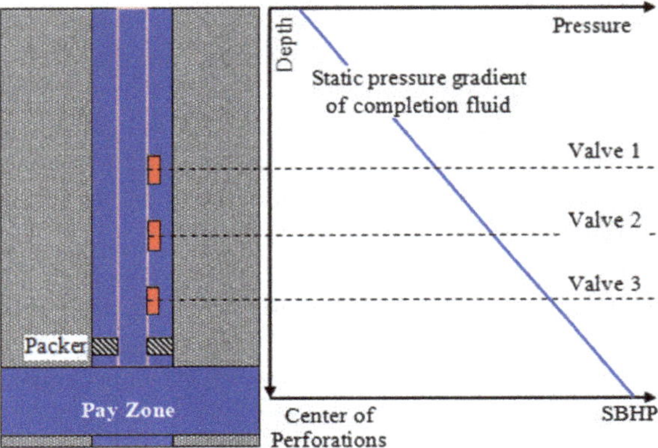

Fig. 2.34 Initial gas lift unloading process

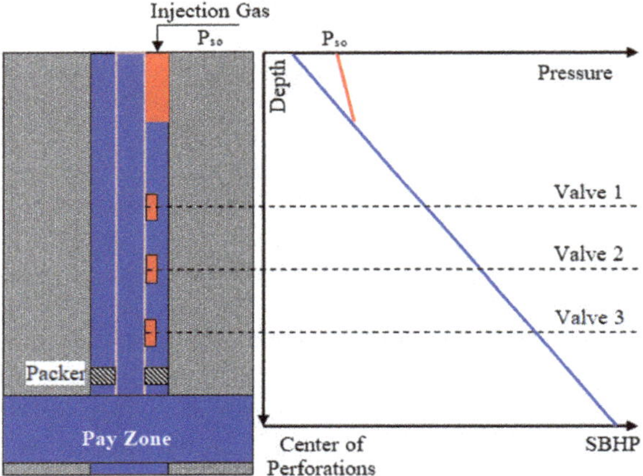

Fig. 2.35 Stage 1 of unloading process

2.5.2 Description of Unloading Process

Let consider a well full of completion fluid inside the tubing and inside the casing annulus. There are three gas lift valves installed in the production tubing. The first two valves are the unloading valves and the deepest valve is the operating valve. Assuming the static liquid levels in the tubing and in the casing are at the surface. All the gas lift valves are in the opening position. The static bottom-hole pressure at the center of the perforations is denoted as SBHP (Fig. 2.34).

The unloading process can be divided into four stages as follows:

Stage 1: Gas is being injected into the casing annulus. The completion fluid is flowing from the annulus into the tubing through the three gas lift valves. As gas travels deeper into the annulus, the injection surface gas operating pressure, P_{so}, is increasing to overcome the difference in densities of the completion fluid and injection gas as described in Eq. (2.5.1). If the friction in the tubing and in the annulus is neglected then the pressure gradient of completion fluid in the tubing remains the same as shown in Fig. 2.35.

As mentioned in the Fundamentals of Unloading Process section, this is a dangerous time for the gas lift valves. If the P_{so} is too high (gas injection rate is too high), the liquid velocity may cut the valve seat causing a communication between the tubing and the annulus when a shut-in occurs. Formation fluid will enter the annulus and hence unloading the well is mandatory before the well can resume production again. In addition, if formation fluid is corrosive, casings can be severely damaged and the well may be abandoned. Therefore, it is critical that operators must allow sufficient time for an unloading process. API RP 11V5 recommends that it should take 10 min for each 50 psi increase in casing pressure up to 400 psi. Then 100 psi increase in casing pressure every 10 min is acceptable until gas injects into the tubing at the first unloading valve (shallowest valve). Following this rule, it takes about 140 min for the injection gas operating pressure to reach 1000 psi.

Stage 2: Gas is being kept injecting into the casing annulus. The injection surface gas operating pressure keeps increasing until gas reaches to the first valve. Injection gas is now entering the production tubing and aerates the liquid column from the first valve to the surface inside the tubing, D_{v1}. The P_{so} now is called the kick-off pressure, P_{ko}. This is the maximum gas injection pressure required at surface to complete the unloading process. Because of the reduction in hydrostatic pressure of this fluid column, the flowing bottom-hole pressure reduces from SBHP to P_{wf1} as illustrated in Fig. 2.36.

Note that the liquid column from the first valve to the bottom of the hole is still the completion fluid and hence the pressure gradient remains the same.

As stated the injection rate during the unloading process is low and hence the frictional pressure losses in the tubing and in the annulus can be neglected. If the pressure loss across the valve 1 is also neglected then the depth of the first valve, D_{v1}, can be achieved by equalizing the two hydrostatic pressures in the tubing and in the annulus at the depth of valve 1.

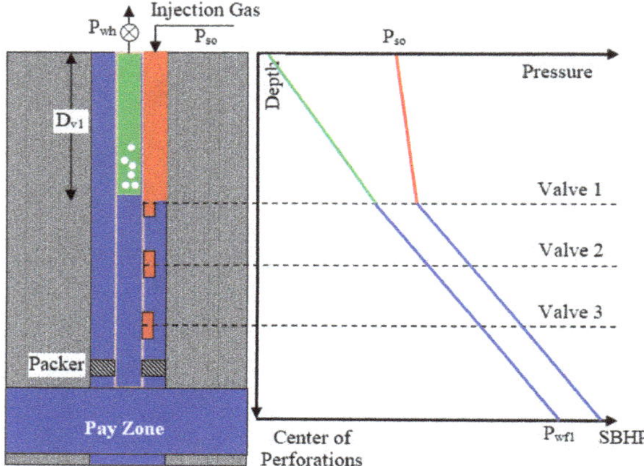

Fig. 2.36 Stage 2 of unloading process

$$P_{wh} + \rho_f g D_{v1} = P_{ko} + \rho_g g D_{v1} \quad (2.5.2)$$

Gas density is normally very small compared to completion fluid density and hence Eq. (2.5.2) can be written as follows:

$$D_{v1} = \frac{P_{ko} - P_{wh}}{g\rho_f} = \frac{P_{ko} - P_{wh}}{G_f} \quad (2.5.3)$$

where P_{wh} and P_{ko} are the well head pressure and the kick of pressure. ρ_f and ρ_g are the completion fluid and gas density. $G_f = g\rho_f$ is the completion fluid pressure gradient in the unit of pressure per unit length. If the pressures are in psi and the pressure gradient is in psi/ft then the depth D_{v1} is in ft.

Equation (2.5.3) can be used to locate the first valve based on the maximum surface injection pressure.

Stage 3: Gas is being kept injecting into the casing annulus. Valves 2 and 3 are still open allowing the completion fluid to u-tube from the annulus into the tubing. The liquid level in the annulus is decreasing from valve depth 1 to valve depth 2. When gas enters the tubing at valve 2, the weight of the fluid column inside the tubing reduces further leading to a reduction in the bottom-hole pressure from P_{wf1} to P_{wf2}. At the same time, when gas first enters the tubing, there is a small pressure decrease in the casing causing valve 1 to close (Fig. 2.37).

In other words, valve 2 is the only one used to inject gas into the production tubing. This will maximize the efficiency of the gas lift unloading. If the current flowing bottom-hole pressure is smaller than the pore pressure, reservoir fluid may flow into the tubing.

As gas is approaching the valve depth 2 (gas has not entered valve 2), a mixture of gas and completion fluid with a density of ρ_m is present inside the tubing from the surface to valve depth 1. The fluid inside the tubing from between valve depth 1 and valve depth 2 is still completion fluid. Assuming the friction in the tubing and

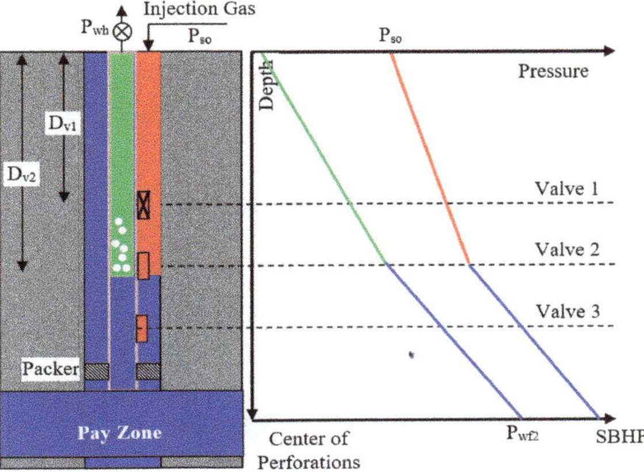

Fig. 2.37 Stage 3 of unloading process

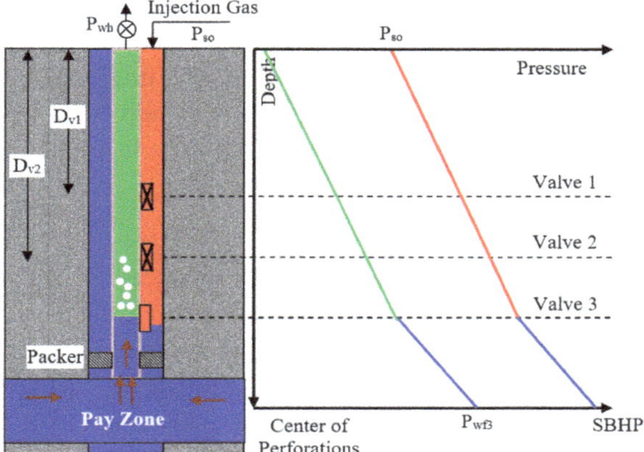

Fig. 2.38 Stage 4 of unloading process

in the annulus is small and neglected due to the fact that the unloading rate is small. The hydrostatic pressures in the tubing and in the annulus at valve depth 2 are the same and given as:

$$P_{wh} + \rho_m g D_{v1} + \rho_f g (D_{v2} - D_{v1}) = P_{ko} + \rho_g g D_{v2} \tag{2.5.4}$$

Neglecting the hydrostatic pressure of gas column inside the annulus, Eq. (2.5.4) becomes

$$D_{v2} = D_{v1} + \frac{P_{ko} - P_{wh} - \rho_m g D_{v1}}{G_f} \tag{2.5.5}$$

In general, the depth of unloading valves can be determined as

$$D_{vi+1} = D_{vi} + \frac{P_{ko} - P_{wh} - G_m D_{vi}}{G_f} \tag{2.5.6}$$

where $G_m = g\rho_m$ is the pressure gradient of the mixture of gas and completion fluid inside the production tubing. In field unit, if the mixture density is pound per gallon (ppg) then $G_m = 0.052\rho_m$ and has the unit of psi/ft. Note that for the first unloading valve, i = 0 and $D_{v0} = 0$, then Eq. (2.5.6) becomes Eq. (2.5.3).

Stage 4: The process of injecting gas to displace the completion fluid inside the annulus is repeated until gas enters the operating valve. The operating valve is normally the deepest valve in the well. Note that as gas enters the next valve, the valve above it needs to be closed to maximize the lift efficiency. In this particular demonstration, the operating valve is the third one. Therefore, as gas enters the third valve, the bottom-hole pressure, P_{wf3}, must be low enough for the well to produce hydrocarbon at the desired rate as the initial design (Fig. 2.38).

The well is now ready for normal production. Gas is injected constantly into the annulus at a designed rate. The operating valve is the sole valve allowing gas to enter the tubing from the annulus. The fluid inside the tubing now is a mixture of hydrocarbon and gas. Gas is normally separated out of the mixture at the surface, filtered, compressed, and reinjected into the well.

2.6 Continuous Gas Lift Design

In this section, we will incorporate the PVT presented in Chap. 1, Gas Lift Valve and the Unloading sections in this chapter to completely design a gas lift system. The main objectives of designing a continuous gas lift system are to determine the following items:

- Volumetric flow rate of injection gas, Q_g, and surface injection operating pressure P_{so}.
- Numbers of valves and the location of unloading valves and operating valve.
- Type and size of gas lift valve.
- Size of valve port (port size).
- Closing and opening pressures for each valve.
- Test rack opening pressure for each valve.

To achieve these objectives, the common following inputs are required:

- Inflow Performance Relationship (IPR) and Outflow Performance Relationship (OPR).
- Well geometry: true vertical depth, measured depth, inclination angle (or dogleg severity), sizes of tubing and casings.
- Pressure rating of surface equipment such as choke, separator, flowline, etc.
- Formation fluid properties: API gravity of oil, water cut, formation producing gas liquid ratio (GLR), oil and gas viscosities, oil and gas densities, and formation volume factor. Some of these parameters can be calculated using PVT correlations presented in Chap. 1.
- Maximum injection gas pressure. This will be used as the kick-off pressure.
- Specific gravity, pressure gradient of the injection gas.
- Surface and bottom-hole temperature and the geothermal gradient.

The design of a continuous gas lift system consists of the determination of the following items (1) IPR and OPR; (2) Location of the operating gas lift valve; (3) Locations of unloading gas lift valves; (4) Volume of gas injection; (5) Selection of gas lift valves; (6) Calibration of selected gas lift valve [1], [6], and [7].

2.6.1 Determination of IPR and OPR

2.6.1.1 Inflow Performance Relationship (IPR)

If the reservoir pressure is greater than the bubble point pressure (undersaturated oil reservoir), Eq. (1.5.5) can be used as the IPR to describe the radial flow in the porous media. On the other hand, if the reservoir pressure is smaller than the bubble point pressure (saturated oil reservoir), two-phase flow exits in the reservoir. Equation (1.5.10) can be used as the IPR to describe the radial flow in the porous media.

To fully characterize the flow in the reservoir using either Eq. (1.5.5) or Eq. (1.5.10), the flowing inputs must be known: reservoir pressure (P_r), flowing bottom-hole pressure (P_{wf}), bubble point pressure (P_b), radius of the reservoir (r_e), radius of the wellbore (r_w), formation permeability (k), payzone thickness (h), oil formation volume factor (B_o), and oil viscosity (μ_o). The P_b, B_o, and μ_o can be obtained from the PVT analysis as presented in Chap. 1, Sect. 1.4.

2.6.1.2 Outflow Performance Relationship (OPR)

The OPR describes the relationship between flow rate (Q) inside the production tubing and the flowing bottom-hole pressure. The flowing bottom-hole pressure is the summation of hydrostatic pressure, separator pressure, wellhead pressure, and frictional pressure loss in the surface flowline and in the production tubing. To have natural flow, the reservoir pressure has to be high enough to overcome the frictional pressure drop inside the reservoir, pressure drop near the wellbore, frictional pressure drop inside the production tubing and in the flowline, hydrostatic pressure, the choke and separator pressure.

$$P_r > \Delta P|_{reservoir}^{friction} + \Delta P|_{near\,wellbore} + \left(\Delta P|_{tubing}^{friction} + \Delta P|_{flowline}^{friction} + \rho_f g TVD + P_{wh} + P_{sep}\right) \quad (2.6.1)$$

The terms in the bracket of Eq. (2.6.1) is the flowing bottom-hole pressure or the back pressure applying to the reservoir at the center of the perforation.

$$P_{wf} = \Delta P|_{tubing}^{friction} + \Delta P|_{flowline}^{friction} + \rho_f g TVD + P_{wh} + P_{sep} \quad (2.6.2)$$

If natural flow does not exist or the natural flow is too low to be considered as a profitable rate, gas needs to be injected into the well to reduce the hydrostatic pressure and hence reduce the flowing bottom-hole pressure. Figure 2.39a shows a well that does not give any natural flow. As gas is being injected into the well at the surface operating pressure of P_{so}, gas liquid ratio is increasing from GLR2 to GLR3

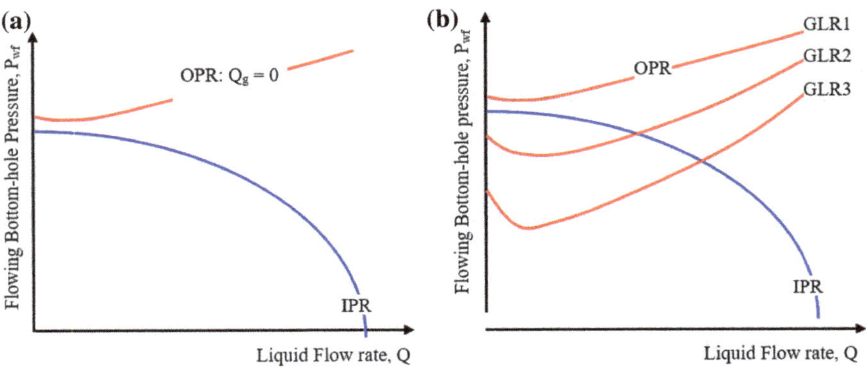

Fig. 2.39 **a** Initial OPR—injection gas rate = 0, **b** OPR with different injection gas rates

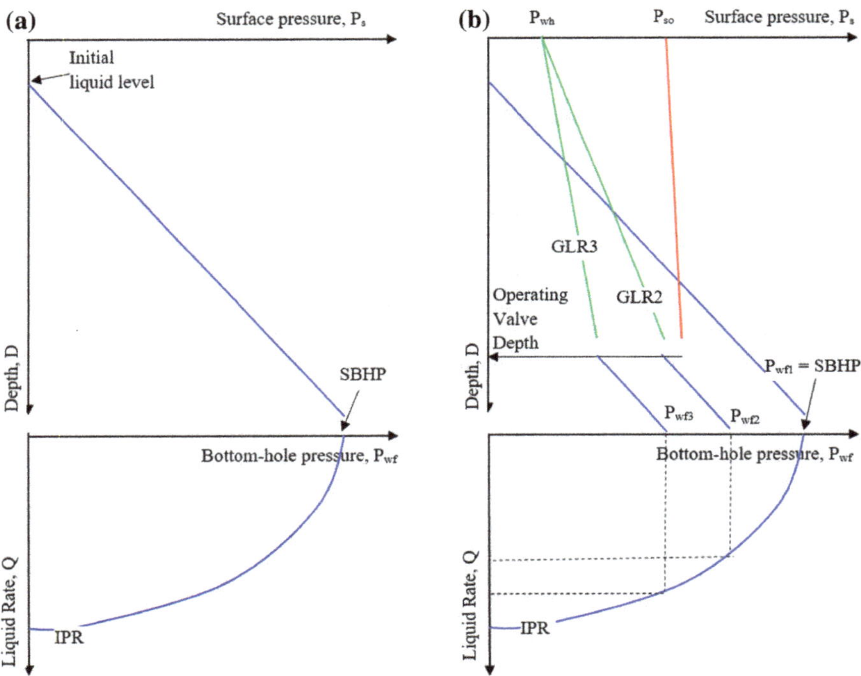

Fig. 2.40 **a** Pressure gradient of formation fluid inside the tubing before gas injection, **b** pressure gradient of fluids inside the tubing after gas injection

as shown in Fig. 2.39b. The well is now producing at liquid rates of Q_2 and Q_3 due to the reduction in the flowing bottom-hole pressure from P_{wf2} to P_{wf3}, respectively.

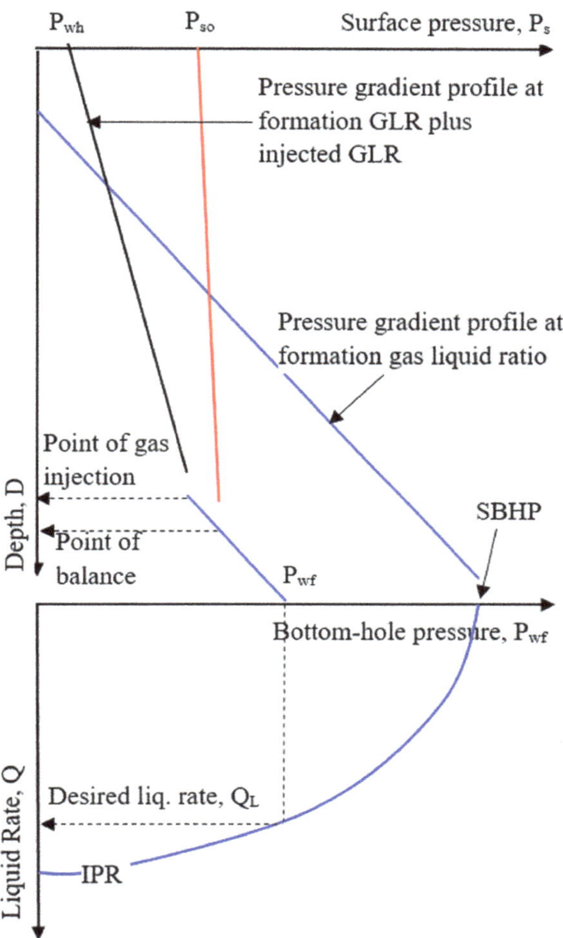

Fig. 2.41 Location of the operating valve

The scenarios described in Fig. 2.39a, b can be illustrated in a different ways as shown in Figs. 2.40a, b, respectively.

2.6.2 Determination of Operating Valve Location

The location of the operating valve needs to be determined in the second step. From the design point of view, the desired total liquid rate, Q_L, is a known parameter. From Fig. 2.39b, the desired flowing bottom-hole pressure, P_{wf}, can be achieved with this Q_L. Note that the static bottom-hole pressure and the formation gas liquid ratio, are the two inputs. Therefore, the pressure gradient profile under static conditions of the formation fluid inside the tubing can be obtained as shown in Fig. 2.41. Assuming the frictional pressure loss inside the tubing is much smaller than the hydrostatic pressure.

Thus, the pressure gradient profile of the formation fluid under flowing conditions will have the same slope (parallel) to the pressure gradient profile of the formation fluid under static condition. From the desired flowing bottom-hole pressure, draw a line parallel to the pressure gradient profile at formation gas liquid ratio. This line intersects with the annular injected gas gradient (red line shown in Fig. 2.41) at a point called point of balance. For gas to flow through the operating gas lift valve, the tubing pressure at the valve depth (pressure at point of gas injection shown in Fig. 2.41) must be smaller than the pressure at the point of balance. This pressure reduction is approximately the same as the pressure drop when gas flows through the operating gas lift valve. This pressure reduction value depends on the port size, gas

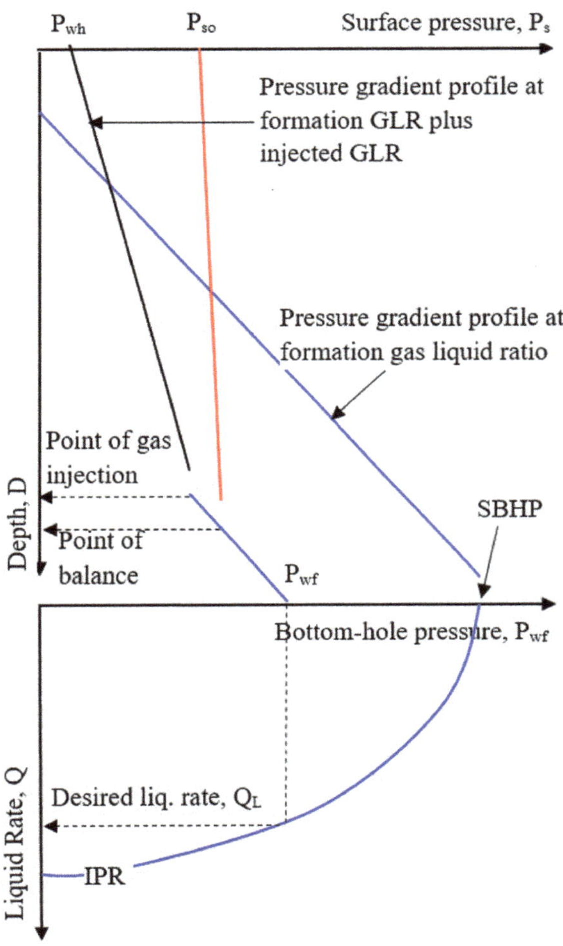

Fig. 2.42 Location of the operating valve

injection rate, and valve geometries. Thornhill Craver equation as presented in Eq. (2.4.4) or (2.4.5) can be used to predict this pressure drop across a sharp-edged orifice gas lift valve. A pressure of 100 psi can be a good approximate value for this pressure drop. The operating valve depth location can be now determined.

2.6.3 Determination of Unloading Valve Location

To determine the number and location of unloading gas lift valves, the design tubing pressure gradient need to be plotted. To have more flexible operations during the unloading process, the designed wellhead pressure is selected as (P_{wh} + $0.2P_{so}$) or (P_{wh} + 200) whichever is greater. The design tubing pressure gradient line can be attained by connecting the designed wellhead pressure and the point of injection as shown in Fig. 2.42.

The location of the first unloading gas lift vale can be achieved by drawing a kill fluid pressure gradient line starting from the operating wellhead pressure. This line intersects with the kick-off pressure at one point as shown in Fig. 2.42. This point is the depth of the first unloading gas lift valve. From the location of valve 1, draw a horizontal line to the left until intersecting the design tubing pressure line. From this intersection, draw a line downward parallel to the previous kill fluid pressure gradient line until intersecting the operating gas injection pressure gradient line. This is the depth of the second valve. Repeat this process until reaching the point of gas injection. According to Fig. 2.42, there are a total of nine gas lift valves including eight unloading valves and one operating valve.

Readers are highly recommended to review the “Multiple Valve Gas Lift Unloading” Sect. 2.5 to understand the true meaning beside this graphical method.

2.6.4 Determination of Injection Gas Rate

The main purpose of injecting gas into the production tubing is to reduce the back pressure applied on the reservoir at the perforation so the well can produce at a desired rate. In other words, how much gas injected into the production tubing mainly depends on the desired production liquid rate and the formation gas liquid ratio.

From the design tubing pressure gradient as shown in Fig. 2.41, the total required GLR can be achieved graphically or analytically using available two-phase flow models presented in Chap. 1, Sect. 1.6.3. The volumetric gas injection rate at surface condition can be now determined as follows:

$$Q_g^s = (TGLR - GLR_F)Q_L \tag{2.6.3}$$

where Q_g^s, Q_L are the volumetric gas injection rate in standard cubic feet per barrel of liquid (scf/B) at surface conditions and the desired liquid rate in barrel of liquid per day (BLPD); TGLR, GLR_F are the total gas liquid ratio in the tubing above the point of injection and the gas liquid ratio of the formation fluids below the point of injection (formation gas liquid ratio).

Note that the volumetric gas injection rate calculated using Eq. (2.6.3) is the gas rate under surface conditions. To convert this rate to the valve depth conditions, the temperature correction factor needs to be taken into consideration. The temperature correction factor can be given as:

$$TCF = 0.0544\sqrt{\gamma_g T_{vd}} \tag{2.6.4}$$

where γ_g is the gas specific gravity; T_{vd} is the temperature at valve depth in Rankin.

The correct volumetric gas injection rate at valve depth is calculated as:

$$Q_g^{vd} = Q_g^s \times TCF \tag{2.6.5}$$

2.6.5 *Selection of Gas Lift Valves*

2.6.5.1 Selection of Port Size

The port size (orifice size) of a gas lift valve can be determined based on the amount of injected gas and the pressure drop across the orifice. Thornhill Craver model can be used to predict the port size if the gas injection rate and the upstream,

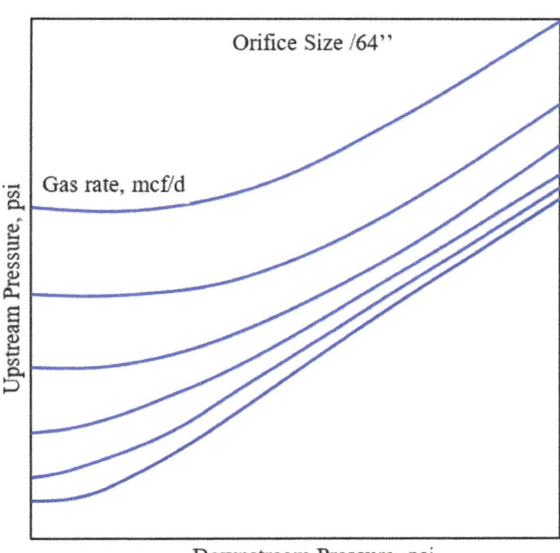

Fig. 2.43 Example of one particular square edge orifice size gas lift valve

downstream pressures are given. Upstream pressure is the pressure of injection gas in the annulus at valve depth and downstream pressure is the pressure of the fluid inside the production tubing at valve depth. If the surface operation gas pressure, P_{so}, is known, the upstream pressure can be calculated using Eq. (1.3.22). The downstream pressure can be estimated based on the design tubing pressure line at valve depth.

Note that the amount of gas injection for the operating valve and for the unloading valves is different. The volumetric gas injection rate for the operating valve is estimated as described in Sect. 2.6.4. The volumetric gas injection rate for the unloading valves should be maintained at low rate to avoid damages to the valves. The recommended injection gas rate for unloading valves is presented in detail in Sect. 2.5.2 "Description of Unloading Process".

Gas lift valve manufacturers also provide charts showing the performance of a specific gas lift valve. These charts are developed by changing the upstream and downstream pressures (pressure drop across the valve) and measuring the gas rate flowing through the valve. If gas lift valve performances (charts) from manufacturers are available, it is recommended to use these charts for selecting the port size instead of using Thornhill Craver or other available models. An example of a particular orifice size gas lift performance is shown in Fig. 2.43.

2.6.5.2 Estimation of Opening Pressure

IPO valves are the most common valves selected in a continuous gas lift well in comparison to PPO valves because they are easy to operate and to control. The opening pressure is the injection gas pressure in the annulus at valve depth maintained to allow gas to flow through the valve. If the surface operation gas injection pressure, P_{so}, is chosen then the opening pressure at valve depth can be calculated using Eq. (1.3.22).

$$P_{vo} = P_{so} e^{\left(\frac{0.01877 \gamma_g D_{vd}}{z T_{ave}}\right)} \tag{1.3.22}$$

2.6.5.3 Calculation of Tubing Pressure

The tubing pressure above the point of injection can always be predicted using the design tubing pressure gradient line as shown in Fig. 2.41. At any given depth, draw a horizontal line until intersecting the design tubing pressure gradient, extend upward to get the tubing pressure at this depth.

Table 2.2 Summary of gas lift valve design parameters

Valve no.	TVD ft	Temp. F	Valve type	Port size	R	P_{so} psi	P_{vo} psi	P_t psi	$P_{vc} = P_d$ psi	P_d (60 °F)	P_{TRO} psi
1			IPO								
2			IPO								
3			IPO								
4			IPO								
5			IPO								

2.6.5.4 Estimation of Closing Pressure (Dome Pressure at Valve Depth)

Recall the equation to calculate the opening pressure for an IPO valve.

$$P_c = P_{vo} = \frac{P_d}{1 - R} - \frac{R}{1 - R} P_t \tag{2.4.12}$$

In Eq. (2.4.12), R is the ratio between the port and the bellow areas. This R value is normally known after selecting the port size. The opening pressure and the tubing pressure are estimated as shown in Sects. 2.6.5.2 and 2.6.5.3. Using Eq. (2.4.12), one can estimate the dome pressure. In addition, dome pressure is the closing pressure for IPO valves.

2.6.5.5 Estimation of Dome Pressure at Surface

If the dome charged pressure at valve depth is known, the dome charged pressure at surface conditions (60 °F) can be obtained using Winkler and Eads [10] equations as shown in Eqs. (2.4.27) or (2.4.28). The test rack opening pressure can also be calculated after obtaining the dome pressures at surface using Eq. (2.4.31).

2.6.6 Discussion on Valve Spacing

Gas lift valve design is very forgiving. As long as gas is injected into the tubing, the flowing bottomhole pressure will reduce leading to an increase in liquid production rate. The heart of designing a gas lift system is to optimize the valve spacing in such a way that the least amount of gas energy is used at surface to obtain the same amount of liquid production rate at surface. Some companies even set the same distance between valves. This practice of course is not recommended because it may restrict production severely when wells, especially high productivity wells, are put into production for a period of time.

When using the geographic method for spacing gas lift valves, if the distance between two valves is less than 200 ft, it is a common practice to space valves 200 ft apart until reaching the operating valve depth. Please note that the minimum value of spacing (200 ft) is totally depending on companies' philosophy.

Cost to complete or to do a work-over of an offshore well is very expensive. Therefore, it is a good practice to run an extra a few more valves (a few more mandrels) in the beginning rather than to take a chance on too few valves. Dummy valves can be used to set into these extra mandrels. If needed, these dummy valves can be replaced by regular IPO valve easily using wireline running tool.

To support the selection of gas lift valves, the information of valves should be summarized in a template as shown in Table 2.2.

2.7 Examples

Example 2.1 Calculate the gas pressure at valve depth in the casing annulus for a gas lift well with the following information: valve depth of 8,000 ft; surface temperature of 80 °F; temperature at valve depth of 160 °F; surface operating injection gas pressure of 800 psi; gas specific gravity of 0.65.

Solution
Assuming this is an ideal gas with Z = 1 and gas temperature in the casing annulus increases linearly. The average gas temperature in the casing annulus is (80 + 160)/2 = 120 °F. The gas pressure in the casing annulus at valve depth can be estimated as follows:

$$P_{8{,}000\,\text{ft}} = P_{so}e^{\left(\frac{0.01877\gamma_g D_{vd}}{ZT_{ave}}\right)} = 800e^{\left(\frac{0.01877\times 0.65\times 8{,}000}{1\times(120+460)}\right)} = 947\,\text{psi}$$

If the compressibility factor Z = 0.95, $P_{8{,}000ft} = 955\,\text{psi}$.

Example 2.2 A gas lift valve has a dome charged pressure of 600 psi at a surface temperature of 60 °F. What is the dome charged pressure of this valve at valve depth of 8,000 ft where the temperature is 140 °F?

Solution
Applying Winkler and Eads model for the case of $P_d^{'} < 1{,}238$ psi gives

$$P_d = P_d^{'} + \left[-0.00226 + 0.001934\left(P_d^{'}\right) + 3.054\times 10^{-7}\left(P_d^{'}\right)^2\right](T_{vd} - 60)$$

$$\begin{aligned} P_d &= 600 + \left[-0.00226 + 0.001934(600) + 3.054\times 10^{-7}(600)^2\right](140 - 60) \\ &= 701.4\,\text{psia} \end{aligned}$$

Using the non-ideal gas law equation, one can estimate the dome pressure at valve depth by rearranging Eq. (2.4.32):

$$P_d = \frac{Z_{vd} T_{vd} P'_d}{520 Z_{60F}}$$

Using the ratio $Z_{60F}/Z_{vd} = 0.985$, the above equation becomes

$$P_d = \frac{T_{vd} P'_d}{520 \times 0.985} = \frac{(140 + 460)600}{520 \times 0.985} = 702.8\,\text{psia}$$

The difference between the Winkler and Eads model and the non-ideal gas law using the ratio $Z_{60F}/Z_{vd} = 0.985$ in this example is about 0.2% which is practically negligible.

Example 2.3 Using Thornhill Craver equation to predict the gas rate in Mscf/D through a gas lift valve with the following inputs: natural gas with the specific heat capacity ratio of 1.3; gas temperature in the casing annulus at valve depth of 120 °F; gas specific gravity of 0.65; orifice size of 16/64″; upstream pressure of 1,100 psi and downstream pressure of 900 psi.

Solution

The ratio $P_d/P_u = 0.818 > 0.546$, applying Thornhill Craver equation for sub-critical flow and for natural gas with the specific heat capacity ratio of 1.3 gives:

$$Q_g^{sc}\left(\frac{MSCF}{D}\right) = 0.43 d_o^2 P_u \sqrt{\frac{1}{\gamma_g (T_u + 460)}} \sqrt{\left(\frac{P_d}{P_u}\right)^{1.52} - \left(\frac{P_d}{P_u}\right)^{1.76}}$$

$$Q_g^{sc} = 0.43(16)^2(1,100 + 14.7)\sqrt{\frac{1}{0.65(120 + 460)}}\sqrt{(0.818)^{1.52} - (0.818)^{1.76}}$$

$$Q_g^{sc} = 1152\,MSCF/D$$

Example 2.4 Using Thornhill Craver equation to predict the orifice size of a gas lift valve with the following inputs: natural gas with the specific heat capacity ratio of 1.3; gas temperature in the casing annulus at valve depth of 176 °F; gas specific gravity of 0.7; gas rate of 401 MSCF/D; upstream pressure of 1,000 psi and downstream pressure of 850 psi.

Solution

The ratio $P_d/P_u = 0.85 > 0.546$, applying Thornhill Craver equation for subcritical flow and for natural gas with the specific heat capacity ratio of 1.3 gives:

$$Q_g^{sc}\left(\frac{MSCF}{D}\right) = 0.43 d_o^2 P_u \sqrt{\frac{1}{\gamma_g (T_u + 460)}} \sqrt{\left(\frac{P_d}{P_u}\right)^{1.52} - \left(\frac{P_d}{P_u}\right)^{1.76}}$$

$$d_o^2 = \frac{Q_g^{sc}}{0.43 P_u \sqrt{\frac{1}{\gamma_g (T_u + 460)}} \sqrt{\left(\frac{P_d}{P_u}\right)^{1.52} - \left(\frac{P_d}{P_u}\right)^{1.76}}}$$

$$d_o^2 = \frac{401}{0.43(1,000 + 14.7)\sqrt{\frac{1}{0.7(176+460)}}\sqrt{(0.85)^{1.52} - (0.85)^{1.76}}}$$

$$d_o^2 = 112.2$$

$$d_o = 10.6$$

The orifice size: $d_o = 10.6/64''$

Example 2.5 An IPO valve is installed at 8,000 ft with a dome charged pressure of 700 psi under the valve depth conditions. The tubing pressure at 8,000 ft is 600 psi. Calculate the valve opening pressure and the spread if the value of R = 0.1.

Solution

The tubing effect factor:

$$TEF = \frac{R}{1 - R} = \frac{0.1}{1 - 0.1} = 0.1111$$

The tubing effect:

$$TE = \frac{R}{1 - R} P_t = 0.1111 \times 600 = 66.67\,\text{psi}$$

For IPO valves, the opening pressure is the gas pressure in the casing annulus at valve depth and expressed as follows:

$$P_c = P_{vo} = \frac{P_d}{1 - R} - \frac{R}{1 - R} P_t$$

$$P_c = P_{vo} = \frac{700}{1 - 0.1} - 66.67 = 711.11\,\text{psi}$$

If the tubing is completely empty with P_t = 0 psi, the valve opening pressure is 777.78 psi. This is the maximum opening pressure. In other words, the tubing pressure helps to reduce the gas pressure in the casing annulus to open the valve. This is the true meaning of the tubing effect.

For IPO valves, the valve closing pressure is the same as the dome pressure: $P_{vc} = P_d = 700$ psi.

Spread $\Delta P = P_{vo} - P_{vc} = 711.11 - 700 = 11.11\,\text{psi}$

If spread is known, the valve opening pressure can also be calculated as:

$$P_{vo} = P_d + \Delta P = 700 + 11.11 = 711.11\,\text{psi}$$

Note that spread can be positive or negative. The positive spread means the injection gas pressure in the casing annulus at valve depth must be greater than the dome pressure at valve depth for the valve to open. The negative spread means the injection gas pressure in the casing annulus at valve depth is smaller than the dome pressure for valve to open.

Example 2.6 A double element PPO valve with the spring tension of 600 psi is being used at the depth of 8,000 ft. The temperature at valve depth and at surface are 180 °F and 90 °F, respectively. The pressure for valve to close at the valve depth is 1,800 psi. The surface operating injection gas pressure is 900 psi. The value of R is 0.0407. Calculate the dome pressure at valve depth and at surface with temperature of 60 °F and calculate the spread of the valve.

Solution

For a PPO valve, pressure to close the valve is the tubing pressure as shown in Eq. (2.4.24)

$$P_t = P_{vc} = P_d + S_t(1 - R)$$

The dome pressure at valve depth:

$$P_d = P_{vc} - S_t(1 - R) = 1,800 - 600(1 - 0.0407) = 1,224.4\,\text{psi}$$

The dome pressure at surface with temperature of 60 °F ($Z_{60F}/Z_{vd} = 0.985$):

$$P_d^{'} = \frac{520 Z_{60F} P_d}{Z_{vd} T_{vd}} = \frac{520 \times 0.985 \times 1224.4}{(180 + 460)} = 980\,\text{psi}$$

The gas pressure in the casing annulus at valve depth with average temperature of 135 °F:

$$P_c = P_{so} e^{\left(\frac{0.01877 \gamma_g D_{vd}}{Z T_{ave}}\right)} = 900 e^{\left(\frac{0.01877 \times 0.7 \times 8,000}{1 \times (135 + 460)}\right)} = 1,074\,\text{psi}$$

The casing effect:

$$CF = \frac{R}{1 - R} P_c = 45.56\,\text{psi}$$

The opening pressure for a PPO valve:

$$P_t = P_{vo} = \frac{P_d}{1-R} - \frac{R}{1-R} P_c + S_t$$

$$P_{vo} = \frac{1,224.4}{1-0.0407} - 45.56 + 600 = 1,830\,\text{psi}$$

The PPO valve spread

$$\Delta P = P_{vo} - P_{vc} = 1,830 - 1,800 = 30\,\text{psi}$$

Example 2.7 The following data are given for a single element nitrogen dome charged IPO valve: valve depth of 8,000 ft; surface operation injection pressure of 800 psi; valve opening pressure at surface of 800 psi; tubing pressure at valve depth of 655 psi; gas specific gravity of 0.7; wellhead and valve depth temperatures of 100 °F and 180 °F, respectively; R ratio of 0.2562. Calculate:

a. Opening and closing pressure at valve depth
b. Spread at valve depth and at surface
c. Test rack opening pressure.

Solution

a. Opening and closing pressure at valve depth
   Because the valve opening pressure at surface is 800 psi, the opening pressure at valve depth can be estimated as follows with the average gas temperature in the casing annulus of 140 °F:

$$P_{vo}^{vd} = P_{so} e^{\left(\frac{0.01877\gamma_g D_{vd}}{ZT_{ave}}\right)} = 800 e^{\left(\frac{0.01877 \times 0.7 \times 8,000}{1 \times (140+460)}\right)} = 953\,\text{psi}$$

As proven, the opening pressure for an IPO valve can also be calculated as

$$P_{vo}^{vd} = \frac{P_d}{1-R} - \frac{R}{1-R} P_t$$

Rearrange this equation gives the dome pressure at valve depth:

$$P_d = (1-R)P_{vo} + RP_t = (1-0.2562)953 + 0.2562 \times 655 = 877\,\text{psi}$$

Because this is a IPO valve and hence the closing pressure is the same as the dome pressure at valve depth:

$$P_{vc}^{vd} = P_d^{vd} = 877\,\text{psi}$$

b. Spread at valve depth and at surface

Spread at valve depth:

$$\Delta P_{vd} = P_{vo}^{vd} - P_{vc}^{vd} = 953 - 877 = 76\,\text{psi}$$

Dome pressure at surface with temperature of 100 °F can be predicted using Winkler and Eads model:

$$P_d = P_d^{'} + \left[-0.00226 + 0.001934\left(P_d^{'}\right) + 3.054 \times 10^{-7}\left(P_d^{'}\right)^2\right](T_{vd} - 100)$$

$$877 = P_d^{'} + \left[-0.00226 + 0.001934\left(P_d^{'}\right) + 3.054 \times 10^{-7}\left(P_d^{'}\right)^2\right](180 - 100)$$

Solving this equation give $P_d^{'} = 747\,\text{psi}$
The valve closing pressure at surface conditions:

$$P_{vc}^{surf} = P_d^{'} = 747\,\text{psi}$$

Spread at surface

$$\Delta P_{surf} = P_{vo}^{surf} - P_{vc}^{surf} = 800 - 747 = 53\,\text{psi}$$

c. Test rack opening pressure
The dome pressure at a temperature of 60 °F:

$$P_d^{'} = \frac{520 Z_{60F} P_d}{Z_{vd} T_{vd}} = \frac{520 \times 0.985 \times 877}{(180 + 460)} = 701\,\text{psi}$$

Test rack opening pressure at 60 °F for the case of $S_t = 0$ psi can be estimated as follows:

$$P_{TRO} = \frac{P_d^{'}}{1 - R} = \frac{701}{1 - 0.2562} = 943\,\text{psi}$$

Example 2.8 Please design a gas lift system for a well with the following information: desired optimum liquid rate; tubing size ID of 2.75 in.; injection gas specific gravity of 0.7; well loaded with kill fluid pressure gradient of 0.45 psi/ft; formation producing gas oil ratio of 400 SCF/STB; average reservoir pressure of 1,920 psi; bubble point pressure of 1,500 psi; formation permeability of 15 mD; reservoir pay thickness of 360 ft; reservoir radius of 1000 ft; skin factor of 0; oil API of 30 °API; wellhead pressure of 120 psi; surface operating gas injection pressure of 600 psi; kickoff pressure of 700 psi; top of perforation depth of 8,000 ft; surface and bottomhole temperature of 105 °F and 170 °F, respectively.

Solution

In this example, we are going to follow Sect. 2.6.

Step 1: Determination of IPR and OPR without Injecting Gas into the Well

Before calculating the IPR, PVT calculation should be done first.

Oil specific gravity:

$$\gamma_o = \frac{141.5}{131.5 + {}^\circ API} = \frac{141.5}{131.5 + 30} = 0.876$$

Applying Standing correlations for estimating the R_s and B_o gives:

$$x = 0.0125API - 0.00091T = 0.0125 \times 30 - 0.00091 \times 170 = 0.22$$

$$R_s = \gamma_g \left[\left(\frac{P}{18.2} + 1.4 \right) 10^x \right]^{1.2048} = 0.7 \left[\left(\frac{1920}{18.2} + 1.4 \right) 10^{0.22} \right]^{1.2048} = 359\, SCF/STBO$$

$$B_o = 0.972 + 0.000147 \left[R_s \left(\frac{\gamma_g}{\gamma_o} \right)^{0.5} + 1.25T \right]^{1.175}$$

$$B_o = 0.972 + 0.000147 \left[359 \left(\frac{0.7}{0.876} \right)^{0.5} + 1.25 \times 170 \right]^{1.175} = 1.2\, bbl/STBO$$

For calculating oil viscosity, Beggs and Robinson model is used.

$$x = \frac{10^{(3.0324 - 0.0202 \times {}^\circ API)}}{T^{1.163}} = \frac{10^{(3.0324 - 0.0202 \times 30)}}{170^{1.163}} = 0.679$$

$$\mu_o = 10^x - 1 = 10^{0.679} - 1 = 3.78\, cp$$

The IPR of this reservoir will consist of two parts: a linear part when the bottomhole pressure, P_{wf}, is higher than the bubble point pressure, P_b of 1,500 psi, and a curve when the P_{wf} is smaller than the P_b.

For P_{wf} = 1,500–1,920 psi, the IPR is a straight line, which the slope is the productivity index, PI. In terms of average reservoir pressure, PI is shown in Eq. (1.5.8) as follows:

$$PI = \frac{q}{\bar{P}_r - p_{wf}} = \frac{kh}{141.2 B_o \mu \left[ln \frac{r_e}{r_w} - 0.5 + S \right]}$$

$$PI = \frac{15 \times 360}{141.2 \times 1.2 \times 3.78 \left[ln \frac{1000}{(0.23)} - 0.5 + 0 \right]} = 1.0 \frac{bbl}{D} /\text{psi}$$

For P_{wf} < 1,500 psi, the IPR is a curve. Using Vogel model shown in Eq. (1.5.10) gives

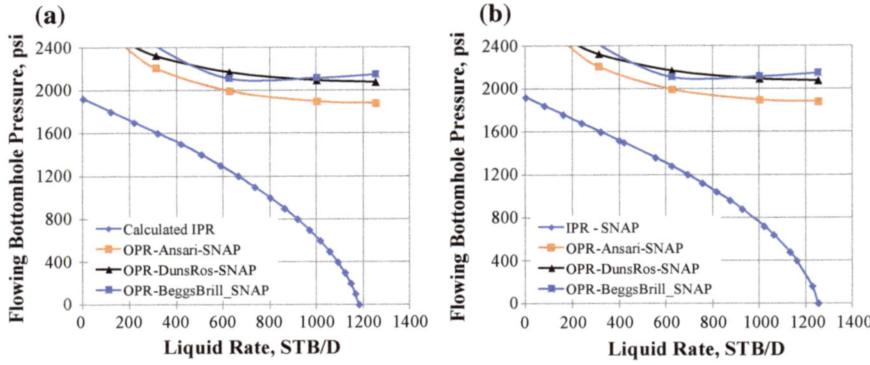

Fig. 2.44 **a** Calculated IPR and OPR using SNAP ID_T = 2.75 in.; P_{wh} = 250 psi; TVD = 8,000 ft; P_r = 1,920 psi; P_b = 1,500 psi; k = 15 mD; h = 360 ft, b: IPR and OPR curves using SNAP ID_T = 2.75 in.; P_{wh} = 250 psi; TVD = 8,000 ft; P_r = 1,920 psi; P_b = 1,500 psi; k = 15 mD; h = 360 ft

$$\frac{q_o}{q_{o,max}} = 1 - 0.2\frac{P_{wf}}{\bar{P}} - 0.8\left(\frac{P_{wf}}{\bar{P}}\right)^2$$

At P_{wf} = 1,500 psi, using PI = 1 bbl/D/psi gives q_o = 419.5 bbl/D and the ratio $\frac{P_{wf}}{\bar{P}} = 0.78$. Therefore, the the maximum oil rate can be estimated using Vogel model as: $q_{o,max} = 1180\,bbl/D$. The complete calculated IPR is shown in Fig. 2.43a

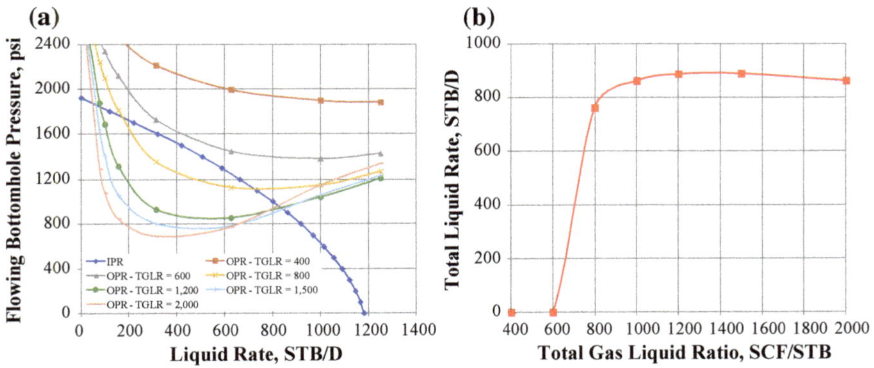

Fig. 2.45 Effect of total gas liquid rate on the performance of the well with D_{Inj} of 7,000 ft

SNAP commercial software was used to obtain the IPR and OPR with the same inputs described in this example. Using Ansari, Duns-Ros, and Beggs-Brill models with the formation producing gas oil ratio of 400 SCF/STB, tubing ID of 2.75 in., wellhead pressure of 250 psi, the OPR curves predicted by SNAP are shown in Fig. 2.44a, b.

Note that the OPR curves in Fig. 2.44a, b are both from the prediction of SNAP using Ansari, Duns-Ros and Beggs-Brill models and hence they are exactly the same. The calculated IPR curve and the IPR predicted using SNAP are slightly different. This difference can be from the accumulation errors while proceeding manually calculations.

The IPR and OPR curves reveal that the well does not give any production under natural flow. Therefore, gas lift is needed to put this well back to production. To determine the optimum liquid rate for this well, a sensitive analysis for the Total Gas Liquid Ratio (TGLR) is needed. Using SNAP, the results of this analysis when the injection depth of 7,000 ft are shown in Fig. 2.45a, b. Note that with the formation producing GLR is 400 SCF/STB, the TGLR of 400 SCF/STB means the surface gas injection rate is zero. As TGLR increases from 400 to 2000 SCF/STB, the OPR curves shift down because of less hydrostatic pressure or less back pressure applied on the reservoir. When the TGLR increases from 1,200 to 1,500

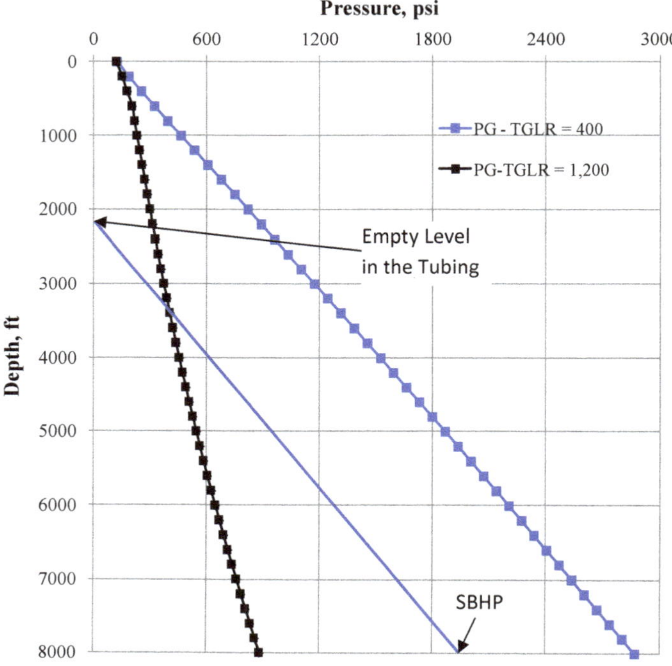

Fig. 2.46 Pressure gradient curves: ID = 2.75 in.; P_{wh} = 120 psi; WC = 0

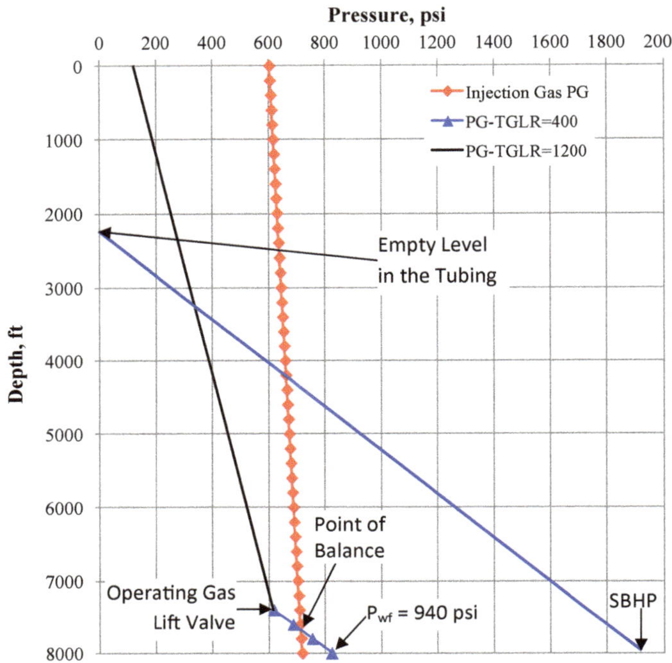

Fig. 2.47 Determination of operating valve location

SCF/STB, the liquid production rate remains more or less the same as around 840 B/D. Keep increasing the TGLR to 2,000 SCF/STB, the liquid production rate gets decreasing. This is because the frictional pressure drop in the tubing is now dominated over the hydrostatic pressure. Therefore, the optimal TGLR in this case is chosen at 1200 SCF/STB and the optimal liquid production rate is expected as about 840 B/D.

Step 2: Determination of Operating Valve Location

Using Ansari two-phase flow model in the tubing with the ID = 2.75 in.; wellhead pressure of 120 psi, WC = 0, the pressure gradient curves for the TGLR of 400 and 1200 SCF/STB are shown in Fig. 2.46. Assuming these two pressure gradients follow a linear relationship. The slopes of the straight lines with TGLR of 400 and TGLR of 1,200 SCF/STB are 0.095 psi/ft and 0.343 psi/ft, respectively. From the SBHP of 1,920, extend a line with a slope of 0.343 (parallel to the 1,200-TGLR pressure gradient line) until intersecting the y-axis at the value of 2,200 ft. The 2,200 ft is the empty level (void level) inside the tubing.

At the optimal liquid rate of 840 STB/D, the optimal flowing bottomhole pressure can be determined from Fig. 2.45 as P_{wf} = 940 psi. From P_{wf} = 940 psi, on Fig. 2.47, draw a line parallel to the pressure gradient line with the

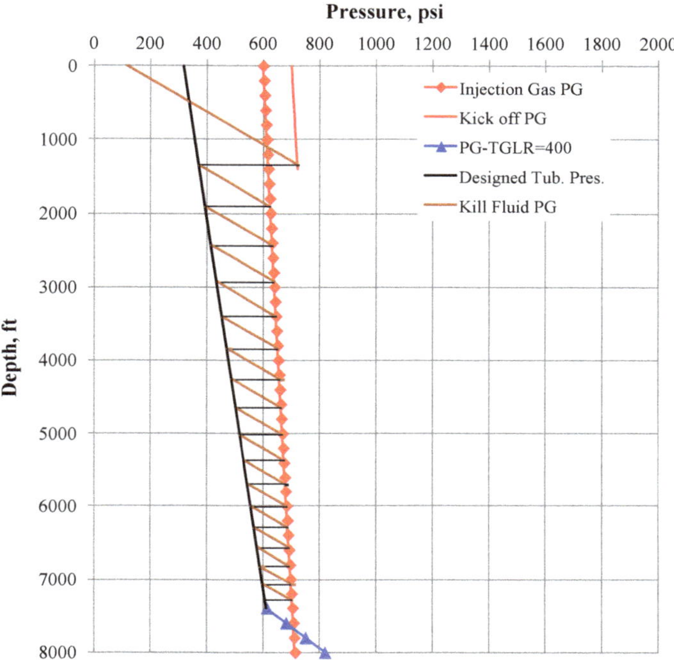

Fig. 2.48 Determination of unloading valve location

TGLR = 1,200 SCF/STB as demonstrated in Fig. 2.45. This line will intersect with the gas injection pressure gradient line at a point called the point of balance where the pressure is about 700 psi. The point of injection (location of the operating gas lift valve) can be determined by subtracting 100 psi from the pressure at the point of balance as shown in Fig. 2.47. The depth of the injection point is 7,400 ft. Connect the point of injection and the wellhead pressure at depth of zero gives the two-phase pressure gradient line with the TGLR of about 1,200 SCF/STB.

Step 3: Determination of Unloading Valve Location

$$P_{wh} + 0.2P_{so} = 120 + 0.2 \times 600 = 240\,\text{psi}$$
$$\text{or } P_{wh} + 200 = 120 + 200 = 320\,\text{psi}$$

It is recommended that the designed wellhead pressure is a greater value between the above two pressures. Therefore, the value of 320 psi is selected as the designed wellhead pressure. Connecting the point of injection and the point at surface with the wellhead pressure of 320 psi gives the designed tubing pressure gradient as shown in Fig. 2.48. From the operating wellhead pressure of 120 psi, extend a line which has a slope the same as that of the kill fluid (0.45 psi/ft) until intersecting the kick-off pressure gradient line of 700 psi. This intersection gives the location of the first unloading valve of 1,350 psi. The depth of the first unloading valve can also be determined analytically using Eq. (2.5.3).

$$D_{v1} = \frac{P_{ko} - P_{wh}}{G_f} = \frac{700 - 120}{0.45} = 1,289\,ft$$

Note that Eq. (2.5.3) neglects the gas hydrostatic pressure. This will cause the valve depth calculated using Eq. (2.5.3) is slightly smaller than that of using the graphical method. If considering the gas hydrostatic pressure, at the depth of 1,350 ft, the kick-off gas pressure is about 720 psi. The first valve depth using Eq. (2.5.3) now is 1,333 ft, which is very similar to that of using the graphical method.
Using the technique described in Sect. 2.6.3, location of the unloading valves is presented in Fig. 2.48. According to the graphical method, the last unloading valve depth is 7,297 ft. As determined above, the depth of the operating valve is 7,400 ft. The spacing between the last unloading valve and the operating valve is 103 ft which is smaller than 200 ft. Therefore, the last unloading valve location is converted to the location of the operating valve. An additional mandrel at the depth of 7,497 ft (200 ft apart from the operating valve depth) should be installed using a dummy valve for safety purpose. Generally speaking, this well should have 16 unloading valves. The seventeenth valve is the operating valve at the depth of 7,297 ft and the eighteenth valve is the dummy one for future operation.

Step 4: Determination of Injection Gas Rate

The optimal TGLR was determined as 1,200 SCF/STB. The formation producing gas liquid ratio (GLR_F) is 400 SCF/STB. The optimal liquid rate is 840 STB/D. Using Eq. (2.6.3) gives the volumetric gas injection rate at surface conditions as follows:

$$\begin{aligned} Q_g^s &= (TGLR - GLR_F)Q_L = (1,200 - 400)840 = 672,000\,SCF/D \\ &= 672\,MSCF/D \end{aligned}$$

To convert the volumetric gas injection rate from surface conditions to valve depth conditions, the temperature correction factor is used as described in Eq. (2.6.4).

$$TCF = 0.0544\sqrt{\gamma_g T_{vd}} = 0.0544\sqrt{0.7(170+460)} = 1.1424$$

The volumetric gas injection rate at valve depth conditions is modified as follows:

$$Q_g^{vd} = TCF \times Q_g^s = 1.1424 \times 672 = 767.7\,MSCF/D$$

Step 5: Selection of Operating Gas Lift Valve

The following calculation demonstration is for the operating valve.

If the surface operating gas injection pressure is maintained as 600 psi, the opening pressure at valve depth of 7,297 ft is estimated as:

$$P_{vo} = P_{so}e^{\left(\frac{0.01877\gamma_g D_{vd}}{zT_{ave}}\right)} = 704\,\text{psi}$$

This is also the upstream pressure of the operating valve. In summary, with the inputs of upstream temperature of 170 °F; gas specific gravity of 0.7; gas rate at valve depth of 767.7 MSCF/D; upstream pressure of 704 psi; downstream pressure of 604.5 psi, using Thornhill Craver equation gives the orifice size (port size) of 18/64″. Readers are recommended to review Example 2.4 in this chapter. With this port size, the R value is selected as 0.25 for this valve.

Table 2.3 Shows the summary of the calculation for all valves

Valve #	Valve type	Depth ft	Temp F	Pvo psi	R	P_t psi	$P_d = P_vc$ psi	Spread psi	P_d @ 60 °F	PTRO psi
1	Unloading	1350	117	722	0.25	374	702.1	20	623	831
2	Unloading	1911	122	627	0.25	396	606.8	20	534	712
3	Unloading	2439	127	634	0.25	417	614.3	20	536	715
4	Unloading	2936	131	641	0.25	437	621.4	20	538	718
5	Unloading	3405	135	648	0.25	455	628.1	20	540	721
6	Unloading	3846	139	654	0.25	473	634.4	20	542	723
7	Unloading	4262	143	660	0.25	489	640.4	20	544	725
8	Unloading	4653	146	666	0.25	505	646.0	20	546	727
9	Unloading	5022	150	671	0.25	519	651.3	20	547	730
10	Unloading	5369	153	676	0.25	533	656.4	20	549	731
11	Unloading	5696	156	681	0.25	546	661.1	20	550	733
12	Unloading	6005	158	686	0.25	559	665.6	20	551	735
13	Unloading	6295	161	690	0.25	570	669.8	20	552	737
14	Unloading	6568	164	694	0.25	581	673.8	20	553	738
15	Unloading	6826	166	698	0.25	591	677.5	20	555	739
16	Unloading	7069	168	701	0.25	601	681.1	20	556	741
17	Operating	7297	170	704	0.25	610	680.8	23.6	553	738
18	Dummy	7497	172	N/A	N/A	N/A	N/A	N/A	N/A	N/A

If IPO valve is chosen for this application, the closing pressure is the same at the dome pressure at valve depth. Recall the equation to calculate the opening pressure for an IPO valve:

$$P_{vo} = \frac{P_d}{1-R} - \frac{R}{1-R}P_t$$
$$P_d = P_{vc} = (1-R)P_{vo} + RP_t = (1-0.25)704.5 + 0.25 \times 610 = 681\,\text{psi}$$

The valve spread

$$\Delta P = P_{vo} - P_{vc} = 704 - 681 = 23\,\text{psi}$$

The dome pressure at 60 °F

$$P_d^{'} = \frac{520 Z_{60F} P_d}{Z_{vd} T_{vd}} = \frac{520 \times 0.985 \times 681}{(170+460)} = 553\,\text{psi}$$

The test rack opening pressure

$$P_{TRO} = \frac{P_d^{'}}{1-R} = \frac{553}{1-0.25} = 738\,\text{psi}$$

The calculations for unloading valves are similar to that of the operating valve. The only different is unloading valve closing pressure. According to API, the unloading valve spread should be maintained about 20 psi. This 20 psi should be sufficient for the upper unloading valve to close when gas enters the next valve. The summary of the calculation for all unloading valves as well as operating valve is shown in Table 2.3.

References

1. Brown K (1980) The technology of artificial lift methods, vol. 2a. Introduction of Artificial Lift Systems, Beam Pumping, and Gas Lift, PennWell Publishing Company
2. Craft BC, Holden WR, Graves ED (1962) Well design: drilling and production. Prentice-Hall, Englewood Cliffs, NJ
3. Ganzarolli M, Altemani C (2010) Nitrogen charge temperature prediction in a gas lift valve. J Braz Soc Mech Sci Eng 17(1)
4. Hepguler G, Schmldt S, Blals R, Doty D (1993) Dynamic model of gas-lift valve performance. J Petrol Technol
5. Potter M, Wiggert D (2001) Mechanics of fluids, 3rd edn. CL Engineering
6. Rouen R (2006) Gas lift proving effective in gas wells. The American Oil & Gas Reporter
7. Schlumberger (2000) Gas Lift Design and Technology.
8. Stojanovic S, Radanovic P (2004) Increase efficiency of intermittent gas-lift by use of plunger lift. ASME/API/ISO Gas Lift Workshop
9. Weatherford (2005) Kickover Tools Series
10. Winkler H, Eads P (1989) Algorithm for more accurately predicting Nitrogen-Charged Gas-Lift valve operation at high pressures and temperatures. In: presented at the SPE production operations symposium, March. Oklahoma, SPE-18871-MS

Chapter 3
Electrical Submersible Pump

3.1 Fundamentals of Electrical Submersible Pump

3.1.1 Introduction and History of ESPs

Electrical Submersible Pump (ESP) in oil and gas industry is a multistage centrifugal pump used to lift moderate or high volumes of fluids from wellbores. In other words, ESP is one of the artificial lift methods that uses a downhole pump to provide an additional energy or additional lift to fluids inside the wellbore and hence improve hydrocarbon production. Electric power is supplied from surface to operate downhole pumps (ESPs) via cables. Since EPSs are multistage centrifugal pumps, their performance is similar to multi-single-stage centrifugal pumps connected in series. When connecting multi-single-stage centrifugal pumps in series, as the pumps are operated at a constant speed, the rate would be similar but pump heads are higher when compared with that of a single-stage centrifugal pump. Theoretical performances of a typical single-stage centrifugal pump and two-stage centrifugal pump are shown in Fig. 3.1a, b.

The advantages of using ESP in comparison with other lift methods are as follows [4]:

- The most important factor that needs to be considered when selecting ESPs is high liquid rate. ESP can be economically designed for both oil and water wells, at production rates ranging from 200 to 60,000 B/D and at depths of up to 15,000 feet.
- ESP can be applied easily in crooked or deviated wells which have dog leg severity of less than 9°/100 ft.
- Surface equipment of an ESP system is quite simple compared to that of other lift methods. It has a relatively small surface footprint and hence is appropriate for use in offshore, urban or other confined locations.
- Generally speaking, ESPs provide low lifting costs for high fluid volumes. In other words, ESPs are best suit for high liquid rate.

T. Nguyen, *Artificial Lift Methods*, Petroleum Engineering,
https://doi.org/10.1007/978-3-030-40720-9_3

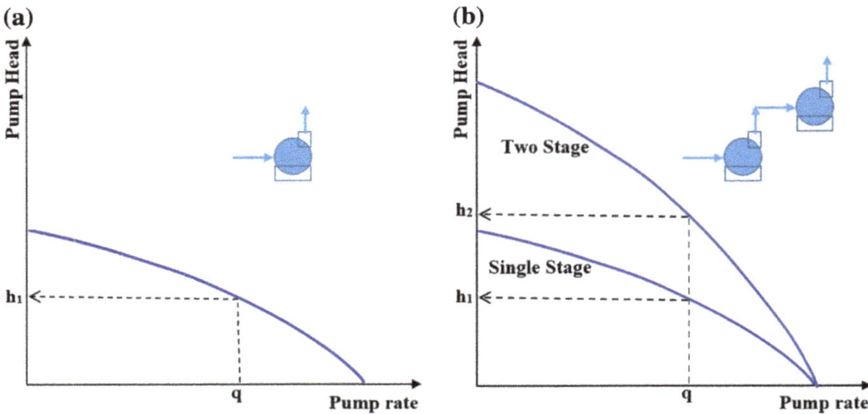

Fig. 3.1 **a** Theoretical performance of a typical single-stage centrifugal pump **b** theoretical performance of a two-stage centrifugal pump connected in series

The disadvantages of using ESP in comparison with other lift methods are as follows:

- For most cases, an ESP system is applicable mainly to single-zone completions. If the production casing is big enough, then the y-tool can be used to have dual-zone completions.
- As always, ESPs requires a stable high-voltage electric power source.
- Running or pulling the tubing string may cause a damage to the power cable installed along the tubing. In addition, cables may deteriorate in high temperature and abrasive conditions. The temperature limit for most of the power cables available on the market is 400 °F (about 200 °C).
- With high Gas Liquid Ratio (GLR) or high solid production, ESPs are not recommended. In the recent years, the industry has improved the efficiency of the downhole gas separator and hence high GLR is not a big concern. However, ESPs still don't work well with the presence of solid due to the solid erosion at surface of impellers and diffuser.

The history of ESPs can be traced back in 1911 when Armais Arutunoff invented the first electrical motor that would operate in water. In 1916, Mr. Arutunoff successfully coupled a centrifugal pump and the motor he developed for dewatering mines and ships. Mr. Arutunoff opened a company named REDA in 1919 when he migrated to Berlin. He then migrated to the United States in 1923 and continued looking for new business in the U.S. Arutunoff finally signed a contract with Phillips Oil Co. to do the field test and to prove the concept that he proposed in the late 1920s. This was the initiation of the new and famous company named REDA Pump Co. established. Camco bought REDA Pump Co. in 1987 and later merged with Schlumberger in 1998 [14].

Another well-known oilfield submersible pump company is Centrilift Baker Hughes Inc. This company was traced back in 1957 when Byron Jackson Pump facility was established in California. The company was moved to Claremore, Oklahoma in 1980 and the company name was changed to Centrilift Inc. Hughes Tool Co. bought Centrilift Inc. in 1980 and later became Centrilift Baker Hughes Inc. until now [14].

3.1.2 A Basic ESP System

A basic ESP system consists of the main following components:

Surface equipment includes transformers, a switchboard, a junction box and a wellhead.

Subsurface equipment includes an electric motor, a protector (or seal), and a pump connected at the end of the production tubing, and cables.

Transformer is used to step-up or step-down the voltage from the primary line to the motor of the submersible pump. Because a range of operating voltages may be used for submersible pump motors, the transformer must be compatible with the selection of the motor voltage.

Switchboard controls the speed of the pump motor. In addition, it provides overload and underload protection. Overload protection is needed to keep the motor from overheated and burning. Underload protection is needed because if liquid rate is too low (underload) there is not enough liquid to cool off the motor (Fig. 3.2).

Junction Box connects the power cable from the switchboard to the power cable from the well. It provides an explosion-free vent to the atmosphere for any gas that might migrate up the power cable from the wellbore.

Wellhead provides means for installing cables with adequate seal. It may include adjustable chokes, bleeding valves.

Electric Motor is used to turn the pump. Depending on the frequency, the motor speed will vary. For example, the motor runs at a relatively constant speed of around 3500 RPM at a frequency of 60 Hz or 2900 RPM at 50 Hz. Voltage, current, and horsepower of a subsurface electric motor may vary from 200 up to 5000 V, 10 up to 200 A, and 12 HP up to 2000 HP at 60 Hz, respectively. Subsurface electric motors are filled with a highly refined mineral oil that provides dielectric strength, lubrication of bearings, and good thermal conductivity. Heat generated by the motor is transferred by the dielectric oil to the motor casing. Produced formation fluids will carry this heat up to the surface. In normal application, the motor is installed above the perforations and the fluids pass along the motor cooling it. In some installations (motor below perforations or pump below motor), it is required to use a shroud to force the fluids to cool the motor.

Protector or Seal serves as a connection between the motor shaft and the pump shaft. The main function of the protector is to support the axial thrust developed by the pump on the seal thrust bearing. It also prevents formation fluids from entering

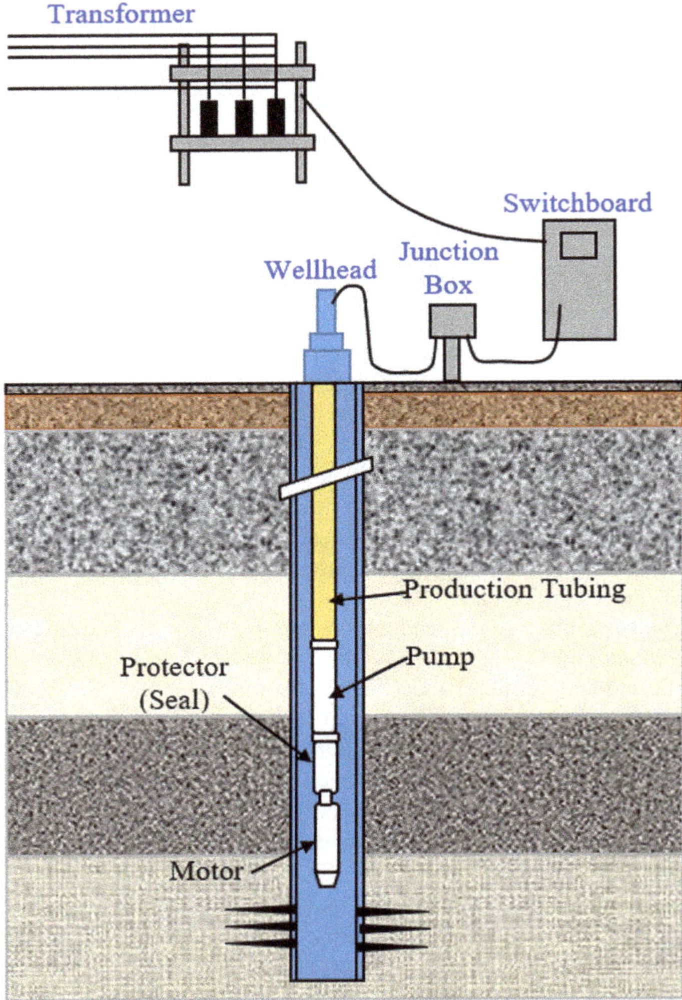

Fig. 3.2 A basic ESP system

the motor and provides an oil reservoir to compensate for expansion and contraction of motor oil due to the change in temperature.

Submersible Pump is a multistage centrifugal pump. Depending on desired liquid rate, pump depth, wellhead pressure, and friction inside the production tubing, the number of pump stages will be different.

Electrical Cables are run from the junction box then through the wellhead and all the way to the bottom to supply power to the pump motor. There are two main functions of the power cable: (1) to transmit the electric power from the surface

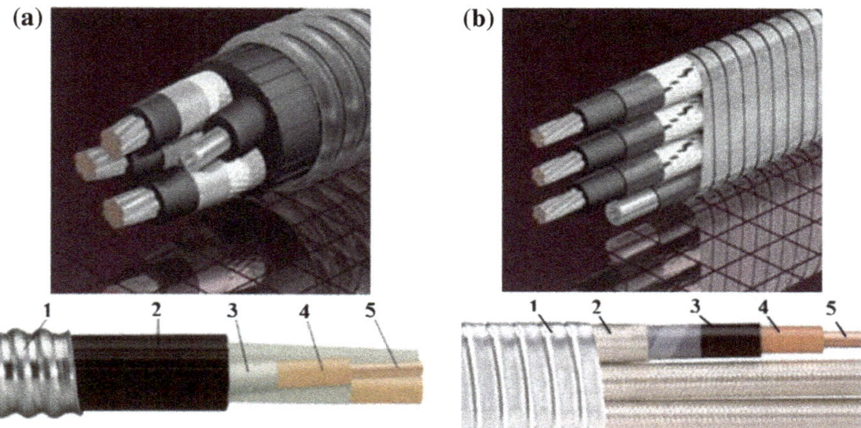

Fig. 3.3 **a** Round cable with jacket and armor **b** flat cable with jacket and armor

down to the pothead in the motor and (2) to transmit signals from downhole to the surface (usually pressure and temperature).

The power cable consists of 3 phase conductors individually insulated with insulation material. The conductors are covered with a protective material and finally protected from abrasion, chemical and mechanical damage with a jacket and an armor as shown in Fig. 3.3a, b.

3.1.3 Working Principle of an ESP

If the reservoir pressure is high enough to overcome the hydrostatic pressure and all frictional pressure losses in the reservoir and in the production system, the well will produce naturally with an equilibrium liquid rate of Q_e^{nf} and a flowing bottomhole pressure of P_{wf}^{nf} as shown in Fig. 3.4. The IPR and OPR without pump will intersect at one point determined by Q_e^{nf} and P_{wf}^{nf}. In other words, in a naturally flowing well without artificial lift equipment, production flowrates higher than the natural flowrate are impossible to be achieved. In order to produce flowrates higher than the natural equilibrium flowrate, an external energy needs to be provided to the well in a form of artificial lift such as pumps.

When an ESP is installed inside the well close to the perforations, the pump provides extra energy (differential pressure across the pump) to lift the liquid column inside the tubing and hence reduces the flowing bottomhole pressure. The discharge pressure of the pump must be equal to the OPR-pressure and the intake pressure of the pump must be equal to the IPR-pressure. These two pressures

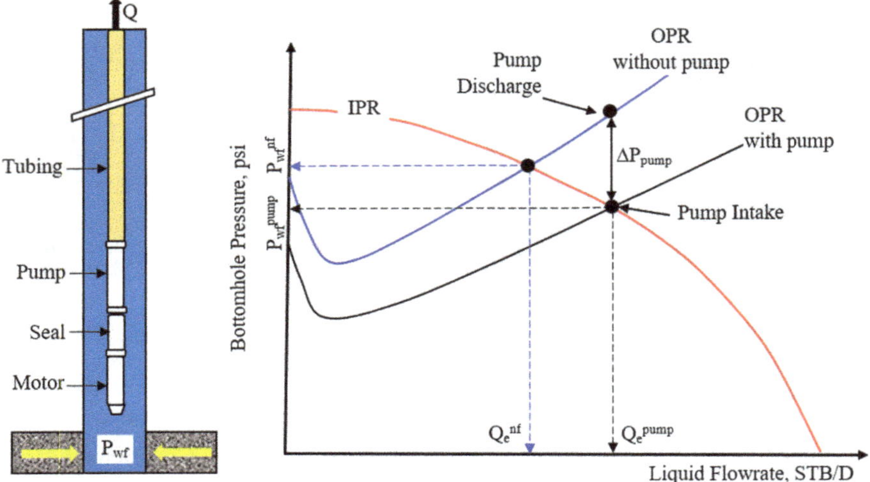

Fig. 3.4 Main working principle of an ESP system

determine the differential pump pressure as shown in Fig. 3.4. The new OPR with pump is plotted and presented in Fig. 3.4. This OPR with pump intersect with the IPR at a higher rate of Q_e^{pump} and lower bottomhole flowing pressure of P_{wf}^{pump}. Please note that the production liquid rate does not really depend on the number of pump stages; It mainly depends on the ESP's performance itself.

3.1.4 ESP Classification

The most common way of classifying an ESP is based on its stage's design: mixed flow and radial flow as shown in Fig. 3.5a, b.

The mixed flow stage design develops pressure/head through centrifugal and axial forces. This design is generally used where a High Flow—High Head installation is required. The angle at which the fluid is discharged at an angle less than 90° from the eye of the impeller. Liquid rates are generally in the range from 1,000 to 50,000 bpd. Pump efficiencies for this design are normally greater than 60% at the best efficiency point. In addition, the mixed flow stage pump is capable of handling gas and solids more efficiently.

The radial flow stage design develops pressure/head through only centrifugal forces. This pump design is generally used where a low flow, high head installation is required. This design is accomplished by allowing the fluid to be discharged at a 90° angle from the eye of the impeller. The liquid rates are in the range of from 100 bpd to 2,000 bpd with pump efficiencies in the order of 60%. The radial stage is a flat stage and is the most efficient design for these lower flow rates.

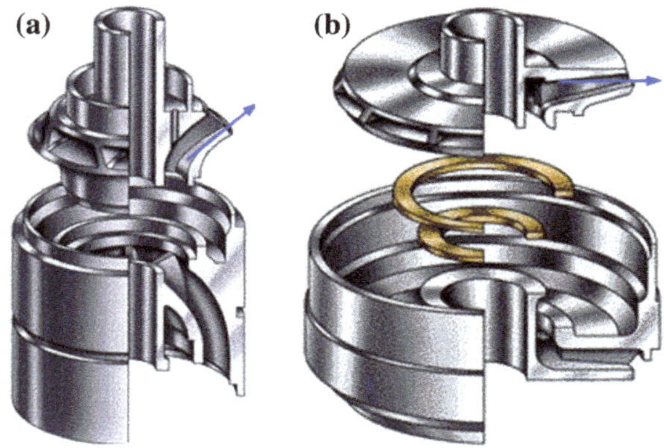

Fig. 3.5 **a** Mixed flow stage **b** radial flow stage

3.1.5 Deploying ESP System

Deploying an ESP system into an onshore oil well can be done easily using threaded pipe joints and a workover rig unit. A downhole ESP system including a sensor unit, a motor, a seal, a separator, a pump, and a discharge head is constructed at the surface. The entire pump system is then run into the hole together with the production tubing using a workover rig unit.

For offshore wells, however, coil tubing and cable suspended are also the two alternative methods of deploying an ESP system into the well beside the use of conventional workover rigs. Coil tubing deployed ESP system allows an ESP to be deployed into a well using standard coiled tubing, eliminating the need for a workover rig and surface equipment. This simplified deployment method is especially beneficial in offshore and deep-water applications, where moving a rig onsite increases costs and delays production. Cable Suspended allows ESPs to be lowered into the well without using a tubing. The main advantage of this installation is the reduction in costs associated with tubing pulling job, especially offshore wells. The pump is suspended from a cable and the power cable is banded to it. A special seating element supports the pump and provides locking to avoid excessive torque on the cable. Differently from the conventional installations, the motor is located above the pump. The system produces through the annular (Fig. 3.6).

3.2 Theoretical Performance of a Centrifugal Pump

A single-stage centrifugal pump consists of an impeller and a diffuser. The diffuser is stationary and the impeller rotates inside the diffuser via a shaft connected with an electric motor. When liquid enters the impeller, the liquid receives kinetic energy from the rotating impeller. When this liquid exits the impeller and enters the diffuser, its kinetic energy is converted to pressure. In other words, as liquid flows

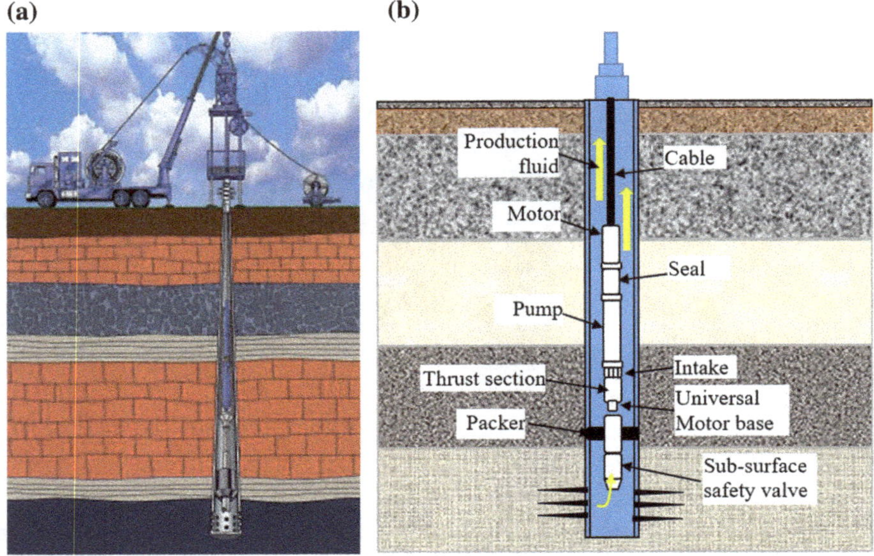

Fig. 3.6 **a** Coil tubing deployed ESP **b** cable suspended deployed ESP

through a single-stage centrifugal pump, the liquid will gain a pressure defined as the difference between discharge and intake pressures. This pressure gain is sometimes called pump pressure.

Pump performance is a relationship between liquid flowrate and pump pressure or pump head at a constant pump speed as shown in Fig. 3.1a. If the liquid rate is zero or maximum, the pump pressure or pump head will be maximum or minimum, respectively. To develop a pump performance curve for a specific centrifugal pump, pump's manufacturers test it with water by varying the discharge pressure and measuring liquid rate. This is the actual pump performance curve. It is sometimes costly and time consuming to obtain actual pump performance curves for every single pump. Therefore, there is a need to develop the theoretical pump performance based on the pump's geometry and operating conditions.

The theoretical pump performance is developed with the following assumptions:

- Two dimensions: radial and tangential direction.
- The impellers are completely filled with the flowing fluid at all time (no void spaces).
- The streamlines have a shape similar to the blade's shape.
- The fluid is incompressible, inviscid, and single phase.
- The velocity profile is symmetric.

Let consider an impeller of a centrifugal pump and point A on a streamline as shown in Fig. 3.7a. The radial direction is defined as a line connecting point A and the center point of the impeller. The tangential direction is the line perpendicular to

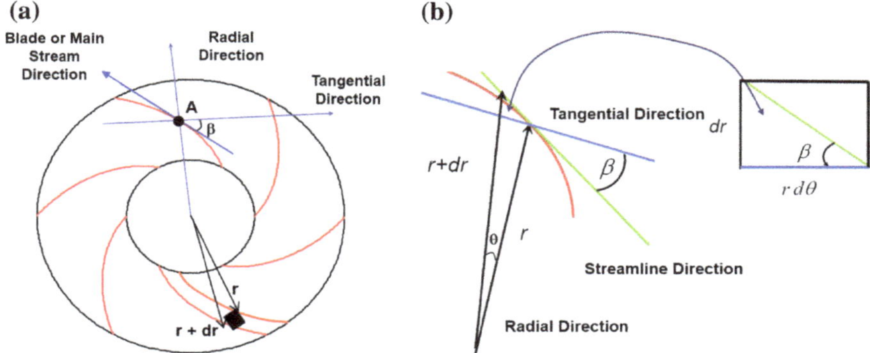

Fig. 3.7 **a** Impeller of a centrifugal pump **b** free body diagram

the radial direction at point A. The main flow direction is the line tangent to the streamline (or blade's shape) at point A. The pump blade angle, β, is the angle between the tangential direction and the direction of the streamline as shown in Fig. 3.7a.

Now let consider a Free Body Diagram (FBD) which is a fluid volume as shown in Fig. 3.7a. This FBD is then zoomed in and shown in Fig. 3.7b. The FBD is rotating at an angular velocity of ω, which is the same as the angular velocity of the impeller. The height of this FBD, h, is the same as the thickness of the impeller. The volume and mass of this FBD are as follows [2]:

$$dV = dA \times h = (dr \times rd\theta)h \tag{3.2.1}$$

$$dm = \rho dV = \rho \times h \times rd\theta \times dr \tag{3.2.2}$$

The centrifugal force acting on this FBD:

$$dF = dm \times r \times \omega^2 = \rho h r^2 \omega^2 d\theta dr \tag{3.2.3}$$

The pressure increase due to this centrifugal force:

$$dP = \frac{dF}{dA} = \frac{\rho h r^2 \omega^2 d\theta dr}{hrd\theta} \tag{3.2.4}$$

Simplifying Eq. (3.2.4) gives

$$dP = \rho r \omega^2 dr \tag{3.2.5}$$

Integrating Eq. (3.2.5) from the intake, R_1, to the discharge, R_2, of the impeller gives:

$$\int_{P_1}^{P_2} dP = \rho\omega^2 \int_{R_1}^{R_2} rdr \tag{3.2.6}$$

$$\Delta P = P_2 - P_1 = \rho\omega^2 \left(\frac{R_2^2 - R_1^2}{2} \right) \tag{3.2.7}$$

The relationship between angular velocity and linear velocity:

$$U = R\omega \tag{3.2.8}$$

Combining Eqs. (3.2.7) and (3.2.8) gives:

$$\Delta P = \rho \left(\frac{U_2^2 - U_1^2}{2} \right) \tag{3.2.9}$$

Equation (3.2.9) tells us that as an incompressible, inviscid fluid enters and exits an impeller rotating at a constant speed of ω, the fluid will receive an energy in the form of pressure gain as described in Eq. (3.2.9) due to only the centrifugal force.

Note that, as the fluid FBD is rotating at the same speed of the impeller ω, it is subjected not only the centrifugal force in the radial direction but also the force due to the difference in velocities in blade's direction. Figure 3.8 shows the velocity triangle of the fluid FBD at point A. When the impeller rotates clockwise at an angular velocity ω, the fluid FBD also moves at the same speed of ω with the direction of AC as shown in Fig. 3.8a. This velocity vector AC can be decomposed of two vectors AB following the blade's direction and AD following the tangential direction. Vector AC can also be decomposed of two vectors DC and AD. Note that vector DC is the same as vector AB.

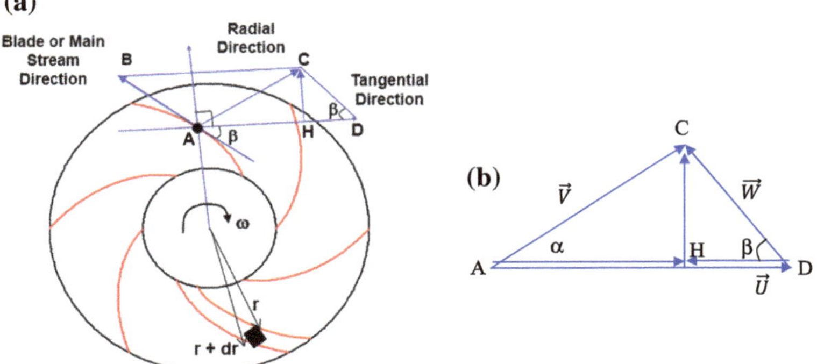

Fig. 3.8 Velocity triangle of the fluid FBD at point A

$$\overrightarrow{AC} = \overrightarrow{AB} + \overrightarrow{AD} = \overrightarrow{DC} + \overrightarrow{AD} \tag{3.2.10}$$

In terms of the magnitude of vector AC:

$$AC^2 = AD^2 + DC^2 - 2AD \times DC \times \cos\beta \tag{3.2.11}$$

$$DC = HC \times \sin\beta \tag{3.2.12}$$

Or

$$DC = \frac{HC}{\sin\alpha} \tag{3.2.13}$$

From Eqs. (3.2.10) to (3.2.13), one can concludes that the magnitude of vector AC (fluid FBD) can be calculated if the angle β and the magnitude of the two vectors AD and HC are known. The magnitude of vector AD can be obtained using the impeller's speed in RPM or radian/second. The magnitude of the vector HC can be achieved using the liquid rate and the pump geometry. Readers should review and exercise Example 3.3 to have better understanding on how to use the triangle vector. Note that if the fluid enters the impeller radially, at the pump intake, vectors AC = HC, vector AH = 0 and α = 90°.

Vector DC (relative velocity) represents the pressure gain of the FBD within the impeller due to the difference in fluid velocities following the blade's direction (streamline direction). This component can be expressed mathematically as follows:

$$\Delta P = \frac{1}{2}\rho \int\limits_{streamline} dW^2 \tag{3.2.14}$$

where W is the relative velocity component of the fluid FBD in the streamline direction as shown in Fig. 3.8b. Solving Eq. (3.2.14) gives:

$$\Delta P = \rho\left(\frac{W_1^2 - W_2^2}{2}\right) \tag{3.2.15}$$

Combining Eqs. (3.2.9) and (3.2.15) gives the total pressure gain of the fluid when it enters and exists the impeller as follows:

$$\Delta P_{impeller} = \rho\left(\frac{U_2^2 - U_1^2}{2} + \frac{W_1^2 - W_2^2}{2}\right) \tag{3.2.16}$$

The pressure gain of the fluid across the impeller is sometimes called the potential pressure. As the fluid exits the impeller, it then enters the pump diffuser and gains an additional pressure due to the fluid velocity, V, described as follows:

$$\Delta P_{diffuser} = \rho\left(\frac{V_2^2 - V_1^2}{2}\right) \tag{3.2.17}$$

The total pressure gain of the fluid when it enters and exits one pump stage (consists of impeller and diffuser) is given as:

$$\Delta P_{stage} = \rho\left(\frac{U_2^2 - U_1^2}{2} + \frac{W_1^2 - W_2^2}{2} + \frac{V_2^2 - V_1^2}{2}\right) \tag{3.2.18}$$

To understand the concept of velocity triangle and to know how to use Eq. (3.2.18), users are recommended to practice Example 3.3 in Sect. 3.9.

Applying the geometrical relationships as shown in Fig. 3.8b gives:

$$V_1^2 = U_1^2 + W_1^2 - 2U_1W_1\cos\beta_1 \tag{3.2.19}$$

$$V_2^2 = U_2^2 + W_2^2 - 2U_2W_2\cos\beta_2 \tag{3.2.20}$$

$$W = \frac{Q}{2\pi rh\sin\beta} \tag{3.2.21}$$

Substituting Eqs. (3.2.8), (3.2.19), (3.2.20) and (3.2.21) into Eq. (3.2.18) gives:

$$\begin{aligned}\Delta P_{stage} &= \rho\omega^2\left(R_2^2 - R_1^2\right) \\ &\quad - \frac{\rho\omega Q}{2\pi h}\left(\frac{1}{\tan\beta_2} - \frac{1}{\tan\beta_1}\right) + \frac{\rho Q^2}{4\pi^2h^2}\left(\frac{1}{R_2^2\sin^2\beta_2} - \frac{1}{R_1^2\sin^2\beta_1}\right)\end{aligned} \tag{3.2.22}$$

Note that the third term in Eq. (3.2.22) is always smaller than the second and the first term and hence practically the third term can be neglected. Equation (3.2.22) reduces to:

$$\Delta P_{stage} = \rho\omega^2\left(R_2^2 - R_1^2\right) - \frac{\rho\omega Q}{2\pi h}\left(\frac{1}{\tan\beta_2} - \frac{1}{\tan\beta_1}\right) \tag{3.2.23}$$

Equation (3.2.23) can be used to describe the theoretical performance of a single stage centrifugal pump in terms of pressure and in SI unit.

3.2.1 Pump Head

Pump head is an indirect measurement of pressure that does not depend on the fluid density. That means for low viscous fluids, the pump performance can be uniquely defined in terms of head. In other words, the pump performance, in pressure, depends on the density of the fluid being pumped, but when this performance is expressed in head, the pump performance is independent of the fluid being pumped. In SI unit, pump head in meter of one pump stage is expressed as follows:

$$H_{stage} = \frac{\Delta P_{stage}}{\rho g} \tag{3.2.24}$$

where ΔP is the pressure gain across the stage in Pa, ρ is the fluid density in kg/m^3, and g is the gravitational acceleration in m/s^2.

In oil field unit, pump head in feet is expressed as:

$$H_{stage} = \frac{\Delta P_{stage}}{0.433\gamma} \tag{3.2.25}$$

where ΔP is the pressure gain across the stage in psi, and γ is the fluid specific gravity which is dimensionless.

To have a better understanding on the pump head, let consider three different fluids: oil, water, and brine with specific gravities of 0.8, 1.0, and 1.2, respectively. To deliver these three fluids to a height of 2,000 ft, applying Eq. (3.2.25), an ESP must have the differential pressures of 693 psi, 866 psi, and 1,039 psi, respectively. The illustration of this example is shown in Fig. 3.9.

Converting Eq. (3.2.23) from pressure to pump head in meter gives:

$$H_{stage} = \frac{\omega^2}{g}\left(R_2^2 - R_1^2\right) - \frac{\omega Q}{2\pi g h}\left(\frac{1}{\tan\beta_2} - \frac{1}{\tan\beta_1}\right) \tag{3.2.26}$$

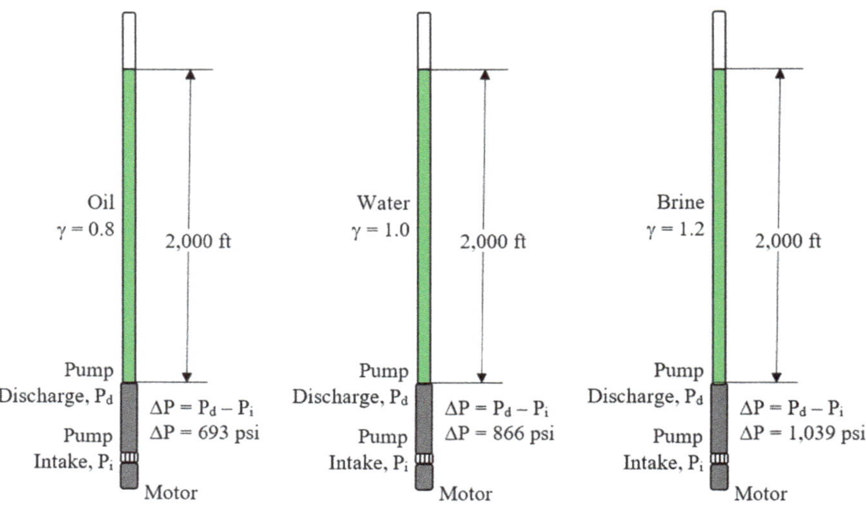

Fig. 3.9 Illustration of pump head

In terms of oil field unit, Eq. (3.2.23) becomes:

$$H_{stage} = 2.3689 \times 10^{-6} N^2 \left(R_2^2 - R_1^2\right) - 2.0398 \times 10^{-6} \frac{NQ}{h} \left(\frac{1}{\tan \beta_2} - \frac{1}{\tan \beta_1}\right) \tag{3.2.27}$$

where N is the rotational speed of the impeller in RPM, R_1 and R_2 are the radii of the impeller at the intake and at the outlet in inches, h is the height of the impeller in inches, Q is the liquid rate in barrel per day (BPD).

As defined, the pump blade angle, β, is the angle between the tangential direction and the direction of the streamline at one point along the streamline as shown in Fig. 3.7a. If the pump blade angle at the outlet, β_2, is smaller than the pump blade angle at the inlet, β_1, than the term $\left(\frac{1}{\tan \beta_2} - \frac{1}{\tan \beta_1}\right)$ is positive. Therefore, the straight line $H_{stage} = f(Q)$ described in Eq. (3.2.27) will have a negative slope. Similarly, if $\beta_2 > \beta_1$ then the straight line $H_{stage} = f(Q)$ will have a positive slope. The effect of the pump blade angle on the pump performance of a centrifugal pump is presented in Fig. 3.10.

3.2.2 *Specific Speed Number*

Applying the theoretical pump performance equation, Eq. (3.2.26), to two different centrifugal pumps, the following equation can be derived:

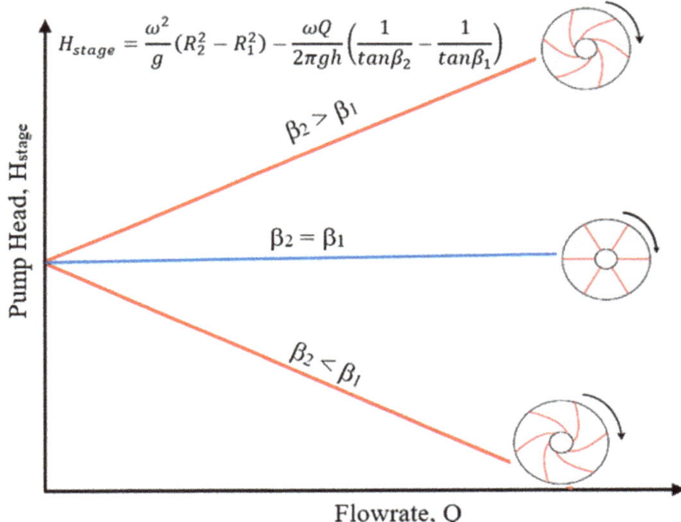

Fig. 3.10 Effect of pump blade angle, β, on a centrifugal pump performance

$$\frac{N_1\sqrt{Q_1}}{H_1^{0.75}g^{0.75}}\sqrt{\frac{r_1}{h_1\tan\beta_1}} = \frac{N_2\sqrt{Q_2}}{H_2^{0.75}g^{0.75}}\sqrt{\frac{r_2}{h_2\tan\beta_2}} \tag{3.2.28}$$

The original specific speed, N_s, is a dimensionless number and defined in SI unit as follows:

$$N_s = \frac{N\sqrt{Q_{BEP}}}{H_{BEP}^{0.75}g^{0.75}} \tag{3.2.29}$$

where N is the impeller speed in revolution per second, Q_{BEP} is the flowrate in m^3/s at the Best Efficiency Point (BEP), H_{BEP} is the pump head at the BEP in m, and g is the gravitational acceleration in m/s^2.

In America, the specific speed sometimes is defined as:

$$N_s = \frac{N\sqrt{Q_{BEP}}}{H_{BEP}^{0.75}} \tag{3.2.30}$$

where N is the impeller speed in Revolution Per Minute (RPM), Q_{BEP} is the flowrate in GPM at the BEP, H_{BEP} is the pump head at the BEP in ft. If Q_{BEP} is in BPD, N is in RPM, and H is in ft, then Eq. (3.2.30) becomes:

$$N_s = 0.171\frac{N\sqrt{Q_{BEP}}}{H_{BEP}^{0.75}} \tag{3.2.31}$$

N_s is used to classify, describe the performance and design of centrifugal pumps. With these units, the specific speed number is in the range of 500–15,000: N_s = 500–1,000 are called radial flow pumps; N_s = 1,500–7,000 are called mixed flow pumps; and N_s > 9,000 are called axial flow pump. This classification is illustrated in Fig. 3.11. In other words, with N_s < 1,000, the pump head is developed mainly due to the centrifugal force and the flow is mainly in radial direction. If Ns > 9,000, the pump head developed is not only due to the centrifugal force but also due to the push of the blades and the flow is a mixed between radial and axial flows.

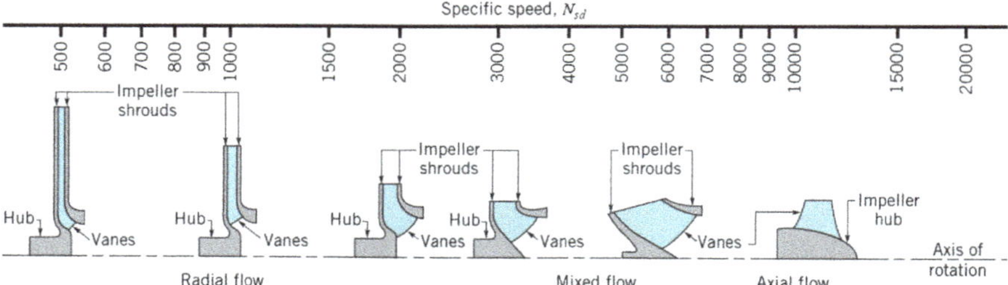

Fig. 3.11 Pump classification based on the specific speed [11]

Based on the definition of the specific speed, one may draw the following conclusions:

- From Eq. (3.2.29), at the BEP, the specific speed number will remain the same when the pump operates at different pump speeds or different flowrates.
- From Eq. (3.2.28), if the two pumps have similar geometry $\sqrt{\frac{r_1}{h_1 \tan \beta_1}} = \sqrt{\frac{r_2}{h_2 \tan \beta_2}}$, the N_s will remain the same along the pump performance curve when pumps are operated at different pump speed.
- Let compare two pumps which have different specific speed number but have the same outside diameter. Equation (3.2.28) tells us at the BEP, the higher specific speed pump will offer higher rate, lower pump head, the blade angle β is higher, and the flow is more mixed-flow type.

3.2.3 Affinity Law

In this section, we will discuss on how pump performance changes for pumps with the same geometric design but different sizes.

The Affinity Law states the similarity between two pumps with the same geometric design but different outside diameter. The Affinity Law reveals that the relationships between flowrate, head, and brake horsepower as a function of pump speed and outside diameter are satisfied by the following equations:

$$\frac{Q_1}{Q_2} = \frac{N_1}{N_2}\left(\frac{r_1}{r_2}\right)^3 \tag{3.2.32}$$

$$\frac{H_1}{H_2} = \left(\frac{N_1}{N_2}\right)^2\left(\frac{r_1}{r_2}\right)^2 \tag{3.2.33}$$

$$\frac{HP_{b1}}{HP_{b2}} = \left(\frac{N_1}{N_2}\right)^3\left(\frac{r_1}{r_2}\right)^5 \tag{3.2.34}$$

If the two pumps are operated at the same speed, $N_1 = N_2$ but have different outside diameter, the affinity law becomes:

$$\frac{Q_1}{Q_2} = \left(\frac{r_1}{r_2}\right)^3 \tag{3.2.35}$$

$$\frac{H_1}{H_2} = \left(\frac{r_1}{r_2}\right)^2 \tag{3.2.36}$$

$$\frac{HP_{b1}}{HP_{b2}} = \left(\frac{r_1}{r_2}\right)^5 \tag{3.2.37}$$

If $r_1 = r_2$ (two pump are the same) but they are operated at different conditions, the Affinity Law predicts as follows:

$$\frac{Q_1}{Q_2} = \frac{N_1}{N_2} \tag{3.2.38}$$

$$\frac{H_1}{H_2} = \left(\frac{N_1}{N_2}\right)^2 \tag{3.2.39}$$

$$\frac{HP_{b1}}{HP_{b2}} = \left(\frac{N_1}{N_2}\right)^3 \tag{3.2.40}$$

3.3 Actual Pump Performance

The actual pump head developed by a centrifugal pump at a given pump rate is always smaller than the theoretical pump head. This discrepancy is first due to the assumptions made when developing the theoretical pump performance. Some of the key assumptions are:

- The impeller are completely filled with the flowing fluid at all time (no void spaces). This is not true if the fluid has the presence of gas.
- The streamlines have a shape similar to the blade's shape. This assumption is only true if the number of pump blades is very high.
- The fluid is incompressible, inviscid, and single phase. With the presence of gas, the fluid is compressible and two phase. Also, most of fluids are viscid and have some levels of viscosity.

In addition to these assumptions, the discrepancy between the actual and theoretical pump performance is also due to recirculation and leakage of fluid inside the impeller, and hydraulic losses. When there is a recirculation of fluid inside the impeller, the streamlines are no longer have the shape which is assumed to be similar to the blade's shape as shown in Fig. 3.12. Fluid inside the impeller can leak through the small gap between the impeller and the diffuser. Also the leakage may also occur at the balancing holes on impellers as shown in Fig. 3.13. The hydraulic losses includes friction loss due to the viscous effect, diffusion loss due to divergence or convergence of the passage, fluid shock loss at the impeller inlet, etc. When considering all the losses, the actual pump performance is not a straight line but it is a curve and presented in Fig. 3.14.

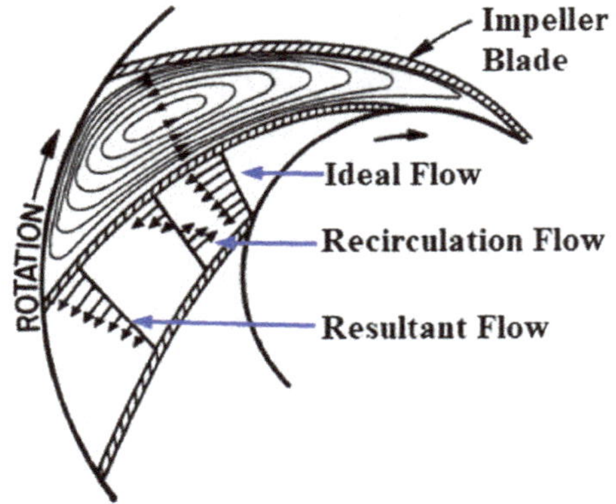

Fig. 3.12 Recirculation flow inside the impeller

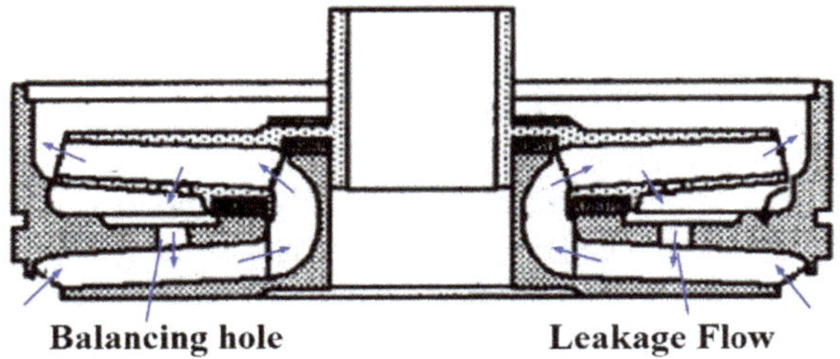

Fig. 3.13 Leakage through balancing holes on impellers

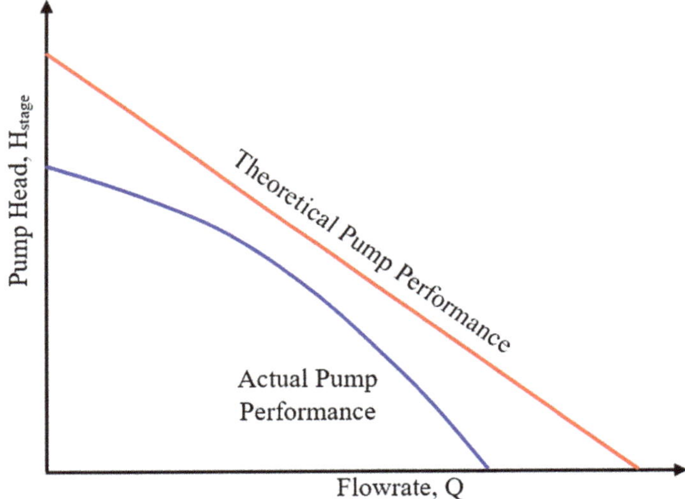

Fig. 3.14 Theoretical and actual pump performance

3.3.1 Hydraulic Horsepower

Hydraulic horsepower or fluid horsepower is the energy transmitted to the fluids by the pump and defined as the power to lift a continuous volume rate of fluid to a specific height. Depending on the units used for pressure and rate, equations for calculating hydraulic horsepower are different.

$$HP_h = 1.67 \Delta P_{pump} Q \tag{3.3.1}$$

where HP_h is the hydraulic horsepower in watts, ΔP_{pump} is the differential pressure at pump discharged and pump intake in bar, and Q is the flowrate in liter/min.

$$HP_h = \frac{\Delta P_{pump} Q}{1714} \tag{3.3.2}$$

where HP_h is the hydraulic horsepower in hp, ΔP_{pump} is the differential pressure at pump discharged and pump intake in psi, and Q is the flowrate in GPM.

$$HP_h = 7.37 \times 10^{-6} \gamma h_{pump} Q \tag{3.3.3}$$

where HP_h is the hydraulic horsepower in hp, h_{pump} is the pump head in ft, Q is the flowrate in BPD, and γ is the fluid specific gravity.

3.3.2 Brake Horsepower

The brake horsepower, HP_b, is the energy required by the pump shaft to turn or the power required to overcome all the losses and provide energy needed to pump the fluid. The brake horsepower is always higher than the hydraulic horsepower. The brake horsepower is normally measured while developing the actual pump performance curve.

3.3.3 Pump Efficiency

The pump efficiency is defined as the ratio between the hydraulic horsepower and the brake horsepower. The value of the pump efficiency depend on the losses including friction loss due to fluid viscosity, mechanical losses at bearings, and washers, turbulent loss, etc.

$$\eta = \frac{HP_h}{HP_b} \tag{3.3.4}$$

3.3.4 Actual Pump Performance Curves

In theory, pump manufacturers test each pump to develop pump curves before selling the products to the market. Water is used to conduct the test. During the test, the pump discharged pressure is varied by controlling the valve on the discharge line. The following parameters are measured: flowrate, intake pressure, discharge pressure, and brake horsepower. The pressure gain is then converted to pump head. The hydraulic horsepower and the pump efficiency are also calculated. The relationships between head versus rate, brake horsepower versus rate, and pump efficiency versus rate are plotted on the same plot. These curves describe the actual performance of the pump under testing conditions. An example of the actual pump performance curves is shown in Fig. 3.15.

Generally speaking, if pump performance curves described in Fig. 3.15 are available from manufacturers (normally water is used as a testing fluid and the pump is tested at 60 Hz or about 3,500 RPM), one can predict the pump performance at different speeds based on the Affinity Law as shown in Fig. 3.16.

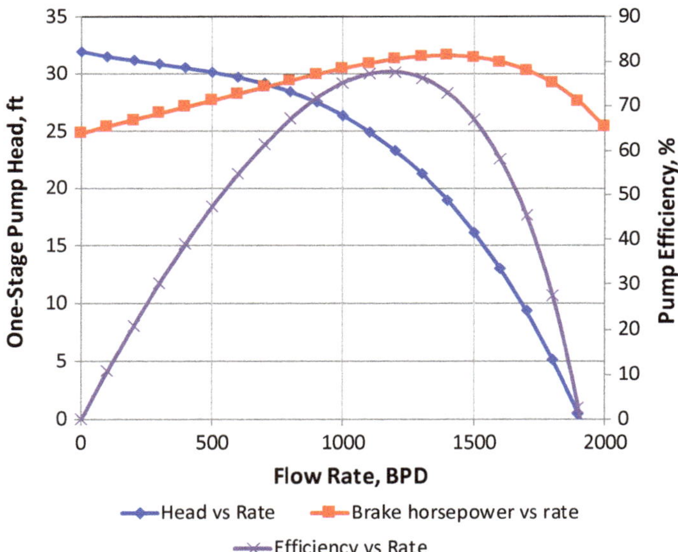

Fig. 3.15 An example of actual pump performance curves

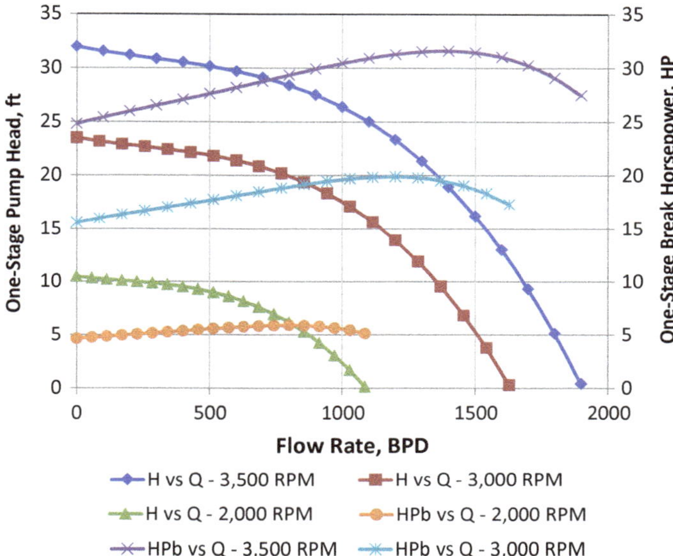

Fig. 3.16 Applying affinity law to predict pump performance curves at different pump speed

3.4 Viscous Effect on Pump Performance

Recall that Eq. (3.2.23) was developed to predict a theoretical performance of a centrifugal pump handling inviscid fluid. We then convert pump pressure to pump head to take care of fluid density. The affinity law can be used to predict the pump performance in different rotational speeds. The actual pump performance for handling inviscid fluids (normally water) is obtained from experimental tests in a laboratory as presented in Sect. 3.3.4. A question rises up is that how can we predict the actual pump performance of an ESP which is handling viscous fluid?

When a centrifugal pump handles viscous liquids, the flowing will be observed: increase in brake horsepower, reduce in head, and decrease in capacity. Stepanoff [16] stated that: "Since centrifugal pump manufacturers' experience has been concerned mostly with the handling of water and because most factory test facilities permit the testing of pumps on water only, the expression of performance when handling viscous oils has been based on a comparison with water performance". In other words, it is very reasonable to predict the actual pump performance of a centrifugal pump handling viscous fluid with an assumption that the actual pump performance of the same pump for handling water is known. Following are the common methods to correct pump performance for viscous applications: (1) Stepanoff (1940–1957); (2) Hydraulic Institute charts [10]; and (3) Turzo correlation [21].

3.4.1 Stepanoff Method

Stepanoff [16–18] performed a number of experiments using conventional design centrifugal pumps for handling water as well as eleven different oils with kinematic viscosities ranging from 1 to 2020 cSt. Stepanoff stated that for a certain pump at a constant speed, the head-capacity decreases as the viscosity increases, in such a way that the specific speed at the Best Efficiency Point (BEP) remains constant. Recall Eq. (3.2.30) for the specific speed gives:

$$N_s = \frac{N\sqrt{Q_{BEP}^{w}}}{\left(H_{BEP}^{w}\right)^{0.75}} = \frac{N\sqrt{Q_{BEP}^{vis}}}{\left(H_{BEP}^{vis}\right)^{0.75}} \tag{3.4.1}$$

The superscripts w and vis are for water and viscous fluids.

Therefore, if pumping performance curves for water is known, only one correction factor, either for the capacity or the for head, is needed to obtain the new performance curve of the pump for handling viscous fluid.

$$\frac{Q_{BEP}^{vis}}{Q_{BEP}^{w}} = \left(\frac{H_{BEP}^{vis}}{H_{BEP}^{w}}\right)^{1.5} \tag{3.4.2}$$

In terms of correction factors, C, Eq. (3.4.2) becomes (Note that Stepanoff used the abbreviation F for correction factors instead of C. For the consistency purposes, C will be used for correction factors.):

$$C_Q = C_H^{1.5} \tag{3.4.3a}$$

Digitizing Fig. 3.17 give a correlation for calculating C_H as follows:

$$C_H = 1 - 417.8\left(R_{Stepanoff} + 9651\right)^{-0.732} \tag{3.4.3b}$$

Based on the experimental results, Stepanoff presented a diagram for head and efficiency correction factors as shown in Fig. 3.17 for several pumps valid at the BEP. Readers should be careful when using the non-linear x-axis scale. The independent variable of the diagram is a Reynolds Number like parameter, defined as Stepanoff Reynolds Number:

$$R_{stepanoff} = 6.0345\frac{NQ_{BEP}^{vis}\gamma}{\mu\sqrt{H_{BEP}^{w}}} \tag{3.4.4}$$

where N is the pump speed in rpm, Q_{BEP}^{vis} is the flowrate of the viscous fluid at the BEP in barrel per day, μ is the fluid dynamic viscosity in cp, and H_{BEP}^{w} is the pump head for water at the BEP in ft. For different set of unit, the Stepanoff number is as follows:

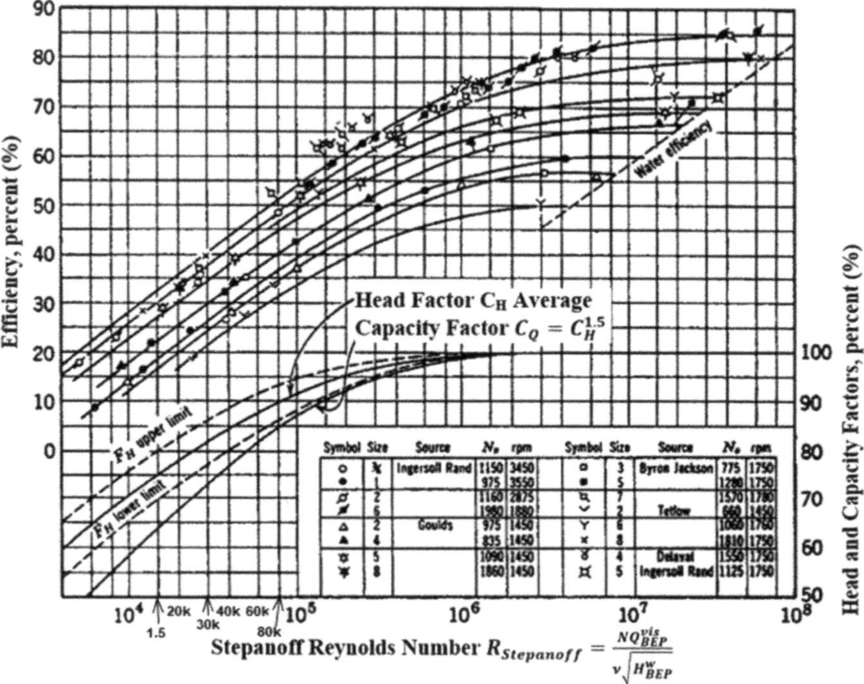

Fig. 3.17 Head and efficiency correction factors for viscous fluids at the BEP

$$R_{stepanoff} = \frac{NQ_{BEP}^{vis}}{v\sqrt{H_{BEP}^{w}}} \tag{3.4.5}$$

where N is the pump speed in rpm, Q_{BEP}^{vis} is the flowrate of the viscous fluid at the BEP in ft^3/s, υ is the fluid kinematic viscosity in ft^2/s, and H_{BEP}^{w} is the pump head for water at the BEP in ft.

To use Stepanoff method and take care of the viscosity effect for a centrifugal pump, an iteration is needed. The procedure of this iteration can be summarized below:

- Step 1: Obtain the head and flowrate at the BEP ***for water*** (H_{BEP}^{w}, Q_{BEP}^{w}) at the desired speed using Affinity Laws.
- Step 2: Guess a value for the viscous flowrate (Q_{BEP}^{vis}), then obtain the viscous head (H_{BEP}^{vis}).
- Step 3: Calculate the Stepanoff Reynolds Number, $R_{stepanoff}$.
- Step 4: Obtain the head correction factor, C_H, from Fig. 3.17 then calculate the C_q using Eq. (3.4.3a).

- Step 5: Calculate the viscous flowrate at the BEP (Q_{BEP}^{vis}) and compare with the flowrate in step 2.
- Step 6: If convergence is obtained, calculate the viscous head at the BEP (H_{BEP}^{vis}). If convergence in not achieved guess a new value for the viscous flowrate and return to step 3.

An example showing detail how to use Stepanoff method is presented in the example section.

3.4.2 Hydraulic Institute Method

The Hydraulic Institute [10] introduced correction charts to predict single-stage pump performance at the BEP when handling viscous fluids using the known water single-stage pump performance. Figure 3.18 is one of the examples proposed by Hydraulic Institute showing how to obtain head, capacity, and efficiency correction factors for capacity less than 10,000 gallons per minute (GPM) at the BEP.

Following is the step-by-step of how to use the Hydraulic Institute chart:

- Step 1: From the pump performance curves for water, obtain the head, H_w, in feet and capacity, Q_w, in GPM at the BEP. Locate this point A on the Hydraulic Institute chart.
- Step 2: From this point, move horizontally until intersecting with the viscosity (centistokes) line. This is point B in Fig. 3.18.
- From point B, draw a vertical line to intersect with the efficiency correction factor curve (C_η), capacity correction factor curve (C_Q), and head correction factor curves (C_H). From these intersections, we can read the C_η, C_Q, and C_H.
- Applying the following equations to calculate the capacity, head and efficiency of pumps handling viscous fluids.

$$Q_{ris} = C_Q Q_w \tag{3.4.6}$$

$$H_{vis} = C_H H_w \tag{3.4.7}$$

$$\eta_{vis} = C_n \eta_w \tag{3.4.8}$$

Since the Hydraulic Institute method is based on empirical correlations, the pump has to be operated within the constraint of the chart to obtain reasonable results. Generally speaking, the Hydraulic Institute method provides a quick, simple way to predict the pump performance with a viscous fluid with reasonable in accuracy.

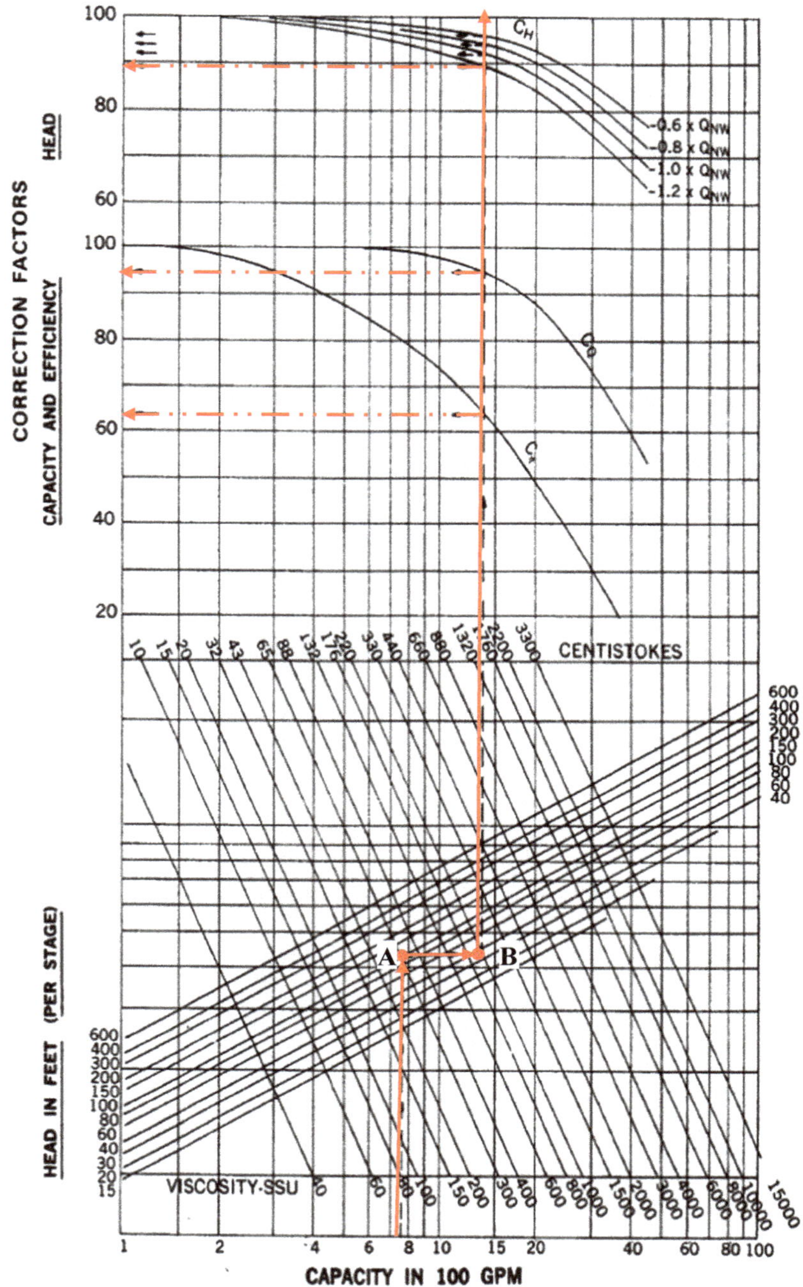

Fig. 3.18 The Hydraulic Institute chart for correcting the effect of fluid viscosity

3.4.3 Turzo et al. Method

Turzo et al. [21] digitized the Hydraulic Institute charts and apply the curve fitting techniques to find the most accurate correlations. To correct for the viscous effect, the capacity and head at the BEP based on the water pump performance curve are identified first. The following two variables, y and q^*, are calculated next.

$$y = -112.1374 + 6.6504\ln\left(H_{BEP}^{w}\right) + 12.8429\ln\left(Q_{BEP}^{w}\right) \tag{3.4.9}$$

$$q^* = \exp\left(\frac{39.5276 + 26.5605\ln(v) - y}{51.6565}\right) \tag{3.4.10}$$

where H_{BEP}^{w} is the head of the water performance at the B.E.P in ft, Q_{BEP}^{w} is the flow rate of the water performance at the BEP in BPD and v is the kinematic viscosity of the viscous fluid in cSt.

From these two variables, the correction factors for capacity (C_Q), efficiency (C_η) and head (C_H) are calculated as shown below:

$$C_Q = 1 - 10^{-4}\left(40.327q^* + 1.724q^{*^2}\right) \tag{3.4.11}$$

$$C_\eta = 1 - 10^{-4}\left(330.75q^* - 2.8875q^{*^2}\right) \tag{3.4.12}$$

$$C_{H1} = 1 - 10^{-5}\left(368q^* + 4.36q^{*^2}\right) \tag{3.4.13}$$

$$C_{H2} = 1 - 10^{-5}\left(447.23q^* + 4.18q^{*^2}\right) \tag{3.4.14}$$

$$C_{H3} = 1 - 10^{-5}\left(700q^* + 1.41q^{*^2}\right) \tag{3.4.15}$$

$$C_{H4} = 1 - 10^{-5}\left(901q^* + 1.31q^{*^2}\right) \tag{3.4.16}$$

where C_{H1}, C_{H2}, C_{H3} and C_{H4} are the correction factors for the pump heads at 60%, 80%, 100% and 120% of the BEP flow rate, respectively. It is obvious that if Turzo method is used then the iteration (Stepanoff method) and charts (Hydraulic Institute) are not needed.

3.4.4 Evdocia Method

To avoid the iteration when using Stepanoff method, Evdocia [6] had digitized the Stepanoff's charts using dimensionless analysis. Since the correction factors and

Stepanoff Reynold number are dimensionless, he proposed that any combination of powers of dimensionless number would also be a dimensionless number. Therefore, the relationship between Stepanoff Reynolds number and Evdocia Reynolds number are as follows:

$$R_{Evdocia} = 6.0345 \frac{NQ^{w}_{BEP}}{\sqrt{H^{w}_{BEP}}\nu} = R_{Stepanoff} \frac{Q^{w}_{BEP}}{Q^{vis}_{BEP}} \tag{3.4.17}$$

where N is the pump speed in RPM, Q^{w}_{BEP} and H^{w}_{BEP} are the capacity in BPD and head in ft at the BEP when pumps are handling water, respectively, Q^{visc}_{BEP} is the capacity in BPD at the BEP of a viscous fluid and v is kinematic viscosity in cSt. Evdocia also arrived to the relationship between the head correction factor and his Reynold number:

$$C_H = 1 - \exp\left(-0.033823 R^{0.367769}_{Evdocia}\right) \tag{3.4.18}$$

The calculation procedure of Evdocia method is summarized as follows:

- Step 1: Obtain the head and flowrate at the BEP ***for water*** $(H^{w}_{BEP}, Q^{w}_{BEP})$.
- Step 2: Calculate Evdocia Reynolds number using Eq. (3.4.17).
- Step 3: Calculate the head correction coefficient C_H using Eq. (3.4.18).
- Step 4: Obtain the capacity correction coefficient C_Q using Eq. (3.4.3a).
- Step 5: Calculate the viscous flowrate and head at the BEP (Q^{vis}_{BEP} and H^{vis}_{BEP}) using the calculated C_H and C_Q from steps 3 and 4.

With this proposed relationship, it is no longer iterative as opposed to Stepanoff's model.

3.5 Gas Effect on Pump Performance

The pump design and pump selection can be wrong if there is a significant volumes of gas not being pumped. The amount of free gas inside the tubing causes the change in the fluid mixture density and hence reduces the hydrostatic pressure or reduces the required pump head. If the pump discharge pressure is too low due to the free gas in the tubing then the pressure inside the pump can be much less than the bubble point pressure. This leads to a phase change inside the impeller and diffusers causing the dissolved gas to come out of the solution and become the fee gas. The entire pump cavities may be occupied by this fee gas instead of liquid; this is called gas-locked. When gas-locked occurs, the pump provide very small energy to the fluid and hence there may not be any production at surface.

Another phenomenon causing a deterioration of pump performance is pressure surging. Surging occurs when there are trapped gas pockets entering a pump at high rate causing the flow to fluctuate and pump to vibrate. When gas pockets enter a centrifugal pump, they accelerate and cause fluids inside the pump cavities to decelerate leading to a sudden increase in pump pressure. This is a common problem when operating centrifugal pumps. Pressure surging can lead to very poor pump performance, vibrations, shorter pump life and even pump failures.

In reality, pump manufacturers provide downhole gas separators to prevent gas entering the pump. The free gas separated by the gas separator is vented up the annulus. With this installation, there is normally no packers installed on the tubing. However, if the gas separator's efficiency is not high, there may still be some volume of gas entering the pump and cause a pump performance degradation. Although handling gas-liquid mixture has gradually become common for ESPs, the physical mechanism of how gas affecting pump performances is not well understood. Zhu and Zhang [23] stated that: "the gas bubble formation, coalescence and breakup mechanisms inside ESPs, which affect the two-phase flow characteristics, are still unclear. Due to the compact and complex geometries of multistage ESPs, the visualization of internal flow structures and bubble movement is very difficult".

To model the performance of an ESP under a two-phase mixture flow, there have been many different approaches. The simplest one is the homogeneous model.

3.5.1 Homogeneous Flow Modeling

Homogeneous model assumes that the centrifugal pump performance of a two-phase liquid-gas mixture would behave the same as that of a homogeneous mixture inside the pump. The model suggests that there is no head degradation between single-phase and two-phase flow inside the pump if *the mixture liquid flow rate verses pump head is used to describe for the pump performance*. In other words, if the two-phase pump performance curve is based on the mixture flow rate and density under non-slip flow, single-phase pump performance and two-phase pump performance would be the same. However, *when the pump performance is only based on single-phase liquid flow rate* and density, head performance degradation can be seen on the homogeneous model prediction.

According the two phase flow concept, the mixture flow rate and the mixture density for the case of no slip flow are defined as follows:

$$Q_m = Q_l + Q_g \tag{3.5.1}$$

$$\rho_m = (1 - \lambda_g)\rho_l + \lambda_g \rho_g \tag{3.5.2}$$

where the subscripts m, l, and g are for mixture, liquid, and gas, respectively. Q, ρ, and λ are flow rate, density, and no-slip gas fraction, respectively.

In this model, the mixture pressure gain across a centrifugal pump from the intake to the discharge is defined as follows:

$$\Delta P_{TP} = \rho_m g H_{TP} \tag{3.5.3}$$

Combining Eqs. (3.5.2) and (3.5.3) gives:

$$\Delta P_{TP} = \left[\left(1 - \lambda_g\right)\rho_l + \lambda_g \rho_g\right] g \mathrm{H}_{TP} \tag{3.5.4}$$

$$\Delta P_{TP} = \rho_l g \mathrm{H}_{TP} - \left(\rho_l - \rho_g\right)\lambda_g g \mathrm{H}_{TP} \tag{3.5.5}$$

One of the assumptions listed above tells us that the two-phase pump head is the same as the single-phase pump head if the mixture flow rate is used to describe the two-phase pump performance. This statement can be written as follows:

$$H_{TP}\left\{Q_l, Q_g\right\} = H_{SP}\{Q_m\} \tag{3.5.6}$$

Combining Eqs. (3.5.5) and (3.5.6) gives:

$$\Delta P_{TP} = \rho_l g H_{SP} - \left(\rho_l - \rho_g\right)\lambda_g g H_{SP} \tag{3.5.7}$$

$$\Delta P_{TP} = \Delta P_{SP}\{Q_m\} - \left(\rho_l - \rho_g\right)\lambda_g g H_{SP} \tag{3.5.8}$$

where the subscripts TP and SP are for two phase and single phase. The second term in the right hand side of Eq. (3.5.8), $\left(\rho_l - \rho_g\right)\lambda_g g H_{SP}$, is the pressure degradation due to the presence of gas. Equation (3.5.8) tells us that the two-phase pump pressure is deteriorated based on the single-phase pump pressure using mixture flow rate with an amount of $\left(\rho_l - \rho_g\right)\lambda_g g H_{SP}$. The theoretical two phase pump performance at different gas fraction, λ, is shown in Fig. 3.19.

3.5.2 *Empirical Correlations*

Turpin et al. [20] presented an empirical correlation using experimental.

$$H_m = H \exp\left[-\frac{Q_g}{Q_l}\left(\frac{346{,}430\, Q_g}{\left(C_1 P_{in}\right)^2 Q_l} - \frac{3410}{C_1 p_{in}}\right)\right] \tag{3.5.9}$$

where, H_m is the pump head with gas/liquid flow (ft); Q_g and Q_l are gas and liquid flow rates in barrels per day, respectively; P_{in} is pump intake pressure in psi; $C_1 = 0.145$ is a unit conversion constant.

Romero [15] introduced another empirical correlation to predict the two-phase pump performance using Cirilo's [3] experimental data.

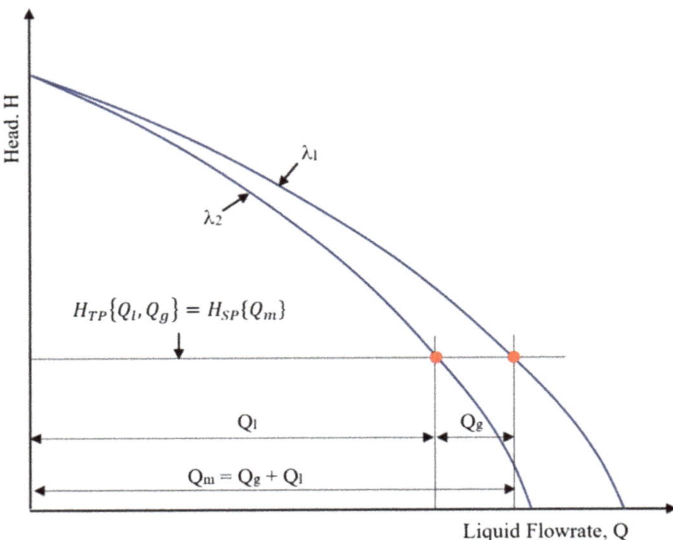

Fig. 3.19 Effect of gas on the two phase performance using homogeneous model prediction

$$\frac{H_m}{H_{\max}} = \left(1 - \frac{Q_{ld}}{Q_{d\max}}\right)\left[a\left(\frac{Q_{ld}}{Q_{d\max}}\right)^2 + \frac{Q_{ld}}{Q_{d\max}} + 1\right] \tag{3.5.10}$$

$$a = 2.902\lambda_g + 0.2751 \tag{3.5.11}$$

$$Q_{d\max} = 1 - 2.2035\lambda_g \tag{3.5.12}$$

where H_{max} is the shut-in pump head; Q_{ld} is the dimensionless liquid flow rate, which is defined as a ratio of the intake liquid flow rate to the open flow rate.

Duran and Prado [5] presented a model for predicting the two-phase pump performance of an ESP. This model is applicable for mild and severe head degradations and for bubbly and elongated bubble flow regimes.

$$\Delta P_m = \begin{cases} (1-\alpha)\rho_l H \frac{Q_l}{1-\alpha} + \alpha\rho_g H \frac{Q_g}{\alpha} \\ -0.47075 - 0.2163 ln\left(\frac{Q_g(1-\alpha)}{Q_{\max}}\right) \end{cases} \tag{3.5.13}$$

where ΔP_m is the single stage ESP pressure gain with gas/liquid flow in psi; α is the in situ gas void fraction.

3.5.3 Experimental Study on Two-Phase Centrifugal Pump Performance

Murakami and Minemura [12] conducted a study on the effect of two-phase flow on a centrifugal pump performance. The authors reported that the pump head decreases as more air is being injected, but each impeller's head remains is very similar. Since then, there have been many researchers focusing on the experimental as well as mathematical works to predict the two-phase performance of a centrifugal pump. Zhu and Zhang [23] presented a thorough review of works related to how gas affects the performance of centrifugal pumps. Table 3.1 shows the summary of works at Tulsa University Artificial Lift Project (TUALP).

A typical plot of a two-phase ESP performance generated at TUALP is shown in Fig. 3.20.

Vo and Nguyen [22] experimentally studied the effects of high viscosity oil and foamy-oil using a single-stage centrifugal pump. The authors concluded that: (1) Up to 37.8% GVF, foamy-oils has similar pump increment pressure performance

Table 3.1 Summary of works at TUALP related to two-phase flow in ESPs

Authors	Study	Pump	Fluid
Cirilo [3]	Compare two-phase flow performance of three different ESPs	GN4000 GN7000	Air/water
Romero [15]	ESP gas-liquid performance with an advanced gas handler installed upstream	GN4000	Air/water
Pessoa [13]	Measure stage-by-stage pump pressure increment of a multistage ESP	GC6100	Air/water
Beltur et al. [1]	Investigate pressure surging in ESP and affecting factors	GC6100	Air/water
Duran (2003)	Correlate experimental data of ESP two-phase performance	GC6100	Air/water
Zapata (2003)	Investigate pump rotational speed effect on ESP two-phase performance	GC6100	Air/water
Barrios (2007)	Visualize the internal flow of a 2nd stage ESP under gas/liquid flow conditions	GC6100	Air/water
Gamboa (2007)	Visualize ESP two-phase flow pattern using a similar pump prototype as Barrios	GC6100	Air/water
Trevisan (2009)	Visualize ESP internal flow under air/ viscous-liquid flow	GC6100	Air/oil visualization
Banjar (2013)	Investigate ESP performance with air/oil flow	DN1750	Air/oil
Salehi (2012)	Investigate ESP gas/liquid performance with various flow conditions	TE2700	Air/water
Croce (2014)	Investigate ESP performance with water/oil emulsion flow	DN1750	Oil/water Emulsion
Zhu (2017)	Investigate ESP gas/liquid flow performance with/ without surfactant injections	TE2700	Air/water Surfactant

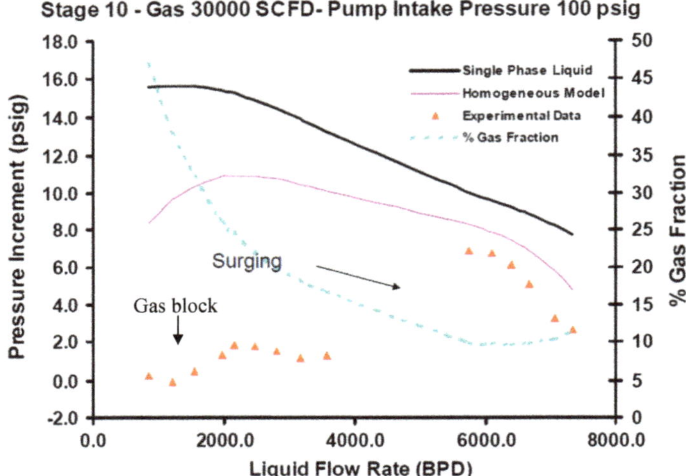

Fig. 3.20 Two-phase performance of an ESP conducted at TUALP

curves as a single-phase oil with up to 8% difference at comparable capacity. As a result, foamy-oil induces a gain in pump head performance; (2) Homogeneous model was unable to predict the pump pressure performance under foamy-oil with acceptable accuracy.

In summary, one can conclude that there is no reliable existing model to predict the two-phase performance of an ESP. The homogeneous model normally over predict due to assumptions. The use of empirical correlations are very limited due to the fact that the correlations were developed based on a specific experimental setup and the testing conditions. Experimental works have revealed to us that the pump performance will be deteriorated with the presence of gas. The higher the gas void fraction, the lower the pump head is. Surging and gas block may occur if the volume of gas is very high. Therefore, it is very critical when design an ESP system, the efficiency of the downhole gas separator must be high to minimize gas coming into the pump.

3.6 Pump Thrust

Pump thrust is defined as the hydraulic axial force acting on the pump shaft. If pump thrust is too high for a long period of time, variety of costly problems may occur such as bearing failure, mechanical seal failure, and impeller's damages. Examples of impeller's damages are shown in Fig. 3.21.

There are two areas where the thrust is generated: at the impellers and at the shaft.

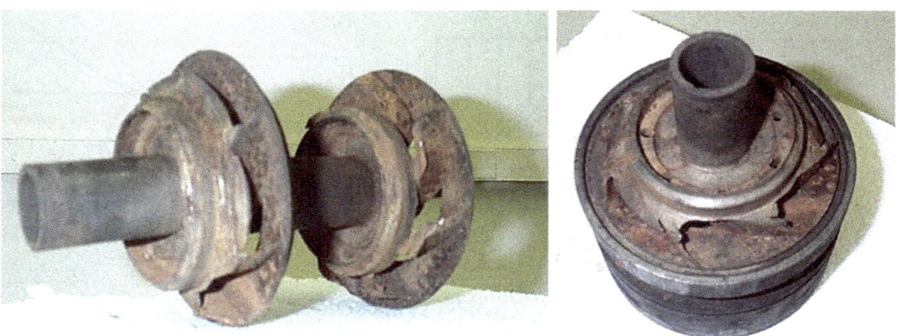

Fig. 3.21 Impeller's damage due to high down thrust of a radial-flow pump

- The thrust generated at the impeller is due to the differential pressure at the intake and outlet of the impeller and also due to the rate of momentum change when the fluid hits the eye of the impeller and changes direction. The total of those forces is called impeller thrust.
- The thrust generated at the shaft is due to the differential pressure between the pump intake and pump discharge. This axial force is called the pump shaft thrust.

3.6.1 Impeller Thrust

At the intake, the impeller wall is exposed to the suction pressure. At the discharge, the impeller wall is exposed to the discharge pressure. The fluid pressure increase from the intake to the discharge of the impeller is mainly due to the centrifugal force and the difference in fluid velocities following the blade's direction as discussed in Sect. 3.2. The magnitude of this pressure increase is in the range of (5–25) psi or equivalent to pump head of (11.5–57.5) ft. This pressure difference times the impeller's area give a force acting downward as shown in Fig. 3.22a.

Another force acting on the impeller is the force due to the change in direction of the fluid when it hits the eye of the impeller. Assuming the turn is 90° and the fluid is incompressible, the resultant force as shown in Fig. 3.22b can be calculated as:

$$F = \frac{\rho U^2}{\sqrt{2}} A_{impeller} \tag{3.6.1}$$

The impeller thrust in Newton in downward direction can now be estimated as follows:

$$F_{impeller} = (P_T - P_B) A_{impeller} - \frac{\rho U^2}{\sqrt{2}} A_{impeller} \tag{3.6.2}$$

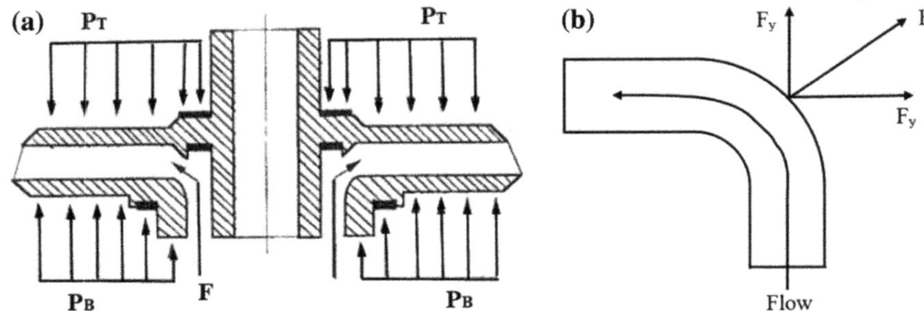

Fig. 3.22 **a** Impeller thrust due to $\Delta P = P_T - P_B$ **b** impeller thrust due to fluid change in direction

where P_T and P_B are the fluid pressure at the discharge and intake of the impeller in Pa, $A_{impeller}$ is the area of the impeller in m^2, ρ is the fluid density in kg/m^3, and U is the fluid velocity inside the impeller in m/s.

3.6.2 Pump Shaft Thrust

The pump discharge pressure is always higher than the pump intake pressure. This pressure difference acts on the cross-sectional area of the pump shaft, A_{shaft}, causing a force acting downward as shown in Fig. 3.23. This force is expressed as:

$$F_{shaft} = (P_d - P_i)A_{shaft} \tag{3.6.3}$$

3.6.3 Total Pump Thrust

The total pump thrust is the summation between the impeller thrust and the pump shaft thrust.

$$F_T = (P_T - P_B)A_{impeller} - \frac{\rho U^2}{\sqrt{2}} A_{impeller} + (P_d - P_i)A_{shaft} \tag{3.6.4}$$

Depending on a specific operating condition, F_T can be negative or positive. If the discharge pump pressure is much higher than the intake pressure (pump is operating under low flowrate and high pump head as shown in Fig. 3.24), the force, F_T, is acting downward or the pump is under down thrust. Similarly, a thrust is called up thrust if the total pump thrust is negative and the force direction is upward. This may happened if the pump is operated at maximum flowrate and very low discharge pressure as shown in Fig. 3.24.

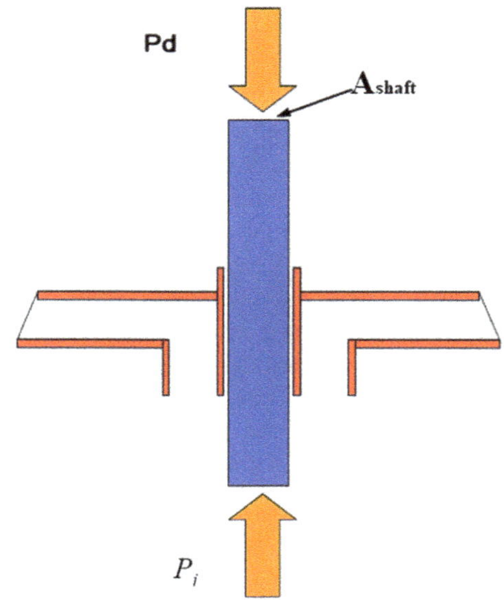

Fig. 3.23 Pump shaft thrust

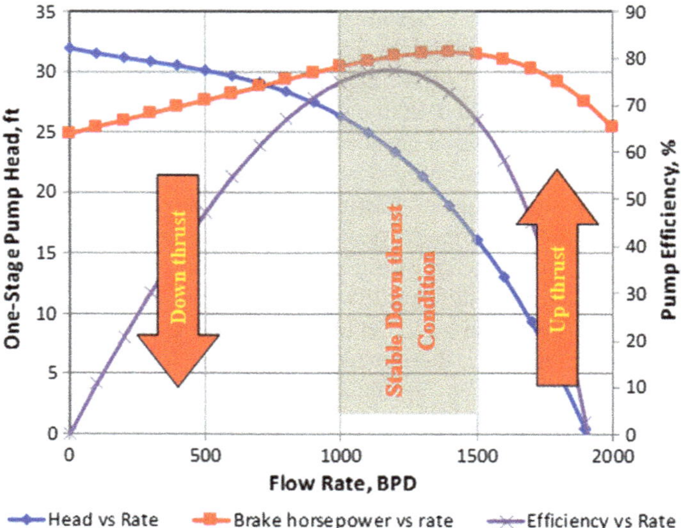

Fig. 3.24 Up thrust and down thrusts operating conditions

3.6.4 Discussion on Pump Trust

To reduce the total axial thrust of an impeller, a pump is normally designed with both front and back wearing rings on the impeller. In addition, pump manufacturers may drill balancing holes through the impeller wall near the inlet to maintain pressure on the back side of the impeller. The volumetric efficiency of those impellers is slightly lower than that of an unbalanced impeller due to the leakage pass the wearing ring into the suction through the balancing holes.

There are two types of impellers: fixed impellers (sometimes called compression impellers) and floating impellers. Fixed impellers are fixed on the shaft, cannot move axially along the shaft and located in the shaft with a certain clearance to the diffusers. The impeller thrust is then transferred to the shaft and hence the total thrust, F_T, will be the summation of the impeller thrust, $F_{impeller}$, and the shaft thrust, F_{shaft}, as shown in Fig. 3.25a. Floating impellers are allowed to move axially along the pump shaft. The impeller touches the thrust surfaces (washers) of the diffuser, and then the diffuser transfers the impeller thrust, $F_{impeller}$, to the pump housing. The pump shaft takes care of only the shaft thrust, F_{shaft}, as shown in Fig. 3.25b.

Note that floating impellers are normally designed to operate in a slightly down thrust condition. In other words, they are not designed to float between the diffuser thrust washers. Under the slightly down thrust condition, the washers act as a seal preventing or reducing the recirculation of fluid from the discharge of the impeller (higher pressure) to the eye of the impeller (lower pressure). This would lead to reduction in efficiency and abrasion. This design point coincides at the BEP for floating impellers. To the left of the BEP the floating impeller is on a down thrust situation and to the right of the BEP the floating impeller is on an up thrust situation as shown in Fig. 3.24. Operation outside the optimum range causes premature wear of the floating impeller and diffuser thrust washers.

For the protector (seal) thrust bearing is not overloaded, the thrust must be calculated correctly. As mentioned, floating Impellers should not operate in an up thrust condition. Fixed or compression impellers transfer all thrust to the protector and should never touch the diffuser. If we want to increase production beyond the operating range, it is better to use fixed impellers, since the thrust is now supported by the protector. In this case, the protector must be design correctly to handle the total thrust.

The reasons for using floating impeller are as follows: (1) each stage handles its own thrust and hence there is no build up of the thrust on the protector bearing; (2) floating stages provide a good seal for mild abrasives (if the pump is working in slightly down thrust condition) and prevent damage to the radial bearing area; (3) assembly of floating impellers is much easier since no shimming is necessary and tolerance stack-up is not a problem.

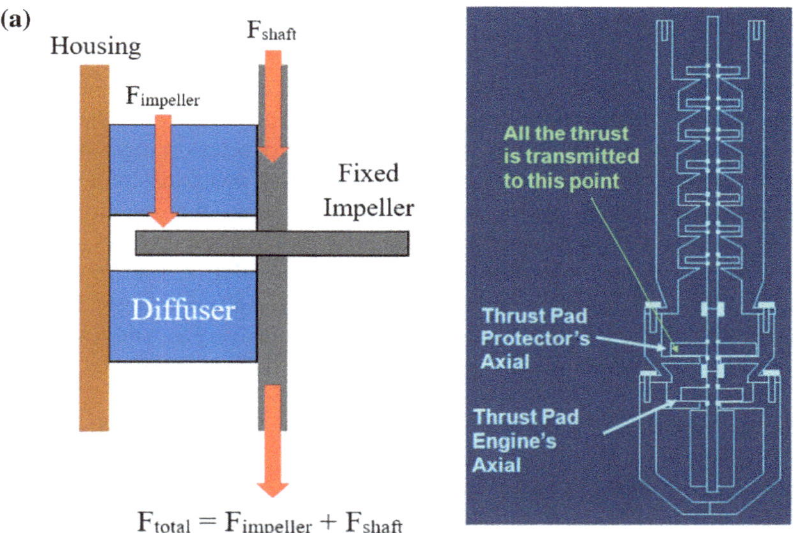

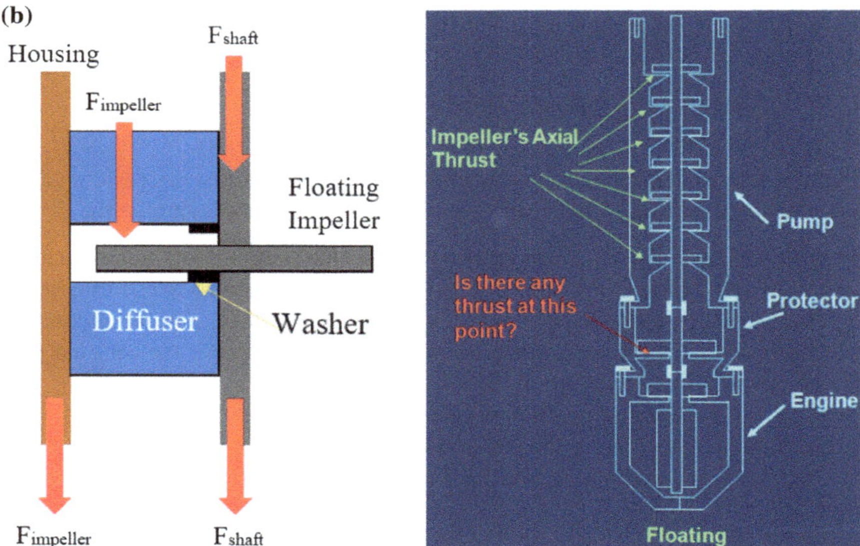

Fig. 3.25 **a** Thrust on fixed impellers **b** thrust on floating impellers

3.7 ESP Design

There are many factors that an engineer has to consider when designing an ESP system. The most two important components in an ESP system are the downhole motor and the downhole pump. Engineers have to select pumps and motors available on the market depending on specific operating conditions of a well. Outer diameter, power, voltage, and current are the four inputs that engineers need to select motors. On the other hand, pump capacity and total pump head at the BEP, pump type, and pump outer diameter are the factors that engineers need when selecting pumps. In addition, inner diameter of production casing, IPR, OPR, formation fluid properties (fluid density and fluid viscosity), gas void fraction, sand concentration, corrosive and abrasive fluids are other factors that greatly impact the pump design.

The procedure of designing an ESP system can be summarized as follows: (1) collecting basic data; (2) selecting pump and motor outer diameter and determining the pump depth; (3) selecting pump depth; (4) analyzing well flow capacity (IPR and OPR); (5) selecting pump; (6) considering effects of gas and viscosity on pump selection; (7) selecting motor; (8) selecting cable; (9) selecting gas separator, and (10) selecting surface equipment. We now examine each step in this procedure.

3.7.1 Collection of Basic Data

The first step to design an ESP system is to collect and verify the basic reliability data used in the design. The basic data may include the following items:

Well data: The well data may include casing size and weight, tubing size and weight, type of completion (cased hole, open hole, perforated, etc.), true vertical depth, measured depth, dog leg severity, azimuth, etc. Well data is important to select pump and motor outer diameter as well as to examine if the selected pump can be run into the well.

Production data: desired production rate, gas oil ratio, water cut, bottom hole and surface fluid temperatures, wellhead pressure, separator pressure, maximum surface pressure, etc. are production data that engineers need when designing an ESP system.

Formation fluids: reservoir pressure, bubble point pressure, specific gravity of oil and gas, oil API, oil viscosity, etc. These parameters will determine shapes of the IPR and OPR.

Other data: checking for possibilities of sand production, deposition, corrosion, paraffin.

3.7.2 Selection of Pump and Motor Diameter

The last casing string run into an oil/gas well is the production casing. The inner diameter of this casing will set a limit for the maximum pump and motor outer diameters. Generally speaking from manufacturing perspectives, pump with larger outer diameters will give lower both initial and operating costs because it is easier for manufacturers to design and manufacture bigger pumps. There are guidelines and recommendations from manufacturers for selecting pumps and motors depending on the inner diameter of production casings. Table 3.2 shows an example of how to select pump and motor outer diameters which depend on production casing inner diameters.

ESP is not considered to be flexible as compared to drillstrings and casings because it is much shorter and more rigid. Different components of a pump assembly are connected using flanges. The flanges can be considered to be the weakest points compared to that of the rest of the unit. As long as the bending stress at the flanges does not exceed the yield strength of the material, the pump can be run and installed in the well. The rule of thumb is that if the DLS is smaller than 3°/100 ft, the pump can be placed into the well [8].

3.7.3 Selection of Pump Depth

Conventionally, ESPs in vertical and horizontal wells are recommended to installed above perforations a few hundred feet and at least 500 ft of fluid over the pump [19]. This recommendation is to make sure that the heat generated from the motor is carried upward by formation fluids to avoid motor overheated.

The pressure gained across the pump (pump pressure), ΔP_{pump}, is defined as the differential pressure between the pump discharge, P_d, and the pump intake, P_i.

$$\Delta P_{pump} = P_d - P_i \tag{3.7.1}$$

The pump discharge pressure, P_d, is to overcome the wellhead pressure, P_{wh}, friction inside the tubing from the pump to the surface, $\Delta P_f^{pump-surf}$, and the hydrostatic pressure from the pump depth, h_{pump}, to the surface.

Table 3.2 Selection of pump outer diameter

Minimum prod. casing OD (in.)	Pump OD (in.)	Motor OD (in.)	Relative cost factor
4 ½	3.750	3.750	2.300
5 ½	4.50–4.560	4.500–4.560	1.440
7	5.400	5.62	1.000
8 5/8	7.250–7.380	7.250–7.380	N/A

$$P_d = P_{wh} + \Delta P_f^{pump-surf} + \rho_f g h_{pump} \quad (3.7.2)$$

Combining Eqs. (3.7.1) and (3.7.2) gives

$$P_i = P_{wh} + \Delta P_f^{pump-surf} + \rho_f g h_{pump} - \Delta P_{pump} \quad (3.7.3)$$

The flowing bottom hole pressure at the midpoint of the perforation

$$P_{wf} = P_i + \rho_f g (TVD - h_{pump}) + \Delta P_f^{perf-pump} \quad (3.7.4)$$

Combing Eqs. (3.7.3) and (3.7.4) gives

$$P_{wf} = P_{wh} + \rho_f g TVD + \Delta P_f^{perf-surf} - \Delta P_{pump} \quad (3.7.5)$$

Applying the definition of pump pressure gives:

$$P_{wf} = P_{wh} + \rho_f g TVD + \Delta P_f^{perf-surf} - (P_d - P_i) \quad (3.7.6)$$

Or

$$P_d = P_{wh} + \rho_f g TVD + \Delta P_f^{perf-surf} - (P_{wf} - P_i) \quad (3.7.7)$$

If the pump is installed close to the perforation then the pump intake is similar to the flowing bottom hole pressure and Eq. (3.7.7) becomes:

$$P_d = P_{wh} + \rho_f g TVD + \Delta P_f^{perf-surf} \quad (3.7.8)$$

where TVD is the true vertical depth of the well, $\Delta P_f^{perf-surf}$, is the frictional pressure drop inside the production tubing from the perforation to the surface.

Equation (3.7.8) tells us that the pump depth (pump position, h_{pump}) does not affect the flowing bottom hole pressure. In other words, changing pump position along the well does not change the performance of the well. For a specific well and a specific pump, varying the pump position does not affect the production rate if the reservoir pressure is assumed to be constant. However, as the pump is installed deeper into the vertical or inclined section of the well, the pump intake pressure is higher leading to a higher in the pump discharge in such a way that the pump pressure ($\Delta P_{pump} = P_d - P_i$) remains the same as shown in Eq. (3.7.8).

In summary, pump depth does impact the performance of vertical (or near vertical) wells. Therefore, it is recommended to install pumps above the perforation from 100 to 500 ft so that formation fluids can carry heat generated from the motor to the surface.

3.7.4 Analyzing Well Flow Capacity

The well flow capacity is mainly dependent on the flow inside the reservoir (IPR) and the flow in the tubing and surface flowlines (OPR). Analyzing the IPR and OPR will help us to determine what ESP we would need to achieve a desired liquid production rate. Figure 3.4 demonstrates the main working principle of an ESP system. Readers are highly recommended to review Sect. 3.1.3.

It is important to note that if reservoir pressure is greater than bubble point pressure, Productivity Index (PI) is a constant and hence the IPR is a straight-line and Darcy law can be applicable. If the reservoir pressure is smaller than the bubble point pressure than the IPR is no longer a straight-line; instead it is a curve. In this case, Vogel's equation (or any other multiphase IPR correlations) can be used to describe for the IPR.

Figure 3.26 illustrates the IPR and OPR of an oil well. Without a pump, the well is producing naturally at an equivalent liquid rate of Q_e^{nf}. For this well to produce at higher liquid rate (desired production rate), an ESP with a total pump head of $H_{pump} = \frac{\Delta P_{pump}}{\rho_f g}$ is needed to overcome the hydrostatic pressure, friction inside the tubing, and wellhead pressures at the surface.

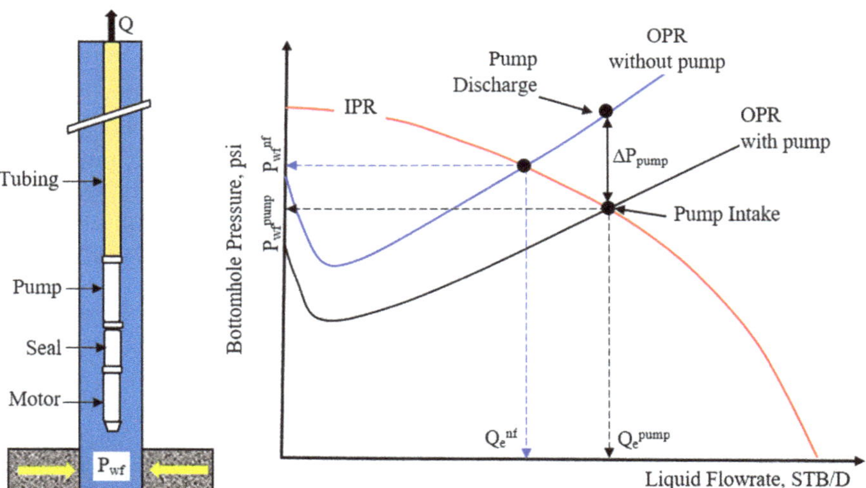

Fig. 3.26 Determination of pump pressure based on IPR and OPR

3.7.5 Selection of ESP

Water pump performance curves are given for each pump series by pump manufacturers. To select an ESP, which is available on the market, we need to know pump outer diameter, pump capacity, pump head, and pump type. Pump outer diameter can be obtained if the production casing size is known using recommended data from pump manufacturers. An example is given in Table 3.2. Pump capacity is basically the desired pump rate which is known when designing an ESP system. Depending on the operating conditions of a specific well, we can be recommended by pump manufacturers if float or fixed impeller type, mixed flow stage pump or radial flow stage pump should be used. The only item left that we need to know to select an ESP is the pump head.

Pump head is defined as follows:

$$H_{pump} = \frac{\Delta P_{pump}}{\rho_f g} = TVD_{pump} + \frac{\Delta P^{tubing}_{friction}}{\rho_f g} + \frac{\Delta P^{flowline}_{friction}}{\rho_f g} + \frac{P_{wh}}{\rho_f g} + \frac{P_{separator}}{\rho_f g} \tag{3.7.9}$$

Or

$$H_{pump} = \frac{P_d - P_i}{\rho_f g} = TVD_{pump} + h^{tubing}_{friction} + h^{flowline}_{friction} + h_{wh} + h_{separator} \tag{3.7.10}$$

where TVD is the true vertical depth from the pump discharge to the wellhead, $\Delta P^{tubing}_{friction}$ and $\Delta P^{flowline}_{friction}$ are the frictional pressure drop inside the production tubing and flowline, respectively, P_{wh} is the wellhead pressure, $P_{separator}$ is the pressure in the separator, P_d and P_i are the pump discharge and intake pressures, respectively. Readers should be careful with units when using Eqs. (3.7.9) and (3.7.10). If pressures are in Pa, fluid density is in kg/m^3, and g = 9.81 m/s^2 then the unit of pump head would be meter. Readers are also recommended to review Sect. 1.6.1 for calculating the frictional pressure losses in tubing and flowline.

The pump head, H_{pump}, is sometimes called the Total Dynamic Head (TDH). From the desired pump rate, pump outer diameter, and pump type, using water performance curves provided by pump manufacturers, the pump head per stage would be achieved. An example is shown in Fig. 3.27 [9]. This is a 400-series-ESP one-stage water pump performance curve. The outer diameter of this pump is 3.38 in. and hence it is used for a minimum casing of 4.5 in. The operating range of this pump is 226–500 BPD. The desired production rate for this example is about 420 BPD at the BEP. Therefore, the one-stage pump head, $h_{one\ stage}$, is found to be 15 ft/stage. The number of pump stage is give as follows:

$$Total\ stages = \frac{H_{pump}}{h_{one\ stage}} = \frac{TDH}{h_{one\ stage}} \tag{3.7.11}$$

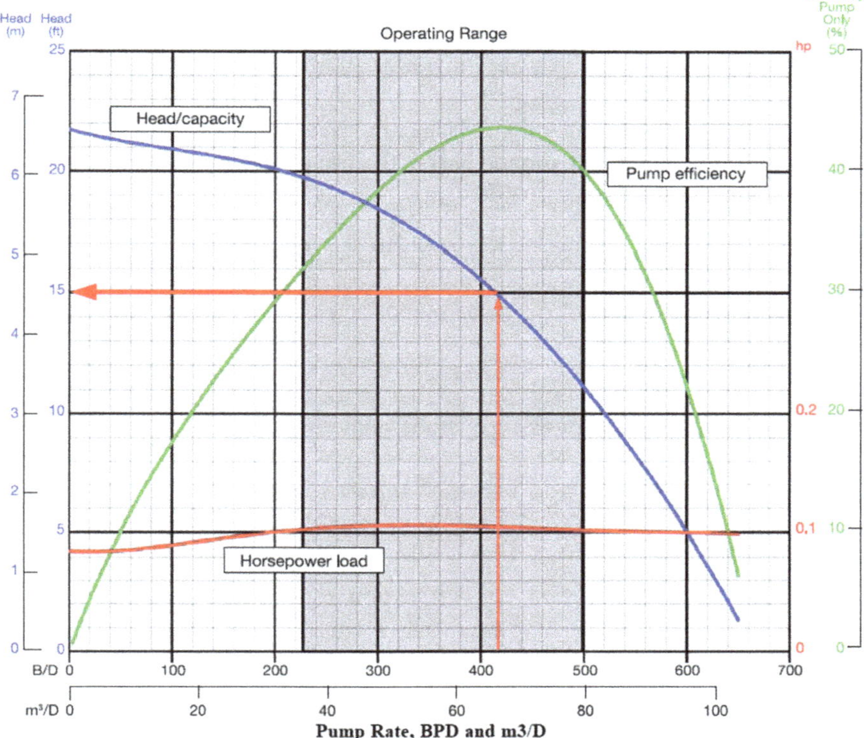

Fig. 3.27 Determination of pump head per stage Halliburton pump—one-stage pump performance—HAL338 400—pump OD = 3.38-in. operating range = (226–500) BPD—minimum casing size = 4.5 in.

The break pump horsepower can also be calculated as

$$HP_b = HP_{one\,stage} \times total\,stages \tag{3.7.12}$$

where the one-stage brake horsepower, $HP_{one\ stage}$, for the pump in Fig. 3.27 is 0.11 hp/stage.

3.7.6 Consideration of the Effect of Gas and Viscosity on Pump Selection

The volume of free gas at the pump intake is also important to know. If the volume of free gas at the intake is high then a gas separator is needed to maintain the pump efficiency. If this free gas is not separated and enters the pump then it may cause a considerably reduction in the required discharge pressure. In order to verify the volume of free gas at the intake, one can proceed the following calculation:

- Determine solution gas oil ratio (dissolved gas oil ratio), R_s, using correlations described in Table 1.2—Chap. 1.
- Determine gas formation volume factor, B_g, using Eq. (1.4.13a).
- Determine oil formation volume factor, B_o, using equations in Table 1.3.
- Determine the total volume of free gas

$$V_{free\,gas} = V_{total\,gas} - V_{solution\,gas} = GOR_{producing} \times Q_o - R_s \times Q_o \tag{3.7.13}$$

- Determine the volume of oil at the pump intake

$$V_{o@intake} = B_o \times Q_o \tag{3.7.14}$$

- Determine the volume of free gas at the pump intake

$$V_{g@intake} = V_{free\,gas} \times B_g \tag{3.7.15}$$

- Total fluid volume

$$V_t = V_g + V_o + V_w \tag{3.7.16}$$

- Percentage of free gas at the pump intake can be defined as:

$$\%\,Free\,gas = \frac{V_{g@intake}}{V_t} \tag{3.7.17}$$

The rule of thumb is that if the % free gas is more than 10%, a gas separator is recommended to maximize the performance the pump. More detail on how gas affect the pump performance can be found in Sect. 3.5.

The amount of free gas and fluid viscosity will change the OPR. Higher free gas reduces the hydrostatic pressure and higher viscosity contributes to higher friction. Readers are recommended to review on how to obtain the OPR in Sect. 1.6. In addition, fluid viscosity has a great impact on the pump performance. Many different methods such as Hydraulic Institute, Stepanoff, etc. showing how to take fluid viscosity into consideration when calculating pump head are presented in Sect. 3.4. As the fluid viscosity is higher, the OPR will shift upward indicating that more pump pressure is needed to maintain the same liquid rate as shown in Fig. 3.28. The new required pump pressure now is the summation of $\Delta P^{viscous}_{pump} = \Delta P^{water}_{pump} + \Delta P^{addition}_{pump}$.

After considering viscous and gas effects, designers should now select a pump to meet specific sets of a well and operating conditions. In other words, designers should know pump rate, pump head per stage, pump efficiency, and pump brake

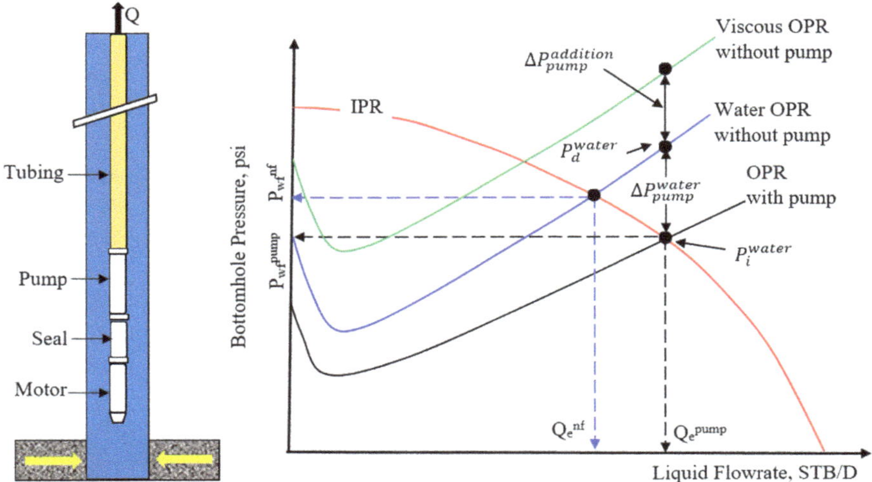

Fig. 3.28 Effect of viscosity on pump pressure

horsepower per stage at the BEP. The total pump brake horsepower can be calculated as follow:

$$HP_b^{total} = HP_b^{stage} \times number\ of\ stages \times \gamma \tag{3.7.18}$$

where the subscript b is for break and γ is for fluid specific gravity.

3.7.7 Selection of Motor

Selection of ESP motors is based mainly on the required power of an ESP system. The power of the motor (or total power required of an ESP system) consists of powers to run the pump, the protector, and the gas separator. Electric power is the product of voltage and amperage and hence there are different combinations between voltage and amperage to achieve a specific ESP motor power. In other words, depending the availabilities of ESP motors in the market as well as the total power, one can select an optimal motor for a specific application. Normally, manufacturer's catalogs recommend motors based on motor outer diameter, brake horsepower, voltage, and amperage. Tables 3.3 and 3.4 show examples of Centrilift® and Halliburton motor catalogs.

For compatibility purposes as well as future pump and motor services, it is highly recommended that the motor and the pump should be selected from one manufacturer. For example, if the pump is from Centrilift® then designers should select the motor from Centrilift®. Most of the time, motors with the largest OD that

Table 3.3 Centrilift® motor catalog: 450 series model—5½ in. OD

Brake horsepower (HP)	Volts/Amps	Length (shipping)	Weight (lbs.)
15	440/22	7′ 2″	269
20	420/31	8′ 5″	332
25	750/22 or 430/46	9′ 9″	396
30	740/27	11′	459
35	430/53 or 960/24	12′ 3″	523
40	420/62 or 965/27	13′ 6″	586
50	1,200/27	16″	713
60	1,270/30	18′ 7″	840
75	1,130/43	22′ 4″	1,031
85	1,290/43	24′ 10″	1,158
100	1,150/56 or 2,080/31	28′ 7″	1,305
120	1,200/64 or 2,080/37	33′ 7″	1,605

Table 3.4 Halliburton motor catalog: 3.75 in. OD—minimum casing size 4.5 in.

60 Hz		50 Hz		Amps	Length		Weight	
HP	Volts	HP	Volts		ft	m	lb	kg
15	550	12.5	458	20.5	6.7	2.0	218	98.9
20	510	16.7	425	30	8.6	2.6	280	127.0
30	420	25	350	54	11.4	3.5	371	168.3
30	535	25	446	42.5	11.4	3.5	371	168.3
40	520	33.3	433	58.5	14.2	4.3	466	211.4
40	785	33.5	654	37.5	14.2	4.3	466	211.4
45	440	37.5	367	76.5	16.1	4.9	524	237.7
45	585	37.5	487	57.5	16.1	4.9	524	237.7
45	975	37.5	812	45	16.1	4.9	524	237.7
50	490	41.7	408	76.5	17.9	5.5	670	303.9
50	650	41.7	541	57.5	17.9	5.5	670	303.9
50	835	41.7	696	45	17.9	5.5	670	303.9
60	585	50.0	487	76.5	20.8	6.3	770	349.3
60	780	50.0	650	57.5	20.8	6.3	770	349.3
60	990	50.0	825	45	20.8	6.3	770	349.3

can be run into the well should be selected due to its lower cost. Due to this reason, motors with difference OD from pump OD are acceptable in reality.

Heat generated from motors is carried up to the surface by formation fluids. Therefore, it is important to maintain adequate fluid velocity passing the motor. A minimum fluid velocity passing the motor of 1 ft/s is recommended when designing an ESP system [7].

3.7.8 Selection of Cable

For an ESP well system, electric cable is run from the wellhead all the way down to the pump depth to supply power to the electric motor. Based on well geometry, desired voltages and amperages, pump designers have to select types of cables (round or flat), materials of cables (copper or aluminum), length of cables, size of cables (cable no. 1, 2, 4, 6, etc.). In addition, designers have to take into considerations of well conditions such as temperature, pressure, corrosive materials when selecting cables. Under high pressure and high temperature, cable constructions need to be stronger compared to shallow cool wells. If the well present corrosive chemicals, attention to jacket and armor selection is a must. If the clearance between production tubings and production casings are small, flat cables instead of round cables should be chosen. To select size of cables, one must take into consideration of voltage drop along the cables. Figures 3.29 and 3.30 show examples of voltage drop per 1,000 ft and current drop depending on fluid temperature of four different cables namely American Wire Gauge cable number 1 (1 AWG), 2 AWG, 4 AWG, and 6 AWG.

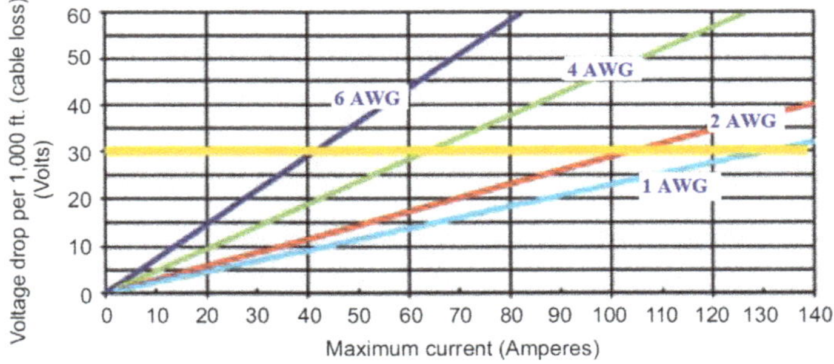

Fig. 3.29 Cable voltage drop—conductor temperature (25 °C) 77 °F Halliburton ESP 2014-catalog

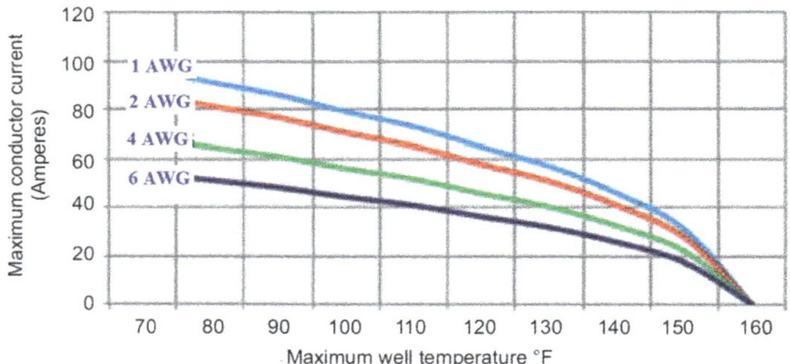

Fig. 3.30 Cable current drop depending on temperature Halliburton ESP 2014-catalog

Table 3.5 presents low temperature flat standard cable size rated 5 kV at the operating temperature up to 205 °F.

3.7.9 Selection of Gas Separator

If the volumetric fraction of free gas at the pump intake is greater than 10%, a gas separator is required to help maintaining the pump performance. The main mechanism of separating gas out of liquid of a downhole gas separator is based on centrifugal force. There are impellers inside the separator rotating at the same speed of the motor to impart centrifugal forces for liquid and gas. Liquid has higher density than gas and hence it has a tendency to move toward the inner wall of the separator. Meanwhile, gas has much lower density than liquid and hence it moves upward toward the center of the separator. With a specific design, separators direct liquid phase into the pump intake and direct gas phase into the annulus between the production tubing and the production casings. Gas is then vented at the surface.

Depending on the well geometry and volumetric free gas, a separator is selected. In reality, a gas separator is chosen based on selections of pump and motor, experience and knowledge of well conditions. There are no clear calculations and guidelines of selecting downhole gas separators. It is heavily dependent on experience and specific well conditions. Table 3.6 presents Halliburton gas separators for three series.

In Table 3.6, the abbreviations SEP is for separator; HAL is for Halliburton; BRG is for bearing; LT is for lower tandem; UT is for upper tandem; HS is for high strength; CS is for carbon steel.

3.7.10 Selection of Surface Equipment

For an ESP system, the surface equipment is composed of transformers, Variable Frequency Drives (VFD) and switchboard, junction box, wellhead, and monitoring and automation.

Transformer: A transformer used in an ESP system is to decrease the high voltages from electric suppliers to lower voltage level where it can be used to supply power for the downhole motor and other surface electric devices. The key when selecting a transformer is to understand the kVA rating. The kVA rating tells us how much apparent power can flow through a transformer without exceeding a temperature rise associated with the rating. kVA is the unit of power in thousand volt-amperes. Let consider a transformer with a rating of 3.5 kVA. This rating means that the transformer has a power limit of 3500 VA. If we divide 3500 by the voltage input, we would get the maximum current output at which the transformer would still work properly. To determine the size of the transformer, you need to calculate the followings:

Table 3.5 Low temperature flat cable 205 °F—Halliburton ESP 2014-catalog

Product key	kV	Conductor size		Conductor diameter		Insulation diameter		Overall dimension		Weight per	
		AWG	mm^2	in. (nom)	mm (nom)	in. (±0.016)	mm (±0.406)	in. (nom)	mm (nom)	lb/ft (nom)	kg/m (nom)
1LTF3015-000	5	1	42.4	0.289	7.3	0.48	12.1	2.00 × 0.75	50.6 × 19.0	1.6	2.4
1LTF3025-000	5	2	33.6	0.258	6.6	0.44	11.3	1.90 × 0.72	48.3 × 18.2	1.4	2.1
1LTF3045-000	5	4	21.1	0.204	5.2	0.39	9.9	1.74 × 0.66	44.2 × 16.8	1.1	1.7
1LTF3065-000	5	6	13.3	0.162	4.1	0.35	8.8	1.61 × 0.62	41.0 × 15.8	0.9	1.4

Table 3.6 Rotary gas separator—Halliburton ESP 2014-catalog

Series	Model	Description	Length		Weight		OD	Shaft OD
			ft	m	lb	kg	in.	in.
338	HAL 13	SEP HAL300 2BRG LT CS	2.5	0.8	40	18.1	3.38	0.875
		SEP HAL300 2BRG LT HS CS	2.5	0.8	40	18.1	3.38	0.875
		SEP HAL300 3BRG LT CS	2.8	0.9	45	20.4	3.38	0.875
		SEP HAL300 3BRG LT HS CS	2.8	0.9	45	20.4	3.38	0.875
400	HAL 14	SEP HAL400 2BRG LT CS	2.6	0.8	55	24.9	4.00	0.875
		SEP HAL400 2BRG LT HS CS	2.6	0.8	55	24.9	4.00	0.875
		SEP HAL400 3BRG LT CS	3.0	0.9	65	29.5	4.00	0.875
		SEP HAL400 3BRG LT HS CS	3.0	0.9	65	29.5	4.00	0.875
		SEP HAL400 2BRG UT CS	2.3	0.7	50	22.7	4.00	0.875
		SEP HAL400 1BRG UT HS	2.3	0.7	50	22.7	4.00	0.875
513	HAL 15	SEP HAL500 3BRG LT CS	3.2	1.0	150	68.0	5.13	1.187
		SEP HAL500 3BRG LT HS CS	3.2	1.0	150	68.0	5.13	1.187
		SEP HAL500 3BRG UT CS	2.6	0.8	120	54.4	5.13	1.187
		SEP HAL500 3BRG UT HS	2.6	0.8	120	54.4	5.13	1.187

- Load voltage (or secondary voltage) is the voltage needed to operate the load;
- Line voltage (or primary voltage) is the voltage from the source;
- Load current is the current needed to operate the load.

For three phase transformers, the power in kVA is determined by:

$$P\,(\text{kVA}) = \frac{u\,(\text{volts}) \times I\,(\text{amps}) \times 1.732}{1000} \tag{3.7.19}$$

For example, if the voltage and current to operate the motor are 480 V and 208 A then the power required from the three phase transformer is 172 kVA. Table 3.7 shows standard Halliburton three phase transformers.

Table 3.7 Halliburton three-phase transformers—Halliburton ESP 2014-catalog

KVA		Primary	Multi-tap secondary		Height		Width		Depth		Weight	
480 V	380 V		Delta	WYE	in.	m	in.	m	in.	m	lb	kg
100	79	480	600/1496	1039/2591	51	1.29	42	1.06	37	0.94	2850	1292
130	103	480	750/1500	1299/2685	51	1.29	42	1.06	37	0.94	2850	1292
150	119	480	750/1500	1299/2685	51	1.29	42	1.06	37	0.94	2950	1337
200	158	480	1100/2200	1905/3811	58	1.47	54	1.37	45	1.14	4210	1909
260	206	480	1100/2200	1905/3811	58	1.47	54	1.37	45	1.14	5125	2324
355	281	480	1100/2200	1905/3811	58	1.47	54	1.37	45	1.14	5350	2426
400	317	480	1100/2200	1905/3811	53	1.35	60	1.52	48	1.22	5210	2362
520	412	480	1100/2200	1905/3811	61	1.54	61	1.54	71	1.80	6500	2948
650	515	480	1250/2500	2165/4330	66	1.68	62	1.57	56	1.42	7150	3243
750	594	480	1250/2500	2165/4330	66	1.68	64	1.63	57	1.45	7650	3470
875	693	480	1400/2800	2425/4850	66	1.68	75	1.91	56	1.42	8525	3867
1000	792	480	1400/2800	2425/4850	66	1.68	76	1.93	57	1.45	9325	4230

Switchboard: is an electric device used to control, protect and monitor a downhole motor of an ESP system. The output power from the transformer is supplied into the switchboard. The output of the switchboard is connected to the downhole motor via. cables (see Fig. 3.2). The switchboard may be configured to constant, variable, or extended torque operations. The selection of a switchboard is mainly dependent on the size of the downhole motor and the power input from the transformer. Table 3.8 presents standard Halliburton switchboards.

Monitoring and Automation: a basic monitoring system of an ESP is installed below the motor. This setup allows us to monitor static and flowing bottomhole pressure (this can be assumed as the intake pump pressure), downhole fluid temperature, motor temperature, and motor vibration. Motor current and motor voltage are recorded on the switchboard. To monitor the fluid temperature and pressure at the discharge of the pump, another ESP gauge must be installed right at the discharge of the pump. In addition, the automation system may allow us to optimize the pump/well performance. Figure 3.31 shows a basic setup of an ESP downhole sensor system. Figure 3.32 gives an example of how the data are recorded for a basic monitoring ESP system.

Table 3.8 Halliburton switchboard—NEMA 4—Halliburton ESP 2014-catalog

Model	Output		Input		Height		Width		Depth		Weight	
	HP	Amps	Volts	Amps	in.	cm	in.	cm	in.	cm	lb	kg
1200-460-50RP	50	65	480	56	48	121.9	24	61.0	20.0	50.8	370	168
1200-460-60RP	60	77	480	67	48	121.9	24	61.0	20.0	50.8	370	168
1200-460-60	60	77	480	66.8	48	121.9	24	61.0	20.0	50.8	370	168
1200-460-75	75	96	480	83.4	48	121.9	24	61.0	20.0	50.8	370	168
1200-460-100	100	124	480	109	48	121.9	24	61.0	20.0	50.8	370	168
1200-460-125	125	156	480	137	48	121.9	24	61.0	20.0	50.8	500	227
1200-460-150	150	180	480	161	60	121.9	24	61.0	20.0	50.8	500	227
1200-460-200	200	240	480	215	60	152.4	36	91.4	36.0	91.4	800	363
1200-460-250	250	300	480	269	60	152.4	36	91.4	36.0	91.4	800	363
1200-460-300	300	360	480	323	60	152.4	36	91.4	36.0	91.4	1000	455
1200-460-400	400	480	480	430	60	152.4	36	91.4	36.0	91.4	1100	500
1200-460-500	500	600	480	538	72	182.9	54	137.2	36.0	91.4	1500	685
1200-460-600	600	720	480	646	72	182.9	54	137.2	36.0	91.4	2100	955
1200-460-800	800	960	480	890	72	182.9	54	137.2	36.0	91.4	2100	955
1200-460-1000	1000	1200	480	1114	72	182.9	54	137.2	36.0	91.4	2100	955

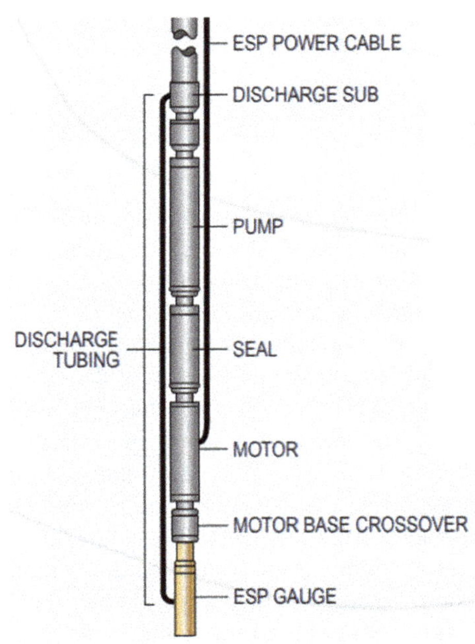

Fig. 3.31 Basic setup of an ESP downhole sensor system

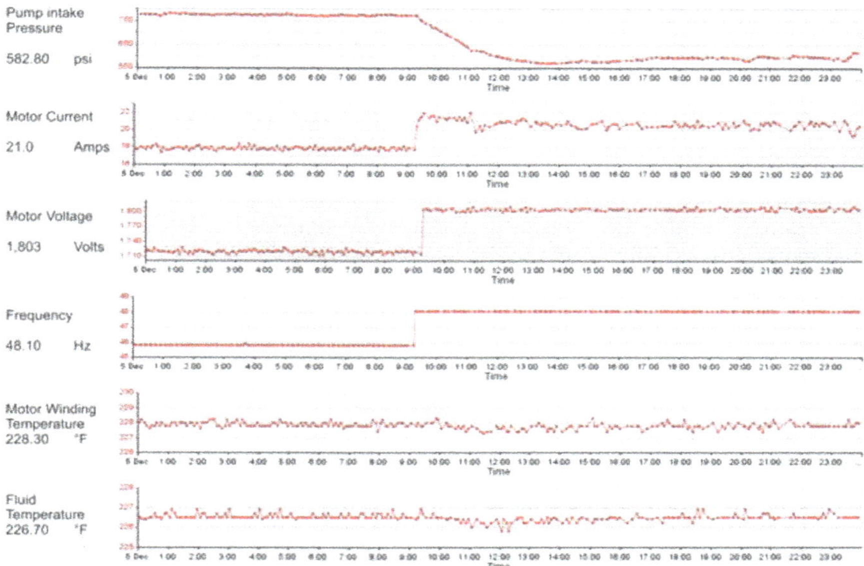

Fig. 3.32 Recorded data of an ESP system—Halliburton ESP 2014-catalog

3.8 ESP Failures

Generally speaking, key components of an ESP system such as pump, motor, separator, and cables are at the bottom of the well and they may expose directly to high pressure, high temperature, and corrosive/erosive environments. Therefore, there are many reasons that can cause an ESP system to fail. These reasons can range from human error, improper design of the ESP for a particular reservoir, severe reservoir or well conditions such as excessive operating temperature, corrosion, or abrasive materials in the produced fluids. Other reasons can be due to mechanical problems such as seal, gasket, bearing, impeller, diffuser, cable damages.

Motor failures due to overloaded or underloaded operating conditions. In addition, deposits of scale, asphaltenes, or paraffin may also cause problems for motor and pump. If the deposits cover the motor housing, the heat transfer from the motor to the fluid is prevented leading to inadequate cooling of the motor. This will cause excessive heat and lead the motor to burn and fail. Overheating motor is one of the leading causes of ESP failures and this problem has been the primary focus of the oil industry in the past decade.

Knowing causes of ESP system failures, one can change pump design, well completion, monitoring system, and operating conditions to improve the life of ESP system.

3.9 Examples

Example 3.1 Derive equations to calculate the pump pressure for the surface and subsurface single-stage centrifugal systems described in Fig. 3.33a and b, respectively.

Solution

For the surface single-stage centrifugal pump (system 1):

Applying Bernoulli equation (momentum equation) at the surfaces of the liquid inside the two reservoirs gives

$$\rho g Z_1 + P_1 + \frac{\rho u_1^2}{2} + \Delta P_{pump} = \rho g Z_2 + P_2 + \frac{\rho u_2^2}{2} + \Delta P_{friction} + \Delta P_{minor} \quad (3.9.1)$$

Assuming the pipe diameter is the same for the entire system and hence $u_1 = u_2$. Also assuming that the minor pressure losses at valves, elbows, etc. are negligible and hence $\Delta P_{minor} = 0$. Equation (3.9.1) becomes:

$$\Delta P_{pump} = \rho g Z + (P_2 - P_1) + \Delta P_{friction} \quad (3.9.2)$$

For the subsurface single-stage centrifugal pump (system 2):

$$P_i + \Delta P_{pump} = \rho g TVD + P_{wh} + \Delta P_{friction} \quad (3.9.3)$$

$$\Delta P_{pump} = \rho g TVD + (P_{wh} - P_i) + \Delta P_{friction} \quad (3.9.4)$$

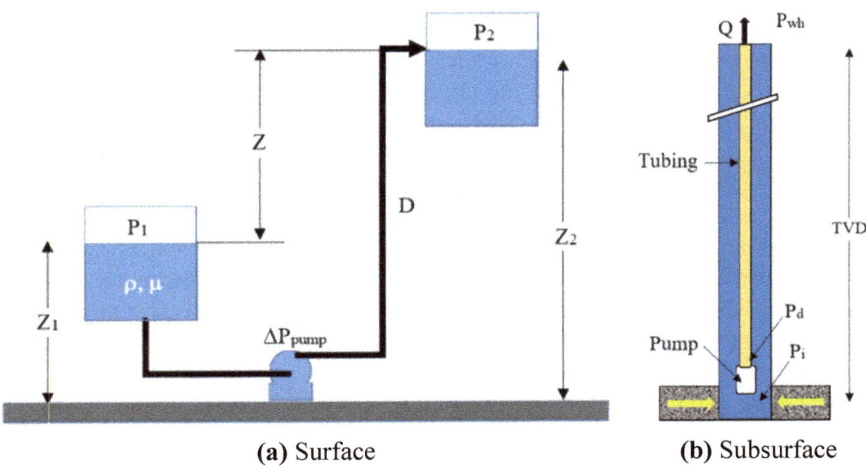

Fig. 3.33 Surface and subsurface centrifugal pump system

Comparing Eqs. (3.9.2) and (3.9.4), one can draw a quick conclusion that the pump pressure (or pump head) does not rely on the pump position. It only depends on the true vertical depth between the two surfaces of the two liquids, the pressure difference between the two surfaces of the two liquids, and the total frictional pressure losses in the system.

Example 3.2 Applying the derived equations from Example 3.1, calculate the pump pressure for the two systems.

a. For system a: D = 2 in.; $P_1 = P_2 = P_{atmosphere}$; Z = 200 ft; liquid flowrate, Q = 500 BPD; fluid specific gravity, $\gamma = 0.8$; fluid viscosity, $\mu = 2$ cp.
b. For system b: D = 2 in.; $P_i = 2000$ psi; $P_{wh} = 500$ psi; TVD = 8,000 ft; liquid flowrate, Q = 500 BPD; fluid specific gravity, $\gamma = 0.8$; fluid viscosity, $\mu = 2$ cp.

Solution

$$\mathrm{D} = 2\,\mathrm{in.} = 0.0508\,\mathrm{m}$$

$$\mathrm{Z} = 200\,\mathrm{ft} = 60.96\,\mathrm{m}$$

a. $\mathrm{Q} = 500\,\mathrm{BPD} = 0.00092\,\mathrm{m^3/s}$

$\rho = 0.8 \times 1000 = 800\,\mathrm{kg/m^3}.$

$\mu = 2\,\mathrm{cp} = 0.002\,\mathrm{PaS}.$

The fluid velocity inside the pipe:

$$u = \frac{\pi D^2}{4} = \frac{\pi \times 0.0508^2}{4} = 0.454\,\frac{\mathrm{m}}{\mathrm{s}}$$

Reynolds number:

$$Re = \frac{\rho u D}{\mu} = 9224$$

The flow inside the pipe is turbulent. The Fanning friction factor (assuming smooth pipe):

$$f_F = \frac{0.0791}{Re^{0.25}} = 0.00807$$

The frictional pressure drop gradient:

$$\left.\frac{dp}{dl}\right|_{friction} = \frac{2 f_F \rho u^2}{D} = \frac{2 \times 0.00807 \times 800 \times 0.454^2}{0.0508} = 52.4\,\frac{\mathrm{Pa}}{\mathrm{m}} = 0.00232\,\frac{\mathrm{psi}}{\mathrm{ft}}$$

The total frictional pressure drop in the system:

$$\Delta P_{friction} = \left.\frac{dp}{dl}\right|_{friction} \times Z = 52.4 \times 60.96 = 3193\,\text{Pa} = 0.463\,\text{psi}$$

Pressure drop due to gravity:

$$\Delta P_{gravity} = \rho g Z = 800 \times 9.81 \times 60.96 = 478{,}414\,\text{Pa} = 69.38\,\text{psi}$$

Required pump pressure:

$$\Delta P_{pump} = \rho g Z + \Delta P_{friction} = 478{,}414 + 3193 = 481{,}607\,\text{Pa} = 69.85\,\text{psi}$$

b. D = 2 in. = 0.0508 m
   TVD = 8000 ft = 2438 m
   Q = 500 BPD = 0.00092 m^3/s
   ρ = 0.8 × 1000 = 800 kg/m^3.
   μ = 2 cp = 0.002 PaS.
   Pi = 2000 psi = 13,790,000 Pa
   Pwh = 500 psi = 3,447,500 Pa
   The fluid velocity inside the pipe, Reynolds number, Fanning friction factor, and the frictional pressure loss gradient are the same as that in question a.
   The total frictional pressure drop in the system:

$$\Delta P_{friction} = \left.\frac{dp}{dl}\right|_{friction} \times TVD = 52.4 \times 2438 = 127{,}730\,\text{Pa} = 18.5\,\text{psi}$$

Pressure drop due to gravity:

$$\Delta P_{gravity} = \rho g TVD = 800 \times 9.81 \times 2438 = 19{,}136{,}563\,\text{Pa} = 2775\,\text{psi}$$

Pressure difference between the top and the bottom of the well:

$$(P_{wh} - P_i) = 3{,}447{,}500 - 13{,}790{,}000 = -10{,}342{,}500\,\text{Pa} = -1500\,\text{psi}$$

Required pump pressure:

$$\Delta P_{pump} = \rho g TVD + (P_{wh} - P_i) + \Delta P_{friction}$$
$$\Delta P_{pump} = 19{,}136{,}563 - 10{,}342{,}500 + 127{,}730 = 8{,}921{,}793\,\text{Pa} = 1294\,\text{psi}$$

From the calculations in the Example 3.2, we can conclude that for the subsurface pump system as described in Fig. 3.33b, there must be multiple centrifugal pumps connecting in series to provide enough pressure (1294 psi) to maintain this liquid rate (500 BPD) in the production tubing. Assuming that each pump provides a pressure of 70 psi as calculated in Example 3.2b, then the number of pumps connected in series required to provide 1294 psi is 1294/70 = 19 pumps. This is the main principle of how an ESP works.

Example 3.3 Water is being pumped using a centrifugal pump operating at a speed of 3000 RPM at a liquid rate of 1714 BPD. The impeller has a uniform blade height of 2 in. with the intake impeller radius of 1.9 in. and the exit impeller radius of 7.0 in., and the exit blade angle β_2 is 23°. Assume ideal flow conditions and that the fluid direction enters the impeller is the same as the radial direction and hence $\alpha_1 = 90°$ as shown in Fig. 3.34. Determine

a. the tangential velocity component at the exit;
b. the ideal pump head;
c. the hydraulic pump power.

Solution
As described in Sect. 3.2, the radial direction is a line connecting point A and the center point of the impeller. The tangential direction is the line perpendicular to the radial direction at a given point (vector AD in Fig. 3.34a). There are two tangential vectors: one at the intake, $\vec{U}_1$ and the other one at the exit of the impeller, $\vec{U}_2$ as shown in Fig. 3.34b. The main flow direction is the line tangent to the streamline (or blade's shape) at a given point along the impeller. Note that the fluid velocities a long the pump impeller are different because of the difference in impeller radius. These velocities are called relative velocity. At the impeller intake and exit, the relative velocity vectors are $\vec{w}_1$, and $\vec{w}_2$, respectively. The pump blade angle, β, is the angle between the tangential direction and the direction of the streamline as shown in Fig. 3.34a.

a. Calculate the tangential velocity component at the exit
   In this example, $\alpha_1 = 90°$ and hence $\vec{V}_{r1} = \vec{V}_1$.
   Q = 1714 BPM = 0.0032 m^3/s.
   r_1 = 1.9 in. = 0.04826 m
   r_2 = 7.0 in. = 0.1778 m
   h = 2 in. = 0.0508 m
   N = 3000 RPM ($\omega = 2\pi N$)
   The linear velocities at the intake and at the exit:

$$U_1 = 2\pi r_1 N = 15.16 \frac{\text{m}}{\text{s}}$$

$$U_2 = 2\pi r_2 N = 55.86 \frac{\text{m}}{\text{s}}$$

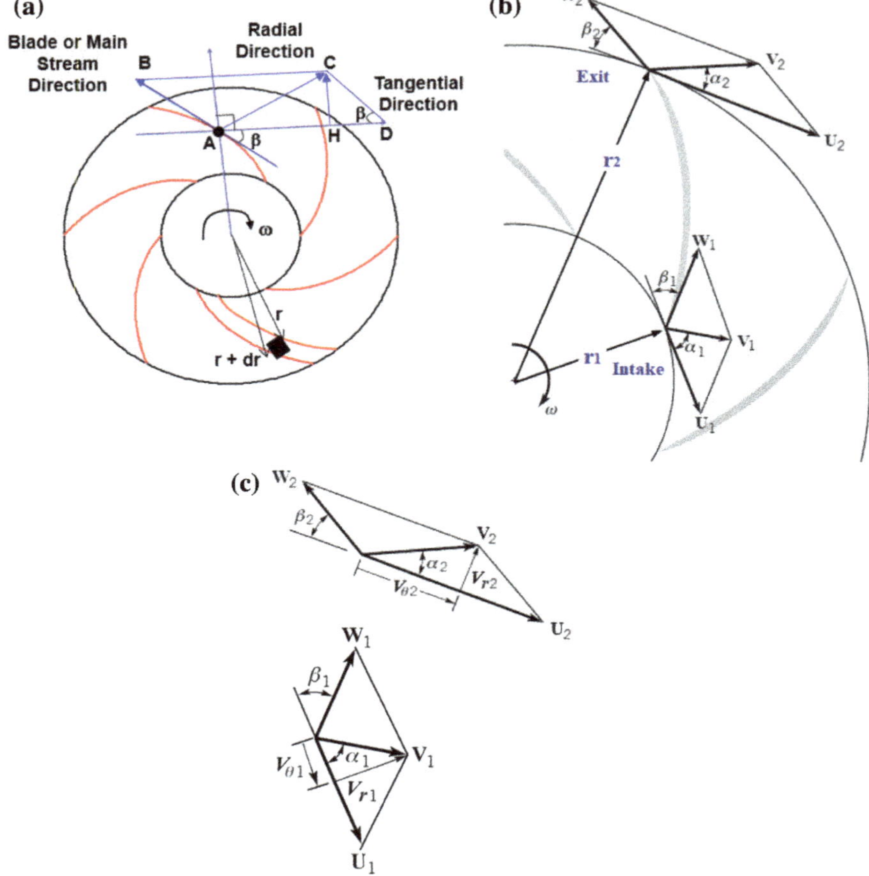

Fig. 3.34 Demonstration of velocity vector for Example 3.3

The radial velocity components at the intake and at the exit:

$$Q = 2\pi r_1 h V_{r1} = 2\pi r_2 h V_{r2}$$

$$V_{r1} = \frac{Q}{2\pi r_1 h} = \frac{0.0032}{2 \times \pi \times 0.04826 \times 0.0508} = 0.2048\,\frac{\mathrm{m}}{\mathrm{s}}$$

$$V_{r2} = \frac{Q}{2\pi r_2 h} = \frac{0.0032}{2 \times \pi \times 0.1778 \times 0.0508} = 0.0556\,\frac{\mathrm{m}}{\mathrm{s}}$$

The tangential velocity components at the exit:

$$\tan(\beta_2) = \frac{V_{r2}}{U_1 - V_{\theta 2}}$$
$$V_{\theta 2} = U_2 - \frac{V_{r2}}{\tan(\beta_2)} = 55.86 - \frac{0.0556}{\tan(23)} = 55.73\ \frac{\text{m}}{\text{s}}$$

b. Calculate the ideal pump head
   The tangential velocities at the intake and at the exit:

$$W_1^2 = V_{r1}^2 + (U_1 - V_{\theta 1})^2$$

Since $\alpha = 90°$ then $V_{\theta 1} = 0$.

$$W_1^2 = V_{r1}^2 + U_1^2 = 0.2048^2 + 15.16^2$$
$$W_1 = 15.16\ \frac{\text{m}}{\text{s}}$$

Similarly,

$$W_2^2 = V_{r2}^2 + (U_2 - V_{\theta 2})^2 = 0.0556^2 + (55.86 - 55.73)^2$$
$$W_2 = 0.14\ \frac{\text{m}}{\text{s}}$$

The main fluid velocity at the intake and at the exit:

$$V_1 = V_{r1} = 0.2048\ \frac{\text{m}}{\text{s}}$$
$$V_2 = \sqrt{V_{r2}^2 + V_{\theta 2}^2} = \sqrt{0.0556^2 + 55.73^2} = 55.73\ \frac{\text{m}}{\text{s}}$$

Pressure gain across the pump can be calculated using Eq. (3.2.18):

$$\Delta P_{pump} = \rho\left(\frac{U_2^2 - U_1^2}{2} + \frac{W_2^2 - W_1^2}{2} + \frac{V_2^2 - V_1^2}{2}\right)$$

The total pump head:

$$h_{pump} = \frac{\Delta P_{pump}}{\rho g} = \frac{\left(U_2^2 - U_1^2\right) + \left(W_1^2 - W_2^2\right) + \left(V_2^2 - V_1^2\right)}{2g}$$

The pump head due to the centrifugal force head

$$\frac{\left(U_2^2 - U_1^2\right)}{2g} + \frac{\left(W_1^2 - W_2^2\right)}{2g} = 147.31 + 11.71 = 159.03\,\text{m}$$

The pump head due to the fluid kinetic change (through the diffuser):

$$\frac{\left(V_2^2 - V_1^2\right)}{2g} = 158.28\,\text{m}$$

The total pump head

$$h_{pump} = \frac{\Delta P_{pump}}{\rho g} = 159.03 + 158.28 = 317.31\,\text{m} = 1041\,\text{ft}$$

The pressure gain across the pump

$$\Delta P_{pump} = \rho g h_{pump} = (1000)(9.81)(317.31) = 3{,}112{,}762\,\text{Pa} = 451\,\text{psig}$$

c. Calculate the hydraulic pump power
   Using Eq. (3.3.3) gives:

$$HP_h = 7.37 \times 10^{-6}\gamma h_{pump} Q = 7.37 \times 10^{-6}(1)(1041)(1714) = 13.15\,\text{HP}$$

Example 3.4 The HAL538 1800 water pump performance at 3500 RPM—60 Hz is given in the Halliburton catalog as shown in Fig. 3.35.

1. Using the Affinity Law, calculate the new pump rates, pump head and pump horsepower at the BEP for handling water if the pump is operated at 2500 RPM.
2. If this pump is used to handle a viscous oil with a viscosity of 40 cp and 25 ° API. The pump speed is 3500 RPM.
   a. Using Hydraulic Institute and Turzo method, please calculate the new pump rate and pump head.
   b. If Stepanoff method is used, what are the values of new pump rate and pump head.

Solution

1. Using the Affinity Law, calculate the new pump rates and horsepower at the BEP for handling water if the pump is operated at 2500 RPM.
   At the pump speed of 3500 RPM and at the BEP, the pump rate, pump head, and pump horsepower are as follows:

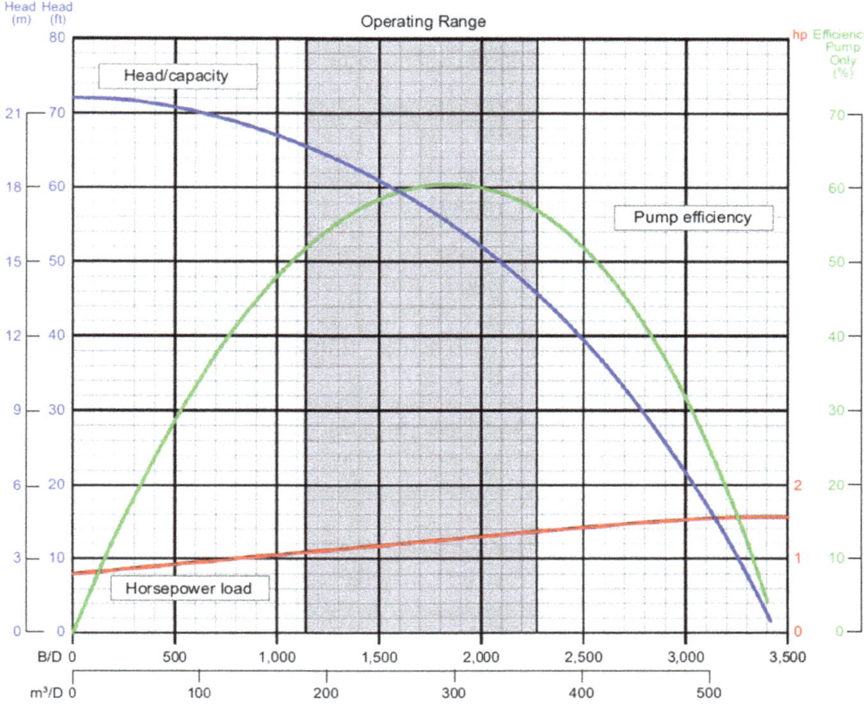

Fig. 3.35 HAL538—1800 BPD—water pump performance—3500 RPM—60 Hz

$$Q_w^{3500} = 1800\,\text{BPD}$$
$$H_w^{B3500} = 56\,\text{ft/stg}$$
$$HP_w^{3500} = 1.22\,\text{HP/stg}$$

Applying Affinity Law:

$$\frac{Q_w^{3500}}{Q_w^{2500}} = \frac{3500}{2500}$$
$$Q_w^{2500} = \frac{2500}{3500} Q_w^{3500} = 1285.71\,\text{BPD}$$

$$\frac{H_w^{3500}}{H_w^{2500}} = \left(\frac{3500}{2500}\right)^2$$

$$H_w^{2500} = \left(\frac{2500}{3500}\right)^2 H_w^{3500} = 28.57\,\text{ft/stg}$$

$$\frac{HP_w^{3500}}{HP_w^{2500}} = \left(\frac{3500}{2500}\right)^3$$

$$HP_w^{2500} = \left(\frac{3500}{2500}\right)^3 HP_w^{3500} = 0.445\,\text{HP/stg}$$

2. If this pump is used to handle a viscous oil with a viscosity of 40 cp and 25 ° API. The pump speed is 3500 RPM.

   a. Using Hydraulic Institute and Turzo method, please calculate the new pump rate and pump head.
      With the API gravity of 25, the oil specific gravity is:

$$\gamma_o = \frac{141.5}{131.5 + {}^\circ\text{API}} = 0.9$$

      The kinematic viscosity of oil:

$$\vartheta = \frac{\mu}{\gamma} = \frac{40}{0.9} = 44.24\,\text{cstk}$$

      Using Hydraulic Institute chart in Fig. 3.18, with $H_w^{BEP} = 56\,\text{ft}$, $Q_w^{BEP} = 1800\,\text{BPD} = 52.5\,\text{GPM}$, the correction factors for capacity and head are: $C_Q = 0.95$ and $C_H = 0.92$.
      The new pump rate and pump head are

$$Q_{vis}^{BEP} = C_Q Q_w^{BEP} = 0.95 \times 1800 = 1710\,\text{BPD}$$
$$H_{vis}^{BEP} = C_H H_w^{BEP} = 0.92 \times 56 = 51.52\,\text{ft/stg}$$

      Using Turzo method in Sect. 3.4.3:

$$y = -112.1374 + 6.6504 \ln\left(H_w^{BEP}\right) + 12.8429 \ln\left(Q_w^{BEP}\right) = 10.49$$

$$q^* = \exp\left(\frac{39.5276 + 26.5605 \ln(v) - y}{51.6565}\right) = 12.31$$

$$C_Q = 1 - 10^{-4}\left(40.327q^* + 1.724q^{*^2}\right) = 0.924$$

$$C_H = 1 - 10^{-5}\left(700q^* + 1.41q^{*^2}\right) = 0.91$$

The new pump rate and pump head are

$$Q_{vis}^{BEP} = C_Q Q_w^{BEP} = 0.924 \times 1800 = 1663\,\text{BPD}$$
$$H_{vis}^{BEP} = C_H H_w^{BEP} = 0.91 \times 56 = 51\,\text{ft/stg}$$

b. If Stepanoff method is used, what are the values of new pump rate and pump head

Assume that the value of viscous flowrate is 1665 BPD. According to Stepanoff, the viscous pump head can be calculated using Eq. (3.4.2) as follows:

$$\frac{Q_{BEP}^{vis}}{Q_{BEP}^{w}} = \left(\frac{H_{BEP}^{vis}}{H_{BEP}^{w}}\right)^{1.5}$$
$$H_{BEP}^{vis} = H_{BEP}^{w}\left(\frac{Q_{BEP}^{vis}}{Q_{BEP}^{w}}\right)^{0.6667} = 56\left(\frac{1665}{1800}\right)^{0.6667} = 53.16\,\text{ft}$$

Stepanoff Reynolds number:

$$R_{stepanoff} = 6.0345\frac{NQ_{BEP}^{vis}\gamma}{\mu\sqrt{H_{BEP}^{w}}}$$
$$R_{stepanoff} = 6.0345\frac{3500 \times 1665 \times 0.9}{40\sqrt{56}} = 106{,}221$$

Using Stepanoff chart gives $C_Q = 0.92$.

$$C_H = C_Q^{0.6667} = 0.946$$

The calculated viscous flowrate and head

$$Q_{vis}^{BEP} = C_Q Q_w^{BEP} = 0.92 \times 1800 = 1656\,\text{BPD}$$
$$H_{vis}^{BEP} = C_H H_w^{BEP} = 0.946 \times 56 = 53\,\text{ft/stg}$$

Comparing the assumed viscous flowrate value and the calculated value, the relative error is about 0.5%. Therefore, no further iteration is needed.

Example 3.5 Select the best pump for the following conditions: production casing OD of 5.5 in., required flowrate of 900 BPD, total dynamic head of 1500 ft, pump speed of 3500 RPM and frequency of 60 Hz.

Solution

Table 3.9 show the summary of the 400 and 338 Halliburton pump series. Since the production casing OD is 5.5 in., we can choose pump OD (housing diameter) of 4.0 in. or 3.38 in. With the required production rate of 900 BPD, there are two options for the 400 pump series and two options for the 338 pump series as highlighted in yellow in Table 3.9. The pump performance of these pumps are presented in Figs. 3.36a, b and 3.37a, b. The performance of these four pumps is summarized and shown in Table 3.10.

A quick look into the pump heads at the required liquid rate of 900 BPM, the HAL400 850A offer the highest head. If the Dog Leg Severity (DLS) is smaller than 8°/100 ft, the 400-series pump should be chosen. HAL400 850A offers higher pump head than HAL400 1000 (31.5 ft compared to 24 ft). However, HAL400 1000 provide larger operating range in comparison to HAL400 850A (600–1250

Table 3.9 Summary of 400-series and 338-series Halliburton pump (Halliburton pump catalog)

Pump type	Speed, RPM and frequency (Hz)	Housing diameter (in.)	Min. casing size (in.)	Optimum flowrate range (BPD)	Opt. head range (ft)	Flowrate at BEP (BPD)	Head at BEP (ft)
HAL400 150	3500-60	4.00	5.50	75–279	23–13	150.0	21.8
HAL400 160A	3500-60	4.00	5.50	75–264	25–15	160.0	22.2
HAL400 330	3500-60	4.00	5.50	216–555	27–12	330.0	25.0
HAL400 400	3500-60	4.00	5.50	300–540	35–15	400.0	24.0
HAL400 500	3500-60	4.00	5.50	339–664	28–16	500.0	25.0
HAL400 650A	3500-60	4.00	5.50	415–867	26–21	650.0	27.5
HAL400 725	3500-60	4.00	5.50	415–867	32–25	725.0	29.5
HAL400 850A	3500-60	4.00	5.50	640–1200	35–25	850.0	30.0
HAL400 1000	3500-60	4.00	5.50	600–1250	28–13	1000.0	21.2
HAL400 1250	3500-60	4.00	5.50	800–1700	27–16	1250.0	23.0
HAL400 1500	3500-60	4.00	5.50	905–2113	32–20	1500.0	28.5
HAL400 1750	3500-60	4.00	5.50	1200–2100	25–15	1750.0	20.5
HAL338 400	3500-60	3.38	4.50	226–500	20–11	400.0	15.4
HAL338 550	3500-60	3.38	4.50	400–660	19–11	550.0	15.5
HAL338 900	3500-60	3.38	4.50	700–980	15–10	900.0	12.5
HAL338 1200	3500-60	3.38	4.50	800–1580	15–9	1200.0	12.5
HAL338 1500	3500-60	3.38	4.50	1000–1880	12–7	1500.0	9.5
HAL338 2500	3500-60	3.38	4.50	1800–3100	16–9	2500.0	13.5

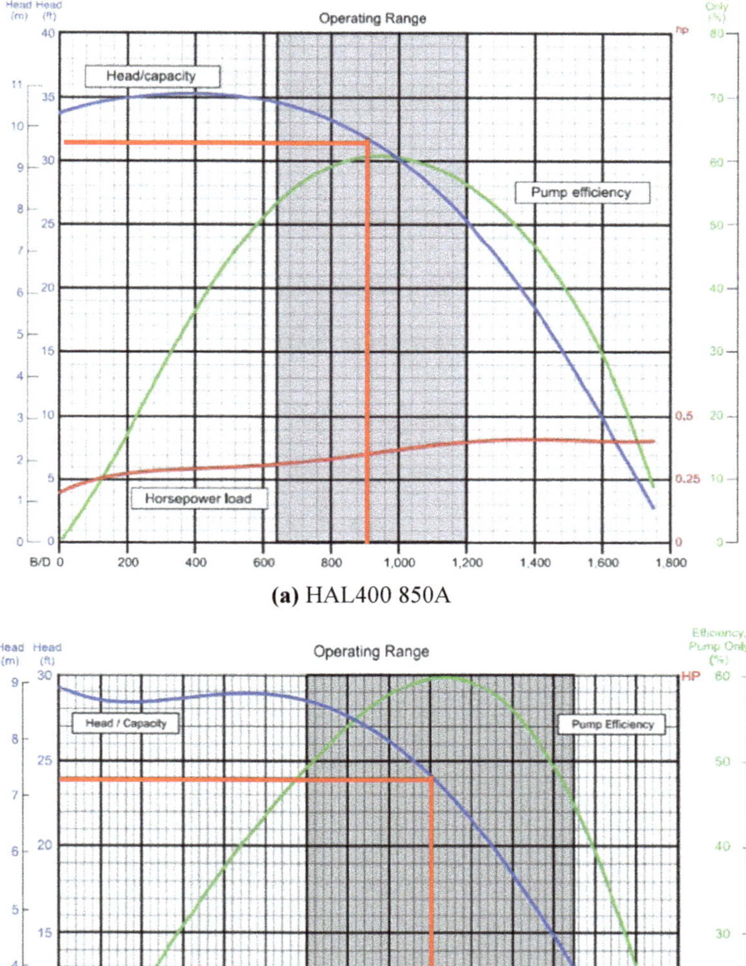

(a) HAL400 850A

(b) HAL400 1000

Fig. 3.36 400-series Halliburton pump performance

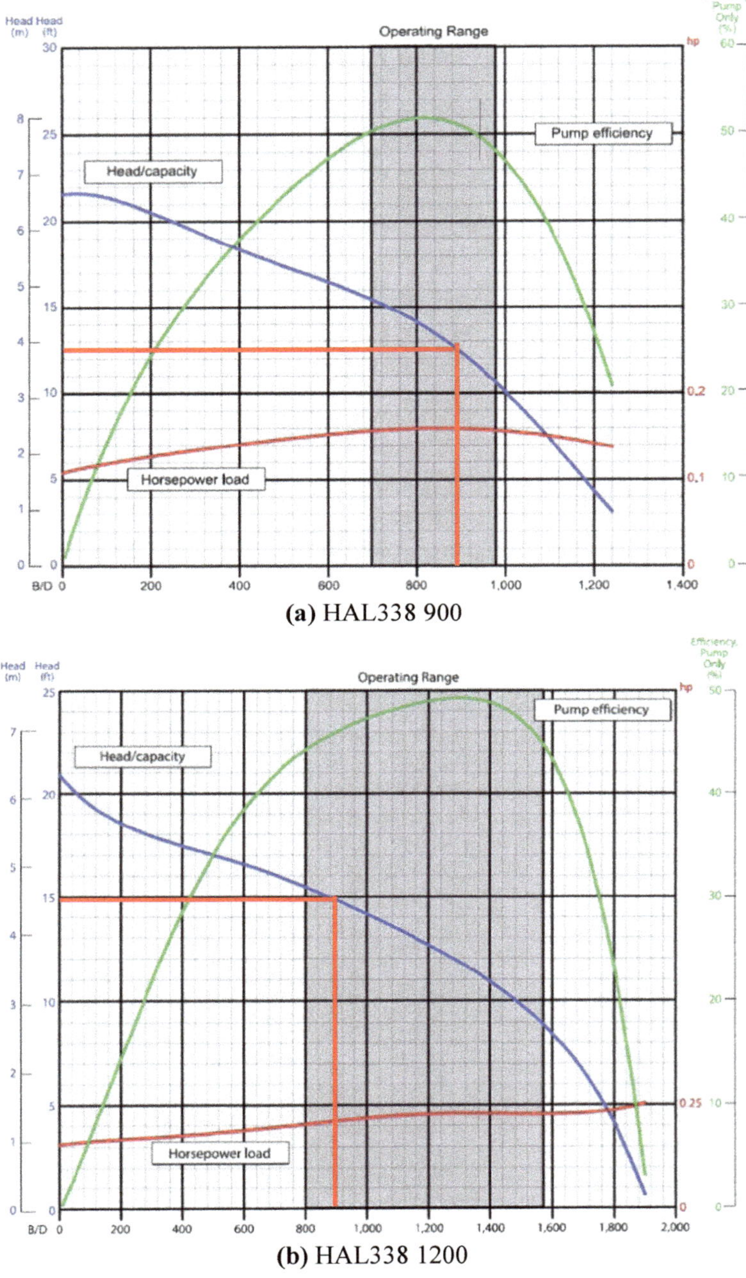

Fig. 3.37 338-series Halliburton pump performance

Table 3.10 Performance summary of four pumps

Pump type	Housing diameter (in.)	Min. casing size (in.)	Optimum flowrate range (BPD)	Opt. head range (ft)	Flowrate at BEP (BPD)	Head at BEP (ft)	Head, ft @ 900 (BPD)
HAL400 850A	4.00	5.50	640–1200	35–25	850.0	30.0	31.5
HAL400 1000	4.00	5.50	600–1250	28–13	1000.0	21.2	24.0
HAL338 900	3.38	4.50	700–980	15–10	900.0	12.5	12.5
HAL338 1200	3.38	4.50	800–1580	15–9	1200.0	12.5	15.0

BPD compared to 640–1200 BPD). Note that the 4.0 in. pump OD is the maximum pump diameter that can be used to run into the 5.5 in. production casing. Selecting HAL400 850 will reduce the pump length and hence minimize the risk of stuck pipe while running the pump into the well. Therefore, it is recommend to select HAL400 850A in this application.

However, if DLS is too high and stuck pipe is a major concern when running the pump into the well than HAL338 120 should be selected because it offers higher pump head and wider operating range.

Example 3.6 Design an ESP system for a vertical well with a desired production liquid rate of 1,200 BPD with the following inputs:

Production casing OD = 5.5 in., casing depth 6000 ft, top of perforation depth = 5900 ft, production tubing ID = 2.441 in.,

Average reservoir pressure = 3500 psi, pay thickness = 100 ft, drainage radius = 3000 ft, wellbore radius = 0.1875 ft, formation permeability = 50 mD, skin factor = 0.

Water cut = 0%, formation producing GOR = 400 scf/B, bottom hole temperature = 170 °F, surface temperature = 80 °F, bubble point pressure = 1800 psi, oil API = 30 °API, gas specific gravity = 0.7, oil specific gravity = 0.75, water specific gravity = 1.02.

Wellhead pressure = 200 psi, primary minimum/maximum voltages = 7200/12,470 V.

Solutions

The solution for this example follows the design procedure presented in Sect. 3.7.

a. **Selection of Pump Diameter**

   With the production casing OD of 5.5 in., users have several options for pump diameter (housing diameter): 4.0, 3.75, and 3.38 in. In this example, we choose pump OD of 4.0 in. because the DLS can be assumed to be zero. Users are highly recommended to review example 3.9.5.

b. **Selection of Pump Depth**
As presented in Sect. 3.7.3, pump depth does impact the performance of vertical (or near vertical) wells. For the formation fluids to carry the heat generated from the motor, we will place the pump 200 ft above the top of the perforation. Therefore, the pump depth is 5700 ft.

c. **Analyzing Well Flow Capacity**
Using Standing correlation described by Eq. (1.4.6) gives the solution GOR:

$$x = 0.0125API - 0.00091T = 0.2203$$

$$R_s = \gamma_g \left[\left(\frac{P}{18.2} + 1.4 \right) 10^x \right]^{1.2048} = 0.7 \left[\left(\frac{1800}{18.2} + 1.4 \right) 10^{0.22} \right]^{1.2048} = 332 \frac{\text{SCF}}{\text{STBO}}$$

Oil formation volume factor using Standing correlation gives:

$$B_o = 0.972 + 0.000147 \left[R_s \left(\frac{\gamma_g}{\gamma_o} \right)^{0.5} + 1.25T \right]^{1.175} = 1.2 \frac{\text{bbl}}{\text{STBO}}$$

Oil viscosity using Begs and Robinson gives:

$$x = \frac{10^{(3.0324 - 0.0202 \times {}^\circ API)}}{T^{1.163}} = 0.6798$$

$$\mu_{o,d} = 10^x - 1 = 3.784\,\text{cp}$$

For simplicity, we assume that if the flowing bottom hole pressure is greater than bubble point pressure, the IPR is a straight line with a constant productivity index (PI). Assuming pseudo steady state flow inside the reservoir, using the average reservoir pressure concept, Eq. (1.5.9) gives:

$$PI = \frac{q}{\bar{p} - p_{wf}} = \frac{kh}{141.2 B_o \mu \left[\ln \frac{r_e}{r_w} - 0.75 + S \right]} = 0.867 \frac{\text{bbl}}{\text{psi}} \Big/ \text{D}$$

If the flowing bottom hole pressure is smaller than the bubble point pressure, there is two-phase flow in the reservoir and the IPR is no longer a straight line; instead it is a parabolic. Using Eq. (1.5.11) with the assumption that B_o and μ_o are constant and does not depend on the average pressure. The IPR is shown in Fig. 3.38.

For the OPR, Ansari two-phase flow model is used. The OPR is presented in Fig. 3.38.

Figure 3.38 reveals that, this well is producing with a natural flow of about 220 BPD at the flowing bottom hole pressure of about 3200 psi. For this well to produce at the desired rate of 1200 BPD, an ESP with a pump pressure of 1500 psi must be installed to lift the fluid inside the tubing as well as to reduce the flowing bottom hole pressure to about 1700 psi. This pressure is just below the

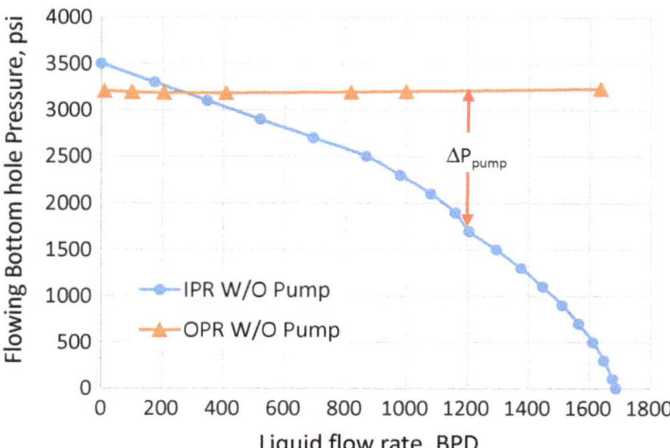

Fig. 3.38 IPR and OPR without pump

bubble point pressure and hence we can assume the flow inside the reservoir is single phase.

d. **Selection of ESP**
Following the principle presented in Example 3.5, with the pump OD of 4.0 in. and the desired liquid rate of 1200 BPD, the HAL400 1250 is selected. This pump has the optimum operating range of 800–1700 BPD; the housing diameter of 4.00 in., shaft diameter of 11/16 in.; housing pressure limit of 5000 psi. The pump performance is shown in Fig. 3.39.
According to the HAL400 1250 pump performance, at the liquid rate of 1200 BPD, the pump head is 23.5 ft/stage and the pump horsepower is 0.35 HP/stage. With the API of 30 °API, the oil specific gravity is 0.86 (oil density of 860 kg/m^3 = 7.14 ppg). As explained in part C of this example, the required total pump pressure is 1500 psi or the total required pump head is:

$$H_{pump} = \frac{\Delta P_{pump}}{0.052 \times \rho} = \frac{1500}{0.052 \times 7.14} = 4040\,\text{ft}$$

The number of pump stages can be determined as follows:

$$Total\,stages = \frac{H_{pump}}{h_{onestage}} = \frac{4040}{23.5} = 172\,\text{stages}$$

The break pump horsepower can also be calculated as

$$HP_b = HP_{onestage} \times totalstages \times \gamma = 0.35 \times 172 \times 0.86 = 51.8\,\text{HP}$$

Reviewing the guideline of Halliburton pump catalog as shown in Table 3.11, one can choose the single pump unit which consists of the housing # 10 with the floater standard stages of 87 and the length of 14.8 ft. To obtain the head of 1500 psi or 4040 ft, two of these pump units are needed with the total stages and length of 174 stages and 29.6 ft, respectively.

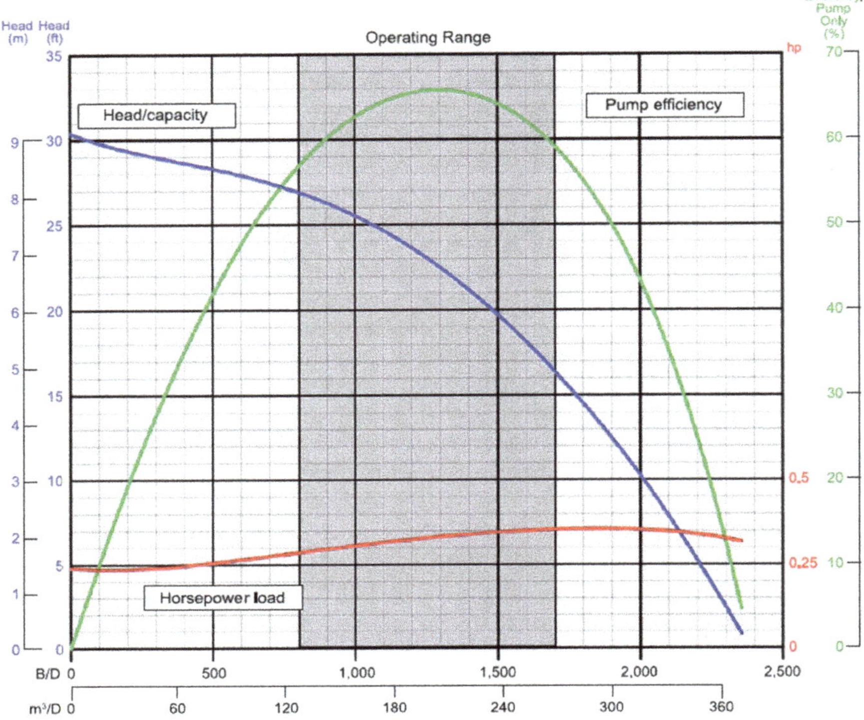

Fig. 3.39 HAL400 1250 pump performance

e. **Consideration of the Effect of Gas and Viscosity on Pump Selection**
The volume of free gas at the pump intake can be calculated following the procedure presented in Sect. 3.7.6:
R_s = 332 SCF/STBO

$$B_g = 0.005\frac{z(T+460)}{p} = 0.005\frac{1(170+460)}{3500} = 0.0009\frac{\text{bbl}}{\text{SCF}}$$

B_o = 1.2 bbl/STBO

$$V_{freegas} = GOR_{producing} \times Q_o - R_s \times Q_o = 400 \times 1200 - 332 \times 1200 = 81{,}600\,\frac{\text{SCF}}{\text{D}}$$

Table 3.11 Standard design of the HAL400 1250 pump

Housing	Standard stages		Abrasion-resistant stages		Length (ft)	Weight (lbm)
	Floater	Compression	Floater	Compression		
1	8	7	8	7	2.1	44
2	16	15	16	15	3.5	73
3	25	24	25	24	4.8	106
4	34	33	34	33	6.3	134
5	43	42	43	42	7.8	165
6	52	51	52	51	9.2	196
7	61	60	61	60	10.6	225
8	70	69	70	69	12.0	258
9	78	77	78	77	13.4	287
10	87	86	87	86	14.8	317
11	96	95	96	95	16.2	348
12	105	104	105	104	17.6	377
13	114	113	114	113	19.0	403
14	123	122	123	122	20.4	438
15	132	131	132	131	21.9	478

The volume of oil at the pump intake:

$$V_{o@intake} = B_o \times Q_o = 1.2 \times 1200 = 1440\,\text{bbl/D}$$

The volume of free gas at the pump intake:

$$V_{g@intake} = V_{freegas} \times B_g = 81{,}600 \times 0.0009 = 73.44\,\frac{\text{bbl}}{\text{D}}$$

Total fluid volume:

$$V_t = V_g + V_o + V_w = 73.44 + 1440 + 0 = 1513\,\frac{\text{bbl}}{\text{D}}$$

Percentage of free gas at the pump intake:

$$\%Freegas = \frac{V_{g@intake}}{V_t} = \frac{73.44}{1513} = 4.8\%$$

Since the percentage of free gas at the pump intake is less than 10%, the impact of gas on the pump performance is minimal and hence installing a gas separator is not highly recommended.

The formation fluid viscosity is 3.78 cp and hence the effect of viscosity on the pump performance is also very minimal. Therefore, all the selections so far are correct. In other words, further modifications on the pump selection are not needed.

Selection of Motor

With the total brake horsepower of 51.8 HP calculated in part d of this example, we choose the HAL450 motor with the specifications as follows: 60 Hz, 60 HP, 640 V, and 59 A. The motor OD is 4.56 in. Note that with the primary min/max power voltages of 7200/12,470 V, a transformer is needed to provide proper voltage to the selected motor.

Readers keep following the design procedure presented in Sect. 3.7 to select cables and surface equipment.

References

1. Beltur R, Duran J, Pessoa R (2003) Analysis of experimental data of ESP performance under two-phase flow conditions. In: SPE production and operations symposium, OK
2. Brown K (1980) The technology of artificial lift methods, vol 2B. Petroleum Publishing Co
3. Cirilo R (1998) Air-Water flow through electric submersible pumps. Master's Thesis, University of Tulsa, Tulsa
4. Clegg J, Buoaram S, Hein N (1993) Recommendations and comparisons for selecting artificial lift methods. J Pet Technol SPE24834-PA
5. Duran J, Prado M (2003) ESP stages air-water two-phase performance modeling and experimental data. SPE-87627-MS
6. Evdocia (2006) TUALP annual report
7. Gabor T (2013) Gas lift manual. PennWell Corporation, USA. ISBN 0-87814-805-1
8. Gallup A, Wilson B, Marshall R (1990) ESP's placed in horizontal lateral increase production. Oil Gas J
9. Halliburton (2014) Artificial lift technology—electrical submersible pump
10. Hydraulic Institute Standards (1955) Determination of pump performance when handling viscous liquid, 10th edn
11. Munson B, Young D, Okiishi T, Huebsch W (2009) Fundamentals of fluid mechanics, 6th edn., John Wiley & Sons, Inc., USA. ISBN 978-0470-26284-9
12. Murakami M, Minemura, K (1974) Effects of entrained air on the performance of a centrifugal pump. Bull JSME, 17:1047–1055
13. Pessoa R, Prado M (2000) Experimental investigation of two-phase flow performance of ESP stages. In: SPE annual technical conference and exhibition, New Orleans
14. PetroWiki—Electrical submersible pump. http://petrowiki.org/Electrical_submersible_pumps
15. Romero M (1999) An evaluation of an electric submersible pumping system for high GOR wells Master's Thesis, University of Tulsa, Tulsa
16. Stepanoff AJ (1940) Pumping viscous oils with centrifugal pump. Oil Gas J 1:26–28
17. Stepanoff AJ (1948) Centrifugal and axial flow pumps. Wiley, New York
18. Stepanoff AJ (1965) Pumps and blowers—two-phase flow. Wiley, New York
19. Takacs G (2013) Detailed design of an esp installation for a geothermal well. Geosci Eng 2 (3):25–36
20. Turpin JL, Lea JF, Bearden JL (1984) Gas-liquid through centrifugal pumps-correlation of data. In Proceedings of the Third International Pump Symposium, Houston, TX, USA, 20–22 May

21. Turzo Z, Takacs G, Zsuga J (2000) Equations correction centrifugal pump curves for viscosity. Oil Gas J 98(22):57–67
22. Vo H, Nguyen T, Al-safran E, Saasen A, and Nes O (2017) An experimental investigation into the effects of high viscosity and foamy oil rheology on a centrifugal pump performance. J Pet Sci Technol http://doi.org/10.22078/JPST.2017.709
23. Zhu J, Zhang H (2018) A review of experiments and modeling of gas-liquid flow in electrical submersible pumps. Energies 11:180

Chapter 4
Progressing Cavity Pump

4.1 Fundamentals of Progressing Cavity Pump (PCP)

Progressing Cavity Pump (PCP), also known as Moineau pump, is an artificial lift method often used for pumping high viscosity and high solids content fluids from producing wells. The Moineau pump was invented by the French, René Moineau, in 1930 [16]. The design of the Moineau pump is composed of two helical gears, one inside of the other as shown in Fig. 4.1. The rotor rotates around its longitudinal axis, which is parallel with the stator axis. The external gear (double helical stator) always has one more tooth (one more lobe) than the internal one (single helical rotor). The rotor is designed so that all the teeth of the rotor are constantly in contact with the stator [3, 27]. As the rotor rotates inside the stator, the cavities move without deformation, thus transferring fluid without pulsation. In other words, the PCP transfers fluid by means of the progress, a sequence of small, fixed shape and discrete cavities, as the rotor turns. This motion mechanism is similar to that of positive displacement pumps. Therefore, PCP has several advantages over other pumps, namely providing a uniform flow rate without any pulsation, capability of pumping very light liquid to very pasty fluids, be able to deliver solids, absence of check valves, and no need for liquid as primer when starting the pump. In oil and gas industry, PCPs are available for a wide range of operating challenges and applications including coal bed methane, medium to light oil, cold production of heavy oil, thermal production of extra heavy oil and bitumen [4–7].

Moineau gear mechanism can be applied to design single-lobe or multi-lobe PCPs as shown in Fig. 4.2. The single-lobe PCP, designated as 1:2, means the rotor has one lobe (one gear or one tooth) and the stator has two lobes (two gears or two teeth). Multi-lobe PCPs are defined when the number of lobes of the rotor is more than one such as 2:3, 5:6, etc. as shown in Fig. 4.2. Detail of single-lobe and multi-lobe pumps and advantages and disadvantages of these pumps will be discussed in the Pump Design Section.

T. Nguyen, *Artificial Lift Methods*, Petroleum Engineering,
https://doi.org/10.1007/978-3-030-40720-9_4

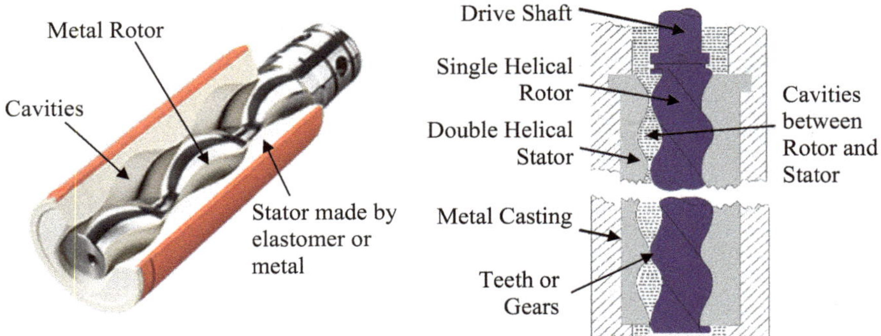

Fig. 4.1 Basic design of a single-lobe PCP (https://www.artemis-kautschuk.de/en/images/Stator.png and https://encrypted-tbn0.gstatic.com/images?q=tbn:ANd9GcQqqfxlUG-YpNalRGPDUwY-F5BsOFVFIg2nMmM8VA-1absAaRxrxg)

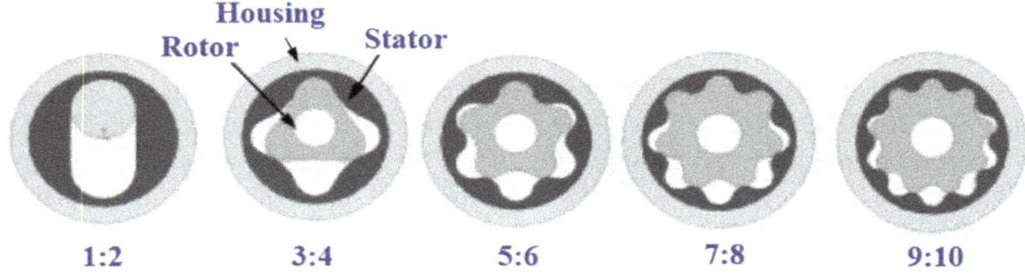

Fig. 4.2 Single-lobe and multi-lobe PCPs

Let's consider a single-lobe PCP and the cross-section "A" as shown in Fig. 4.3. In the beginning, the tooth of the rotor at the cross-section "A" is in contact with the upper part of the stator. The longitudinal axis of the rotor is above the longitudinal axis of the stator a distance defined as the pump eccentricity, e. This is the maximum distance from the center of the stator that the rotor travels up and down inside the stator at one specific cross section along the pump. The total cross-sectional flowing area is the same as that of one cavity at the lower end of the stator as shown in Fig. 4.3a.

As the rotor turns 90° along its axis, the distance between the rotor axis and stator axis equals to zero. There exist two cavities: upper one and lower one as shown in Fig. 4.3b. The total cross-sectional flowing area in this case equals to the summation of the cross-sectional areas of the upper and lower cavities. When the rotor turns from 0 to 90°, the number of cavities at the cross section "A" changes from one cavity to two cavities. However, the total cross-sectional flowing area of these two cases remains the same.

As the rotor turns 180°, the rotor's axis is off from the stator's axis a distance that is the same as the eccentricity, e. Another tooth of the rotor is in contact with the

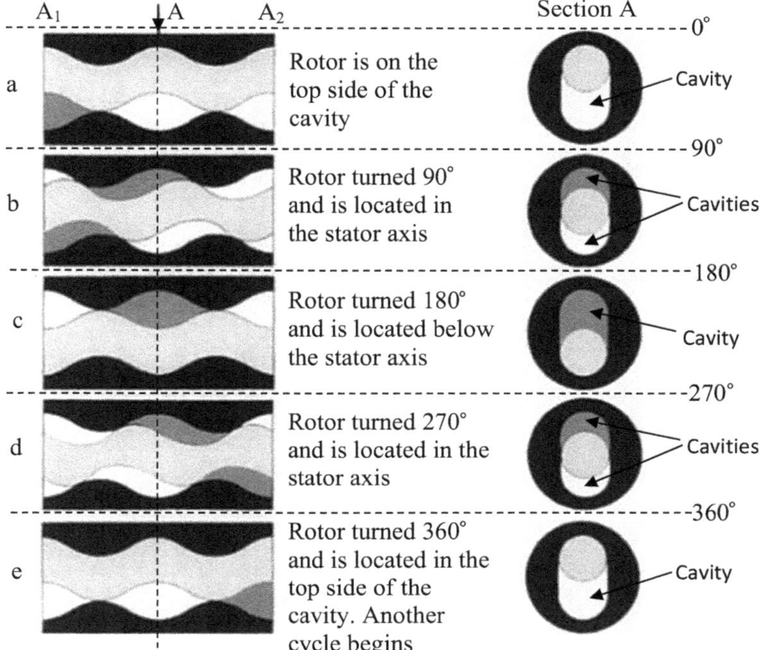

Fig. 4.3 Positions of the rotor inside the stator at a fixed location along the pump when the rotor turns a complete cycle of 360°. *Note* A_1–A_2 is the stator pitch length

lower side of the stator forming an upper cavity as shown in Fig. 4.3c. The cross-sectional flowing area of this cavity at section "A" is the same as the total cross-sectional flowing area in the previous two cases in Fig. 4.3a, b.

As the rotor turns 270°, the offset at section "A" between the rotor axis and stator axis is zero. The rotor is in the middle of the stator creating two cavities: the upper one and lower one. The total cross-sectional flowing area at the section "A" remains the same as compared to that of previous cases shown in Fig. 4.3a, b, c.

As the rotor turns 360°, another cycle begins as described in the section where the rotor is in the zero degree position.

In summary, considering one fixed location along the pump such as the cross section "A", as the rotor turns from 0 to 360° around its own axis, the following changes are observed:

- The distance between the rotor axis and stator axis varies from the pump eccentricity, e, to zero and to e.
- The number of cavities at section "A" varies from one to two. The maximum number of cavities is the same as the stator lobe number. For single-lobe PCP, the stator lobe number is two and hence the maximum number of cavities are two.

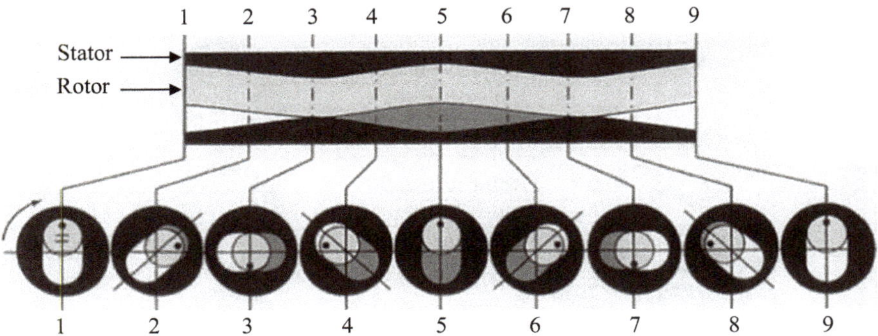

Fig. 4.4 Positions of the rotor inside the stator at various locations along the pump when the rotor turns a complete cycle of 360°. *Note* 1–9 is the stator pitch length

- The individual cross-sectional areas of the cavities at section "A" vary but the total cross-sectional flowing areas remain the same regardless to the location along the pump.
- Fluids inside the cavities travel a longitudinal distance the same as the rotor pitch length (A–A2) or half of the stator pitch length (A1–A2) as shown in Fig. 4.3.

The positions of the rotor at different locations in one complete stator pitch length along the pump can be seen in Fig. 4.4. Note that fixing any locations from 1

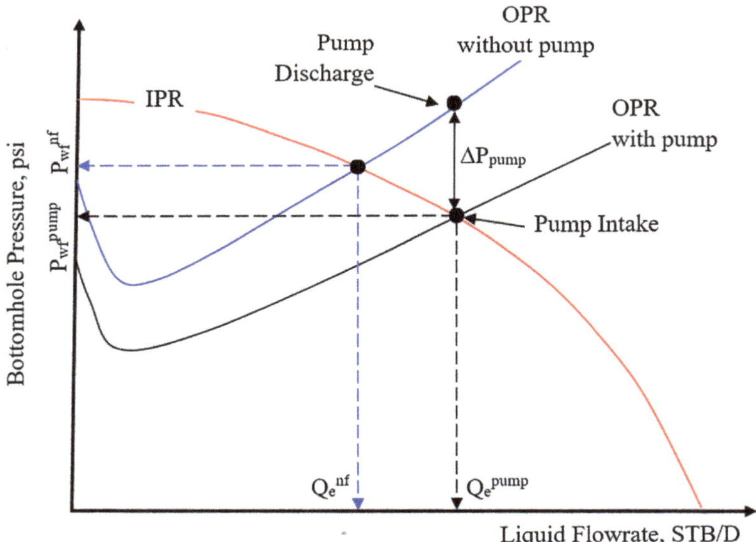

Fig. 4.5 Main Principle of PCPs in an oil well

to 9 in Fig. 4.4 and turning the rotor from 0 to 360° will reproduce the observations in Fig. 4.3.

When a PCP is installed and operated in an oil well, the pump will convert electric energy provided from the surface to hydraulic energy in the form of pressure. The formation fluid receives this hydraulic energy from the pump to overcome more friction inside the tubing and hence to move the fluid upward faster. Like any submersible pumps, PCP does not change the flow inside the reservoir or IPR; instead it changes the flow inside the tubing or the OPR. When the pump is operated, it reduces the flowing bottom hole pressure and hence increases the liquid production rate as presented in Fig. 4.5.

4.2 History of PCP

PCP is a relatively "young" artificial lift technology when compared to other artificial lift methods. The oil and gas industry really adapted PCP as an artificial lift method in late 1970s. The original invention of this pump can be dated back to 1929 when Moineau describe his new invention in the thesis "Une Nouveau Capsulisme". Moineau then filed two patents namely "Gear Mechanism" to the United States Patent Office in April 1931 and in April 1935. The main inventions in these two patents were the adaption of the gear mechanism to be used as a pump, prime mover, compressor, motor, or simple transmission device.

From 1957, Russia studied the opportunity to use the Moineau system for drilling and production but they failed to do so. In the 70s, Russian companies used Moineau pumps for the in situ production of heavy oil and showed some positive signs.

In the late 1970s, Canadian companies started applying PCP to pump heavy oil from shallow wells. They saw it as a better way to lift oil from the Lloydminster area, where billions of barrels of crude oil were found in cold shallow (10–20 °C) and low pressure reservoirs where crude was mixed with a lot of sand. In addition, a large viscous oil reserve was found in Alberta fields, which pump jacks were not the best solution to lift up the oil. The Canadian companies then recognized PCP was a much better lift method than rod pumps.

With the exploration of heavy oil in areas such as Canada, Venezuela, Russia and China, Kuwait, Oman, etc., this method is having more success and has gained more attention from the industry in recent years.

4.3 Applications of PCPs

Because of its unique mechanism, gear pumps are preferred in applications where accurate dosing or high pressure output is required. The discharge pressure has a minimal impact on the pump performance, which is similar to that of positive displacement pumps [21]. Generally speaking, gear pumps are commonly used in the following industries [11, 13, 14]:

Food industry: pumping corn syrup, mayonnaise, peanut butter, chocolate, vegetable fats, vegetable oils, yogurt, fruit cans, etc.
Chemical industry: pumping polymer, chemical mixing and blending, acids, alcohols, solvents, drugs, paints, inks, surfactants, soaps, lube oils, hydraulic oils, asphalt, etc.
Oil and gas industry: pumping heavy oil and bitumen, oil with high sand content and high gas oil ratio. Applying to wells with low flowing bottom hole pressures and high temperature.

4.3.1 Application of PCP to Pump Heavy Oil and Bitumen

Heavy oil is normally defined and accepted as oil having an API gravity of less than 2° API. However, API gravity alone sometimes does not fully represent for the properties of the crude. Some crudes may have low API gravity but low viscosity when comparing to some light oils. Therefore, the industry sometimes uses viscosity to characterize crudes as heavy oils, e.g. 100 cp or greater. Bitumen is defined as having API gravity of 10° or less (viscosity of 10,000 cp or more). In this chapter, the term heavy oil is used interchangeably with bitumen.

As of 2012, the total heavy oil reserve in the world has been estimated at about 4.6 trillion barrels. Another study estimated that about 50% of the world's liquid hydrocarbon has API gravity of less than 20° API. This reveals to us that heavy-oil resources are important for long term supplies of petroleum. The largest heavy-oil deposits are located in Canada, Venezuela, and the Soviet Union. The total reserve from these three countries represents over 90% of the known heavy oil in place in the world.

To produce heavy oils, beam pumping systems have traditionally been chosen as the first option. However, due to high viscosity and corrosive fluids, and high solid contents, conventional rods, couplings, and production tubing have faced a lot of problems. These problems are as follows: abrasive couplings and tubing; rod and coupling fatigue failures; tubing wears; rod buckling during the down stroke leading to high mechanical friction [32]. In addition, the traveling valve positions, which allow for fluid flow, are delayed depending on the fluid viscosity causing a low pump efficiency and pump performance overall. PCPs are operated based on rod rotation; not rod reciprocation and therefore can avoid most of the major problems that beam pump systems have. PCP performance is usually improved (higher pump efficiency) when pumping viscous fluids due to less fluid slippage (less leakage). PCPs are also known to handle solids well as its performance is similar to positive displacement pumps. Because of these benefits, primary heavy oil and bitumen applications today are almost exclusively produced with PC pumping systems.

If fluid viscosity is too high, a method of thermal recovery may be required. There are three common thermal recovery methods being applied to produce heavy oil: cyclic steam injection, steam flooding, and SAGD [22, 23]. Cyclic steam injection is a method where a well is injected with steam, then shut in for a period of time, and finally put back on production. Steam flooding is a method where steam is injected into the reservoir through injection wells to heat up the crude oil and reduce its viscosity. After transferring heat to the formation fluids, the steam becomes condensate and is used to sweep the oil toward the producing wells. SAGD is a method in which a pair of horizontal wells are drilled into the oil reservoir with a vertical separation of a few meters one above another. Steam from the surface is injected into the upper horizontal wellbore to heat up the viscous oil and reduce its viscosity. This thinner viscous oil drains into the lower horizontal wellbore and is pumped up to the surface. Due to high fluid temperature when a thermal recovery method is applied, equal wall PCPs and metallic stator PCPs stand out to be the best artificial lift methods for this application. These two types of PCPs will be discussed later in this chapter.

4.3.2 Application of PCP to Pump High Solid Contents Fluids

Heavy oils are normally found at shallow depths ranging from 1000 to 4000 ft (305–1220 m) and contain high asphaltene, sulfur, and metal contents. The formations where heavy oils are found normally are very unconsolidated and hence sand production in these producing wells is a big concern. If a rod pump system is applied to produce this viscous oil along with solids, the solids in a vortex condition around each coupling cause an increase in the risk of abrasive coupling and tubing failures. The reciprocal motions in vertical direction of the rod with the presence of solids make the rods wear out faster. In addition, the small flowing areas around the couplings (high flow restrictions) decreases the pump's ability to effectively produce solids together with liquid to surface. Lastly, the traveling valves inside the pump can be worn out easily with the presence of solids. PCPs seem to be a better option than beam pump systems because of the following reasons:

- Rod and coupling failures can be minimized because of its rotational motion instead of reciprocal motions.
- Tubing wear can also be mitigated due to the same reason as mentioned above.
- There are no traveling valves inside PCPs and hence reduces the chance of pump failure due to solids.

4.4 PCP System

Most of PCP systems are rod driven with the pump run into the well at the bottom of the production tubing and the rotor connected to the bottom of the rod string as shown in Fig. 4.6a, b [26, 27].

From the surface to the bottom of the hole, following are the main components of a PCP system: wellhead drive unit (drive head), rotating stuffing box, polished rods, pony rods, couplings and centralizers, rotor, stator, stop bushing, torque anchor, and gas separator.

4.4.1 Drive Head

In general, there are two types of drive head: hydraulic or electric drive head as shown in Fig. 4.7. These two types are designed for PCP installations in heavy oil, gas dewatering, and thermal operations. The main functions of PCP drive heads are to drive the rotation of the rotor, control backspin rotation, support the weight of the rod string, and seal around the polished rod. Examples of drive heads are shown in Fig. 4.7a–c.

Like any electric or hydraulic engine, the selection of a drive head for a particular PCP well relies on the mechanical horsepower, pump speed, and torque. If

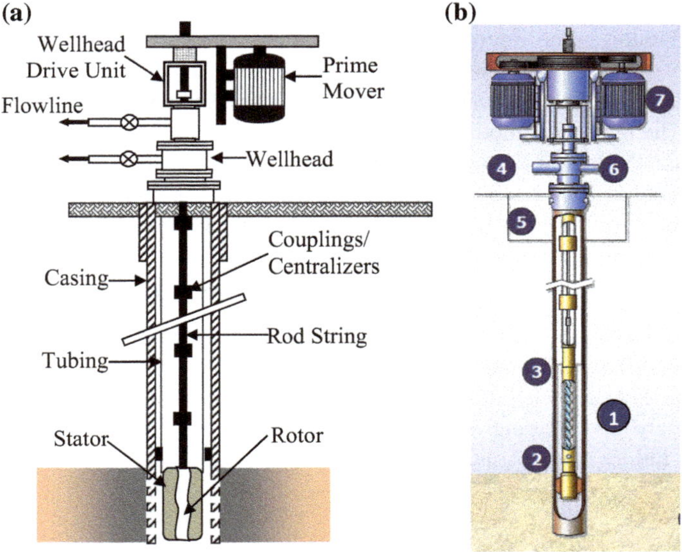

Fig. 4.6 Main components of a PCP system 1: PC pump; 2: Torque anchor; 3: Coupling; 4: Flowline; 5: Polished rod; 6: Wellhead; 7: Prime mover

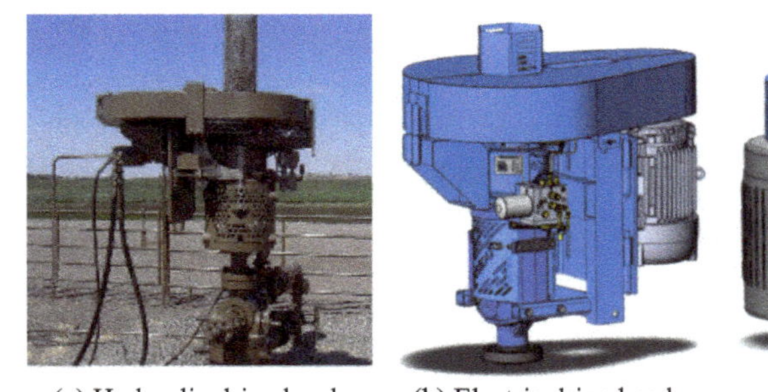

(a) Hydraulic drive head (b) Electric drive head One prime mover

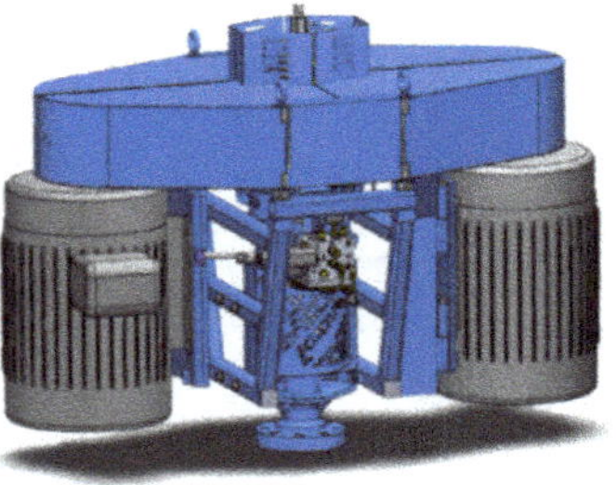

(c) Electric drive head Two prime movers

Fig. 4.7 Hydraulic and electric drive head for PCP systems

the mechanical horsepower and speed are available, torque can be calculated. When selecting a drive head for a PCP system, the maximum speed is used to assure that the drive head's power is enough if the well is operated at a maximum rate. The mechanical horsepower, HP_m, can be estimated using the hydraulic horsepower, HP_h, and the efficiency, η, as follows:

$$HP_m = \eta \times HP_h \tag{4.4.1}$$

The hydraulic horsepower depends on the liquid rate, pressure difference at the pump intake and pump discharge, and the pump geometry. Equation (1) will be discussed in detail in the modeling section. The value of the efficiency relies on the pump design, fluid viscosity, solid (sand) content, formation fluid temperature, and mechanical contacts between centralizers and the production tubing. Table 4.1 shows examples of specifications of different drive head manufactured by Schlumberger KUDU.

4.4.2 Rotating Stuffing Box

Rotating stuffing box is normally incorporated into the drive head as shown in Fig. 4.8. The main functions of a stuffing box in a PCP system are to seal on the rotating polished rod, and to control fluid leakage from the production string and wellhead.

The conventional stuffing box function is similar to those used with beam-pump systems. It uses a special packing material compressed against the polished rod or an inner sleeve that seals against the polished rod. The performance of the stuffing box is strongly dependent on the quality of the polished rod. If the polished rod is bent or scratched, the chance for the stuffing box to leak is quite high.

Table 4.1 Specification of different KUDU drive head

KUDU drive head specifications				
	VHGH—9.3T	VH 60HP 9.3T	VH 100HP 13.7/22.3T	VH 200HP 22.8T
Drivehead speed ratio	4.1:1	1:1	1:1	1:1
Max. torque, ft.lbf [N.m]	1,600 [2,169]	1,250 [1,695]	1,772 [2,403]	3,544 [4,805]
Input shaft	Vertical	Vertical	Vertical	Vertical
Shaft type	Hollow shaft	Hollow shaft	Hollow shaft	Hollow shaft
Axial load capacity	9.3T	9.3T	13.7 or 22.8T	22.8T
Maximum speed, rpm	500	500	500	500
Horsepower range, hp [kW]	Up to 150 [Up to 110]	5–60 [4–45]	15–100 [11–75]	30–200 [22–145]
Polished rod diameter, in [mm]	1.25 [31.8] 1.5 [38.1]	1.25 [31.8]	1.25 [31.8] 1.5 [38.1] 2 [50.8]	1.25 [31.8] 1.5 [38.1] 2 [50.8]

Fig. 4.8 Rotating stuffing box

4.4.3 Polished Rod

Polished rod is the first component of a rod string of a PCP system from the surface as shown in Fig. 4.9a. The main function of the polished rod is to transfer torque from the drive head to the rod string to rotate the pump rotor. Polished rod is finished in such a way that it provides an appropriate sealing in the rotating stuffing box. Conventional polished rod passes through the entire drive head inside a hollow shaft which is an integral part of the drive head. The polished rod is suspended by a clamp seating on the top of the drive head frame as shown in Fig. 4.9.

The conventional design with a hollow shaft provides an easy access to reposition the rod string without removal of the wellhead. This can be done easily by

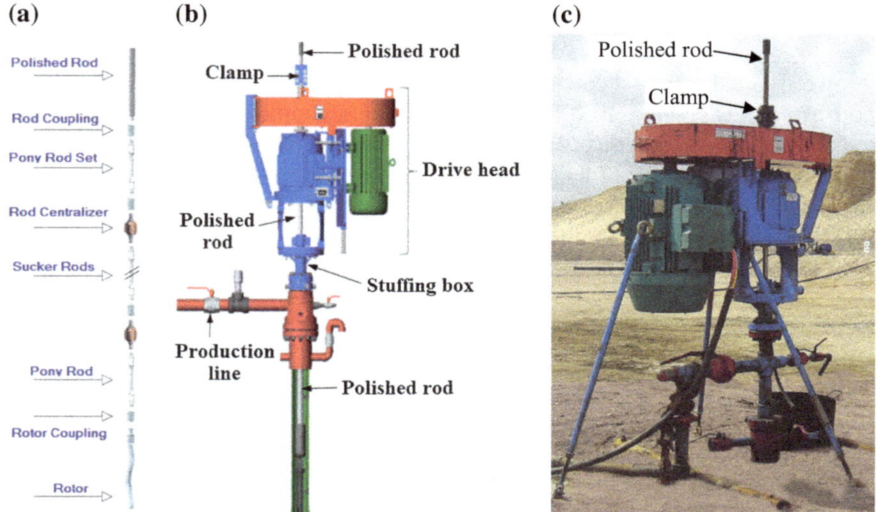

Fig. 4.9 **a** Rod string components; **b** surface equipment; **c** a picture of surface equipment

loosening the polished rod clamp on the top of the drive head, raising or lowering the polished rod as required, and then tightening the clamp.

Common sizes of polished rod are presented in Table 4.2.

Table 4.2 Common size of polished rod in PCP application

Size, inch	Length, ft	Size, inch	Length, ft
1 1/8	8	1 1/2	11
	11		16
	16		22
	22		26
	26		30
	30		36
1 1/4	11		
	16		
	22		
	26		
	30		

4.4.4 Pony Rod

Pony rods are the second component in a rod string of a PCP system. The end of the polished rod is connected to the pony rod using a coupling. Pony rods are sucker rods which are shorter than 25 ft, and vary in length. The main function of pony rods is to provide ways of handling the rod string due to its shapes in cross-section as shown in Fig. 4.10. Another reason of using pony rods is to prevent the string from falling downhole if the polished-rod clamp slips. Pony Rod lengths are available in 2, 4, 6, 8, 10, and 12 ft.

4.4.5 Rod Centralizer

Rod centralizers are installed along the rod string to centralize the rod string and reduce friction between the string and the tubing and hence minimize wear on rod coupling, rod string and production tubing. In addition, centralizers also help to reduce torque in deviated wells and therefore lower workover frequency. Common designs of centralizers are shown in Fig. 4.11.

4.4.6 Centralized Torque Anchor

The centralized torque anchor (CTA) is the last component from the surface of a rod string. The CTA is designed to anchor the tubing string and the PCP within the wellbore. The CTA holds and centralizes the tubing string and the PCP by using rigid slips that guarantee the anchor stays concentric within the wellbore casing as shown in Fig. 4.12.

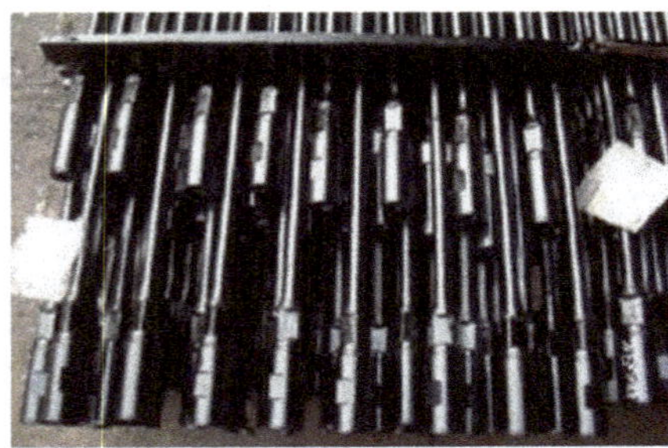

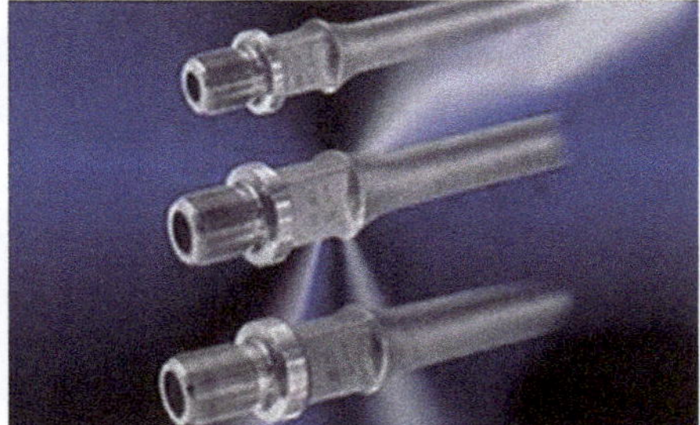

Fig. 4.10 Pony rods

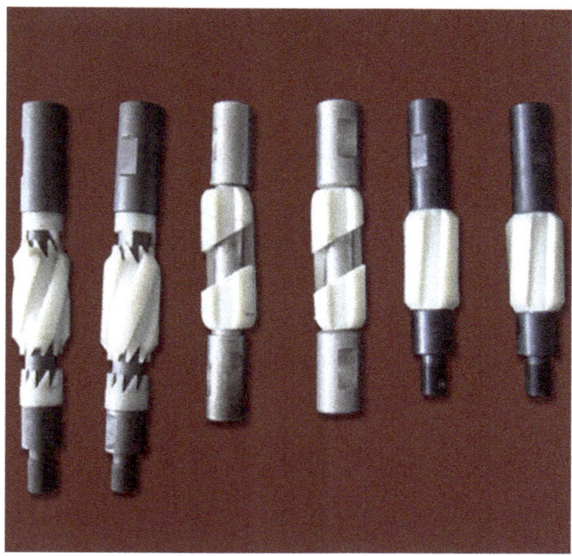

Fig. 4.11 Centralizers used in PCP systems

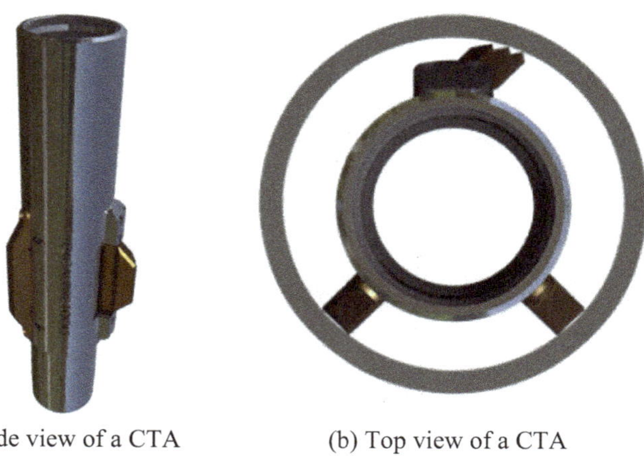

(a) Side view of a CTA (b) Top view of a CTA

Fig. 4.12 Centralized torque anchor

4.5 Pump Design

This section will present the design of a single-lobe and multi-lobe PCP, which is based on two theories: 3-D vector and Hypocycloid theories. Both theories will be applied on PCP geometry in subsequent sections. This design approach was presented by Nguyen et al. [17, 19, 20].

4.5.1 Review of 3-D Vector Theory

One point in a three dimensional space can be determined by a position vector P(x, y, z) in a rectangular coordinate (x, y, z) as described in the Fig. 4.13. The position vector $\vec{r}(s)$ is specified by two components; namely the magnitude (the distance from the origin O of the coordinates to the point P) and the direction (from the origin O to point P). The arc length between the O and P is defined as (s).

In a 3-D space, the position vector of this curve can be expressed as [15, 30, 31]:

$$\vec{r}(s) = x(s)\vec{i} + y(s)\vec{j} + z(s)\vec{k} \tag{4.5.1}$$

The first derivative of position vector with respect to the arc length (s) is defined as tangent vector, which is expressed as:

$$\vec{T}(s) = \vec{r'}(s) = \frac{dx(s)}{ds}\vec{i} + \frac{dy(s)}{ds}\vec{j} + \frac{dz(s)}{ds}\vec{k} \tag{4.5.2}$$

The tangent unit vector is defined as:

$$\vec{t}(s) = \frac{\vec{T}(s)}{\left\|\vec{T}(s)\right\|} \tag{4.5.3}$$

where $\left\|\vec{T}(s)\right\|$ is the magnitude of $\vec{T}(s)$, and assuming the curve is smooth, i.e. $\vec{r'}(s) \neq 0$, the second derivative of the position vector with respect to (s) is defined as the curvature vector. In other words, the curvature vector is the first derivative of the tangent vector with respect to (s), which measures the speed of the curve in changing its direction at a given point and given as:

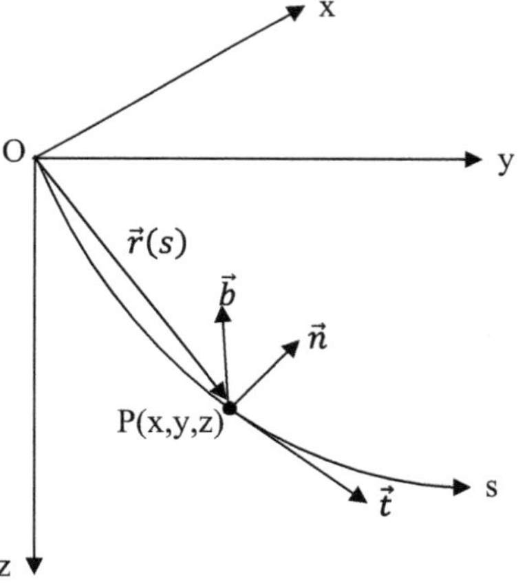

Fig. 4.13 A three dimensional space curve

$$\vec{\kappa} = \vec{r''} = \vec{T'} = \frac{d^2x}{ds^2}\vec{i} + \frac{d^2y}{ds^2}\vec{j} + \frac{d^2z}{ds^2}\vec{k} \tag{4.5.4}$$

The magnitude of curvature vector can then be expressed as:

$$\kappa(\mathrm{s}) = |\vec{K}| = \sqrt{\left(\frac{d^2x}{ds^2}\right)^2 + \left(\frac{d^2y}{ds^2}\right)^2 + \left(\frac{d^2z}{ds^2}\right)^2} \tag{4.5.5}$$

The reciprocal of the magnitude of curvature, $\kappa(s)$, is defined as the radius of curvature. The curvature of a circle in 2D is a constant and equals the inverse of its radius. If the curvature in 3D is constant then the curve is changing direction at the same rate at every point, resulting in a helix geometry. The unit normal vector, which is perpendicular to the tangent vector, can be expressed as follows:

$$\vec{n}(s) = \frac{\vec{\kappa}(s)}{|\vec{\kappa}|} \tag{4.5.6}$$

The definition of the unit tangent vector and the unit normal vector will be applied to design PCP pumps.

4.5.2 Review of Hypocycloid Theory

The principle of the hypocycloid is the original contour of cross section of a PCP pump. In geometry, a hypocycloid is a special plane curve generated by the trace of a fixed point on a small circle (generator circle), that rolls within a larger circle (base circle). Unlike the cycloid, the hypocycloid rolls within a circle instead of along a line. If the smaller circle has a radius of r, and the larger circle has a radius of $R = Kr$, then the parametric equations for the curve in 2D can be given as:

$$x(\theta) = r(K-1)\cos\theta + r\cos[(K-1)\theta] \tag{4.5.7}$$

$$y(\theta) = r(K-1)\sin\theta - r\sin[(K-1)\theta] \tag{4.5.8}$$

As shown in Fig. 4.14, θ is the angle between the x-axis and the line connecting two center points of the two circles (generator and base circles). If K is an integer, then the curve is closed. Figure 4.15 shows a hypocycloid of different K values of 4, 3, 2, and 1, traced as the smaller circle rolls around inside the larger circle. If K = 1, the diameter of the small circle is the same as that of the large circle and the hypocycloid is a circle called 1-lobe hypocycloid. Similarly, if K = 2, the diameter of the small circle is half of the diameter of the large circle and the hypocycloid is a line with its length equals to the diameter of the large circle, it is called 2-lobe hypocycloid. In general, when the value of K is an integer, it is equal to the number of lobes of the hypocycloid.

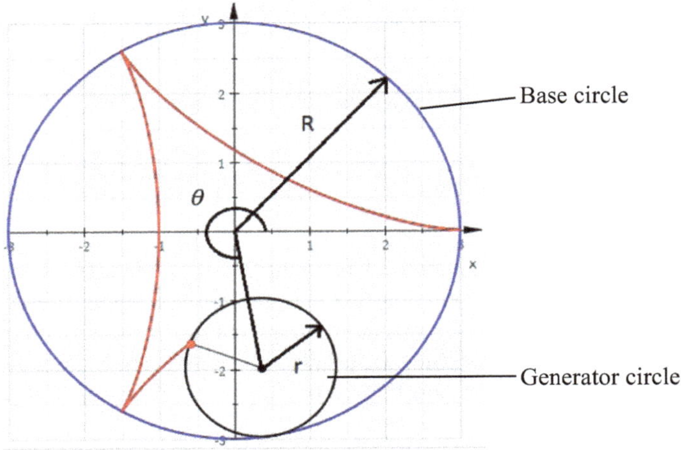

Fig. 4.14 Method of generating a hypocycloid

4.5.3 *Modeling the Design and Theoretical Performance of a Multi-lobe PCP*

4.5.3.1 Design of the Modified Hypocycloid

The main design principle of PCP is that the stator has one lobe (one gear or one tooth) more than the rotor and every lobe of the rotor must always be in contact with the inner surface of the stator. Consider a 4:3 Moineau pump which means that the stator has 4 lobes and the rotor has 3 lobes as shown in Fig. 4.16. In order to design the cross-sectional area of the stator, we will start with a 4-lobe hypocycloid, i.e. K = 4. Any given point on this hypocycloid can be described by using the two parametric equations Eqs. (4.5.7) and (4.5.8) where θ changes from 0 to 360° and x and y are the two components of the position vector $\vec{r}(\theta)$ in 2D.

Applying the vector theory presented in Sect. 4.5.1 gives the tangent vector which is the first derivative of $\vec{r}(\theta)$:

$$\vec{T}(\theta) = \vec{r'}(\theta) = x'(\theta)\vec{i} + y'(\theta)\vec{j} \tag{4.5.9}$$

where

$$x'(\theta) = \frac{dx}{d\theta} = -r(K-1)[\sin\theta + \sin((K-1)\theta)] \tag{4.5.10}$$

$$y'(\theta) = \frac{dy}{d\theta} = r(K-1)[\cos\theta - \cos((K-1)\theta)] \tag{4.5.11}$$

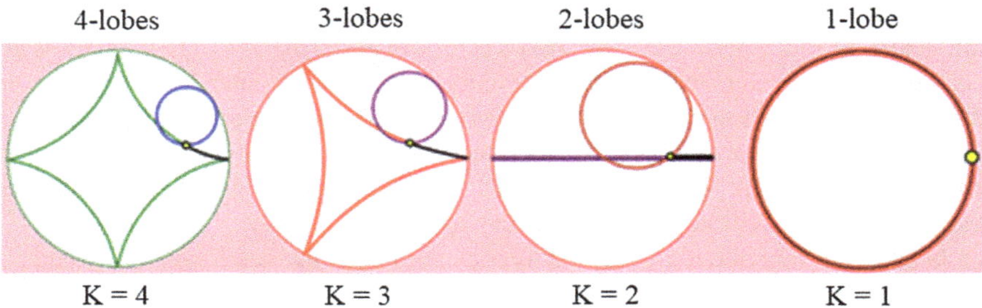

Fig. 4.15 Hypocycloids of different values of K [29]

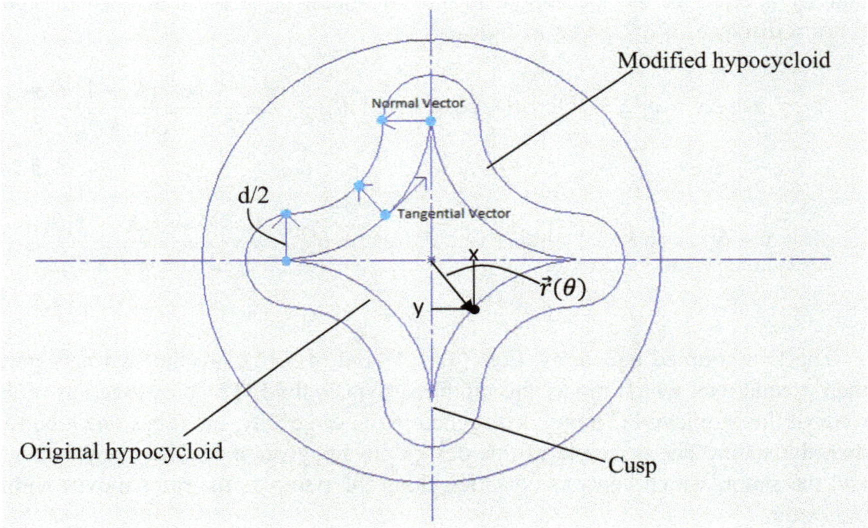

Fig. 4.16 Principle to generate the contour of a 4:3 Moineau pump

The unit tangent vector of any point on the hypocycloid can be determined as follows:

$$\vec{t} = -\frac{[\sin\theta + \sin((K-1)\theta)]}{\sqrt{2}\sqrt{1-\cos K\theta}}\vec{i} + \frac{[\cos\theta - \cos((K-1)\theta)]}{\sqrt{2}\sqrt{1-\cos K\theta}}\vec{j} \tag{4.5.12}$$

The normal unit vector is defined based on Eq. (4.5.6) and given as:

$$\vec{n} = \frac{[\cos\theta - \cos((K-1)\theta)]}{\sqrt{2}\sqrt{1-\cos K\theta}}\vec{i} + \frac{[\sin\theta + \sin((K-1)\theta)]}{\sqrt{2}\sqrt{1-\cos K\theta}}\vec{j} \tag{4.5.13}$$

Note that since the unit tangent vector is perpendicular to the unit normal vector, the scalar product between these two vectors is equal to zero ($\vec{t} \cdot \vec{n} = 0$). When designing a Moineau or PC pump, the cusps (the corners where the curve is not differentiable as shown in Fig. 4.16) on the original hypocycloid need to be modified in such a way that there will be semicircles with diameters "d" at each cusp, as shown in Fig. 4.16. Changing the "d" value will change the performance of the pump, which will be discussed in the application section.

In order to generate the modified hypocycloid with the semicircles (diameter d) on each cusp, each point on the original hypocycloid (x, y) described in Eqs. (4.5.7) and (4.5.8) will be moved with a distance of d/2, following the direction of the unit normal vector. Therefore, each point on the modified hypocycloid will have two new components (x_n, y_n), where $x_n = x + \Delta x$ and $y_n = y + \Delta y$. The magnitude of Δx and Δy is equal to the x-component and y-component of the unit normal vector times a distance of d/2 given as follows:

$$x_n = x + \Delta x = r[(K-1)\cos\theta + \cos((K-1)\theta)] + \frac{\cos\theta - \cos((K-1)\theta)}{\sqrt{2[1-\cos(K\theta)]}}\frac{d}{2} \tag{4.5.14}$$

$$y_n = y + \Delta y = r[(K-1)\sin\theta - \sin((K-1)\times\theta)] + \frac{\sin\theta + \sin((K-1)\theta)}{\sqrt{2[1-\cos(K\theta)]}}\frac{d}{2} \tag{4.5.15}$$

The two coupled equations, Eqs. (4.5.14) and (4.5.15), are the two new parametric equations which model the modified hypocycloid. The cross section of the rotor of the Moineau/PC pump is designed in the same way, but it has one lobe less than the stator. The principle of this design always gives a seal between the rotor and the stator, which generates cavities along the pump as the rotor moves within the stator.

4.5.3.2 Basic Pump Parameters

Recall that the variable r in Eqs. (4.5.14) and (4.5.15) is the radius of the generator circle when generating a hypocycloid. The radius of the base circle used to generate the K-lobe hypocycloid is the product (Kr). For example, the radii of the base circles of the 5-lobe and 4-lobe hypocycloid are 5r and 4r, respectively. Therefore, the eccentricity "e" of a PCP, defined as the difference between the radius of the stator and the radius of the rotor, equals to "r". In other words, the eccentricity of a PCP is equal to the radius of the generator circle.

Let's consider a K:(K − 1) multi-lobe PCP where K is the number of lobe of the stator. This pump is generated by two hypocycloids, namely K lobes and (K − 1) lobes. Assuming the diameter of the base circle of the K-lobe hypocycloid is D_{bs} then the diameter of the generator circle is D_{bs}/K as presented in Fig. 4.17a, b.

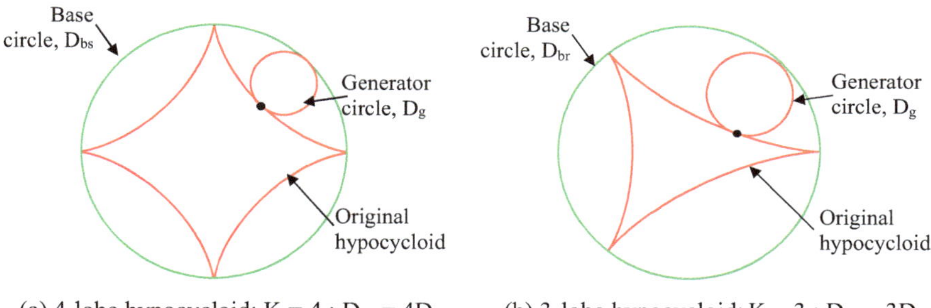

(a) 4-lobe hypocycloid: K = 4 ; $D_{bs} = 4D_g$ (b) 3-lobe hypocycloid: K = 3 ; $D_{br} = 3D_g$

Fig. 4.17 Cross-section of a Moineau/PC pump

Therefore, the eccentricity of this pump, which is the radius of the generator circle, is expressed as follows:

$$e = \frac{D_{bs}}{2K} \tag{4.5.16}$$

The diameter of the stator, as shown in Fig. 4.18a, is defined as the sum of the diameter of the base circle and the diameter of the semicircle at the cusp as,

$$D_s = D_{bs} + d = 2eK + d \tag{4.5.17}$$

The diameter of the modified rotor as shown in Fig. 4.18b, which has one lobe less than the diameter of the stator, is defined as:

$$D_r = D_{br} + d = 2e(K - 1) + d \tag{4.5.18}$$

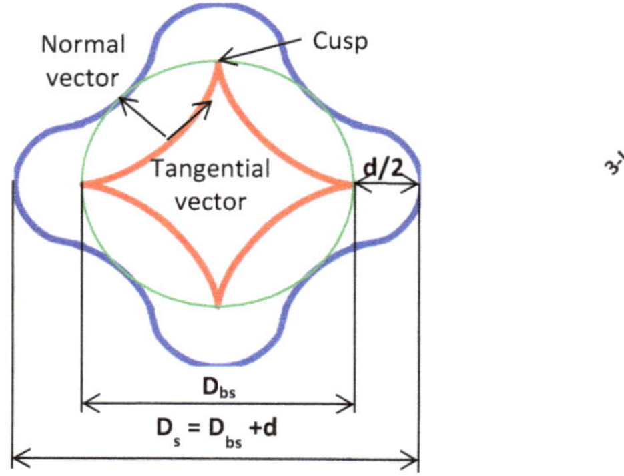

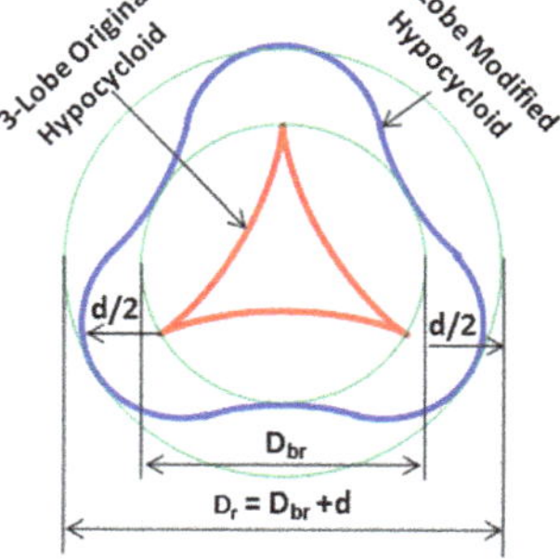

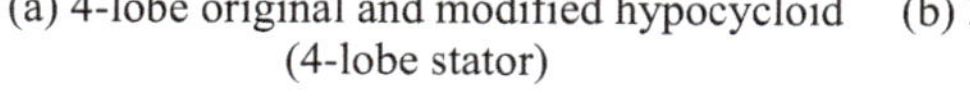
(a) 4-lobe original and modified hypocycloid (4-lobe stator)

(b) 3-lobe original and modified hypocycloid (3-lobe rotor)

Fig. 4.18 Flowing cross-sectional area of a 3:4 (4-lobe) Moineau/PC Pump

where D_{bs} and D_{br} are defined as the diameters of the stator base and rotor base, respectively.

For the original hypocycloid, the relationship between the base diameters of stator and rotor can be expressed as (same generator circle diameter):

$$D_{br} = \frac{K-1}{K} D_{bs} \tag{4.5.19}$$

Combining Eqs. (4.5.18) and (4.5.19) gives an equation to calculate for the rotor diameter as follows:

$$D_r = D_{br} + d = \frac{K-1}{K} D_{bs} + d \tag{4.5.20}$$

Combining Eqs. (4.5.17), (4.5.18) and (4.5.20) gives a relationship between stator and rotor diameters for a multi-lobe Moineau/PC pump as follows:

$$D_s - D_r = 2e = \frac{D_r - d}{K-1} = \frac{D_s - d}{K} \tag{4.5.21}$$

Equation (4.5.21) is only applicable for multi-lobe PCP. For a single lobe 2:1-PCP, the number of lobe of the rotor is one. For single-lobe hypocycloid, the parametric equations give a single point and the modified hypocycloid is a circle with a diameter the same as the semicircle diameter, d as shown in Eq. (4.5.18) for $K = 1$. In other words, for single-lobe PCP, the stator diameter $D_s = 2eK + d$ and the rotor diameter, $D_r = d$. Therefore, the relationship between stator and rotor diameters for a single-lobe PCP is given as:

$$D_s - D_r = 2eK = 4e \tag{4.5.22}$$

4.5.3.3 Estimation of the Cross-Sectional Area of the Modified Hypocycloid

Due to symmetry of a Moineau pump's stator, the cross-sectional area can be obtained by multiplying the area of one section to the number of lobes of the stator (K). The total area of one section includes three parts, namely A_1, A_2, and A_3 as defined in Fig. 4.19. Because the diameters of the base hypocycloid and the semicircle are known, the areas for A_2 and A_3 can be obtained, as follows:

$$A_2 = \frac{Ked}{2} \tag{4.5.23}$$

$$A_3 = \frac{\pi d^2}{8} \tag{4.5.24}$$

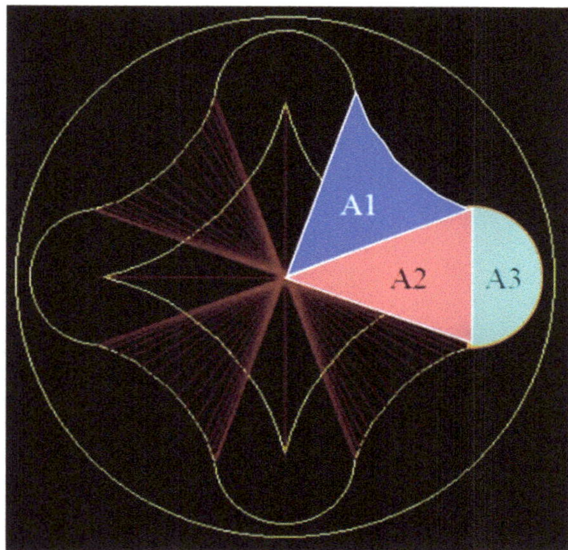

Fig. 4.19 Cross-sectional area of a 4-lobe modified hypocycloid (After [29])

The area A_1 is obtained by integrating the two coupled-parametric Eqs. (4.5.14) and (4.5.15) with respect to θ, which is given as:

$$A_1 = \left(\frac{K^2 - 3K + 2}{K}\right)\pi e^2 + \left(\frac{2 - K}{K}\right)\frac{\pi d^2}{8} + \left(\frac{-K^2 + 8K - 8}{2K}\right)de \tag{4.5.25}$$

Therefore, the total cross-sectional area of a K-lobe modified hypocycloid, $A_K = K(A_1 + A_2 + A_3)$, can now be obtained as:

$$A_1 = \left(\frac{K^2 - 3K + 2}{K}\right)\pi e^2 + \left(\frac{2 - K}{K}\right)\frac{\pi d^2}{8} + \left(\frac{-K^2 + 8K - 8}{2K}\right)de \tag{4.5.26}$$

4.5.3.4 Theoretical Pump Factor and Pump Rate for Multi-lobe PCPs

In general, a PCP will include a K-lobe modified hypocycloid stator and a (K − 1)-lobe modified hypocycloid rotor. In addition, there will be (K − 1) free spaces between the rotor and the stator, where the fluid flows. For example, the PCP design shown in Fig. 4.20 has three free spaces, namely S_1, S_2, and S_3, which change as the rotor rotates within the stator, but the total area ($A = A_1 + A_2 + A_3$) remains constant. Therefore, the flow area of a multi-lobe PCP is ($A_K - A_{K-1}$) and can be calculated using Eq. (4.5.26) as follows:

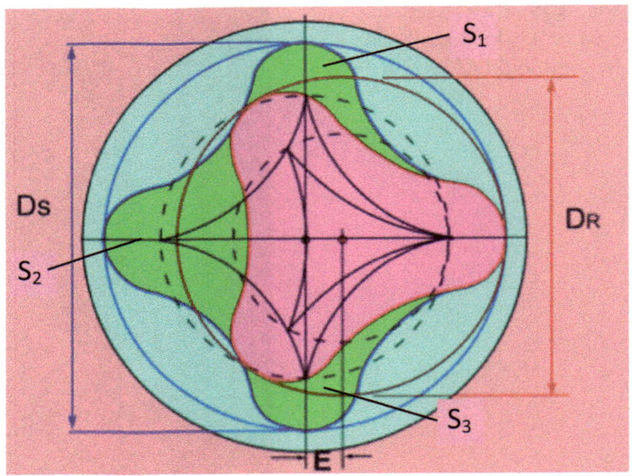

Fig. 4.20 A 4:3 ratio Moineau/PC Pump (After [29])

$$A_F = A_K - A_{K-1} = 2\pi e^2(K-2) + 4de \tag{4.5.27}$$

This new equation can be used to calculate the flow area of any given PCP, regardless of single or multi-lobe.

If there is a clearance (a gap), w, between the rotor and stator, Eq. (4.5.27) becomes

$$A_F = 2\pi e^2(K-2) + 4de + 8(K-1)ew + \pi\left(wd + w^2\right) \tag{4.5.27a}$$

The stator pitch length, P_s, is defined as a length of a 360° rotation of the crest trace of one helix lobe as shown in Fig. 4.21. The relationship between the pitch length of the rotor and the pitch length of the stator is given as:

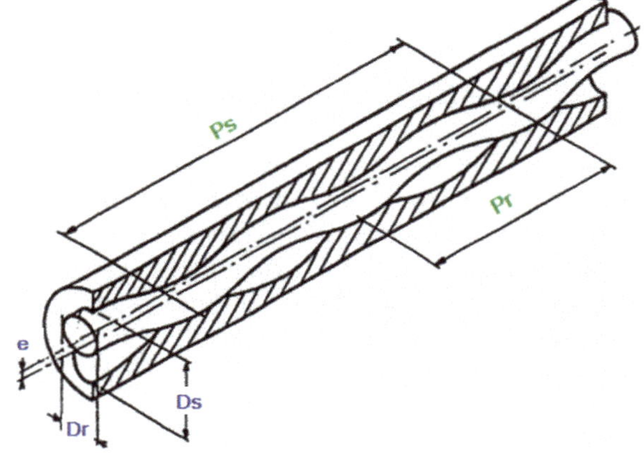

Fig. 4.21 Rotor and stator pitch length of 1:2 PCP

$$P_s = \frac{K}{K-1} P_r \tag{4.5.28}$$

Note that K is the number of lobe of the stator. For a single lobe PCP, K = 2 and $P_s = 2P_r$.

As shown in Fig. 4.20, the number of cavities (green shaded) of a K-lobe PCP is (K − 1); e.g. for the 3:4 PDM, the number of cavities at any point along the pump is three. As the rotor of a K-lobe PDM turns 360/K degrees, the fluid in all cavities along single pitch length is displaced and moved completely to the cavities of the adjacent rotor pitch length. In other words, as the rotor moves from the very top of the stator cusp to the adjacent cusp, the fluid volume passes the motor is the product of flowing cross-sectional area, A_F, and rotor pitch length, P_r. The theoretical pump factor, which is the total fluid volume that the pump can discharge as the rotor turns one cycle, which can be expressed as:

$$F_p = K \times A_F \times P_r \tag{4.5.29}$$

Combining with Eq. (4.5.28) gives:

$$F_p = (K-1)A_F \times P_s \tag{4.5.30}$$

The theoretical pump capacity (idea pump rate when the pump efficiency is 100%) when the rotor turns with a rotational speed of N is given as:

$$Q_{theo} = F_p \times N = (K-1)A_F \times P_s \times N \tag{4.5.31}$$

Combining Eqs. (4.5.27) and (4.5.31) gives:

$$Q_{theo} = \left[2\pi e^2(K-2) + 4de\right](K-1)P_s N \tag{4.5.32}$$

Equation (4.5.32) can be used to estimate the theoretical pump flow rate of any multi-lobe PCP. If the rotor speed is in rev/min or RPM, eccentricity, semicircle diameter, and stator pitch length are in meter then the flow rate calculated using Eq. (4.5.32) would have the unit of m^3/minute.

The flow rate will take the unit of gallons/minute or GPM if the following conversion factor is used:

$$Q_{theo} = 0.00433\left[2\pi e^2(K-2) + 4de\right](K-1)P_s N \tag{4.5.33}$$

In Eq. (4.5.33), the eccentricity, semicircle diameter, and stator pitch length are in the unit of inch.

4.5.3.5 Maximum Flow Area of Multi-lobe PCPs

Let's consider a K:(K − 1) multi-lobe PCP where the flow area can be determined by using Eq. (4.5.27). The maximum flow area (A_{max}) is defined as the cross-sectional area of the stator with K-lobe. Using Eq. (4.5.17), the A_{max} can be computed as follows:

$$A_{\max} = \frac{\pi}{4}(2Ke + d)^2 \tag{4.5.34}$$

In order to optimize the flow area for different pump ratios, a dimensionless flow area (A^*) is defined as the ratio between the flow area (A_F), and the maximum flow area (A_{max}) and given as:

$$A^* = \frac{2\pi(K-2) + 4\phi}{\frac{\pi}{4}(\phi + 2K)^2} \tag{4.5.35}$$

where $\phi = d/e$ is defined as the dimensionless diameter. Differentiating Eq. (4.5.35) with respect to ϕ and setting it equal to zero gives:

$$\frac{\partial A^*}{\partial \phi} = \frac{-16[\pi(K-2) + \phi - 2K]}{\pi(2K + \phi)^3} = 0 \tag{4.5.36}$$

Based on Eq. (4.5.36), the dimensionless flow area (A^*) reaches a maximum value when,

$$\frac{\partial A^*}{\partial \phi} = \frac{-16[\pi(K-2) + \phi - 2K]}{\pi(2K + \phi)^3} = 0 \tag{4.5.37}$$

For a single-lobe PCP of K = 2, Eq. (4.5.37) gives $\phi = 2\ K = d/e = 4$. Thus, A^* reaches a maximum value when d = 4e, indicating that single-lobe PCPs should be designed with d = 4e to have a maximum flow area.

For a multi-lobe PCP of $K \geq 5$, Eq. (4.5.37) gives negative value of ϕ, which is physically impossible. Therefore, A^* reaches the maximum value when $\phi = 0$ for the case of $K \geq 5$. Figure 4.22 is the plot of the dimensionless flow area and the dimensionless diameter for different lobe numbers. The results clearly show that A^* reaches its maximum value at $\phi = 4$ if K = 2, which is the case for single-lobe PCP. In addition, Eq. (4.5.37) points out that A^* is maximum at $\phi = 2.86$ and $\phi = 1.72$, when K = 3 and K = 4, respectively. The results also indicate that as the dimensionless diameter increases, the dimensionless flow area reduces when $K \geq 5$. This is because the rotor diameter increases, which reduces the flow area.

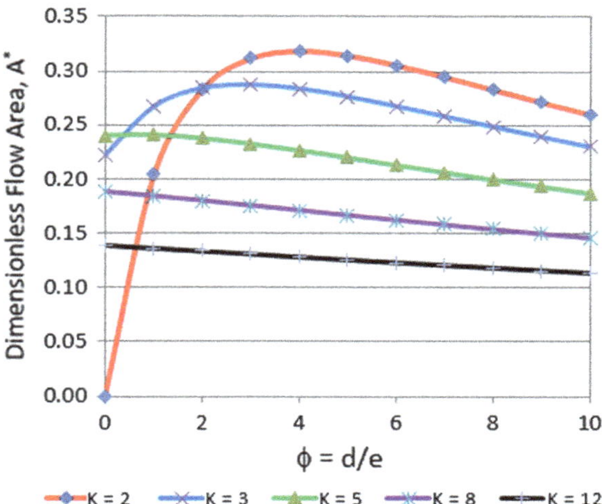

Fig. 4.22 Relationship between A^* and ϕ for different lobe numbers

4.5.4 Estimation of Actual Multi-lobe PCP Performance

Recall Eq. (4.5.33) shows that the theoretical pump rate is only a function of pump geometry and rotor speed. The theoretical pump rate does not take into account the effects of pump intake and discharge pressures, fluid temperature, fluid viscosity, and pump clearance between rotor and stator. The actual pump performance depends strongly on these parameters and will be modeled based on the total internal slippage.

Total internal slippage or slippage is defined as the difference between the theoretical and actual pump rate. If the slippage is zero, this means that the theoretical pump rate is the same as the actual pump rate. Under downhole conditions, at which the discharge pump pressure is much higher than the pump intake pressure, there always exists a reverse flow causing the actual pump rate to be smaller than the theoretical pump rate.

The total internal slippage model for multi-lobe PCPs was developed and extended by six different papers in a period from 1995 to 2016. The authors modeled the total internal slippage for a single-lobe PCP by dividing it into two components: longitudinal slip and transversal slip as shown in Fig. 4.23. According to Gamboa et al. [9, 10], the transversal slip takes place through the sealing lines formed at straight sections of the stator. The longitudinal slip takes place through the sealing lines at fixed positions of the rotor, where this element is located at semi-circular sections of the stator. In this model, the geometry of the internal slip flow is assumed to be equivalent to a flow through a channel at which the cross sectional-area is a rectangle. The two equivalent rectangular shapes in determining the longitudinal slip and transversal slip are defined in Fig. 4.24.

In Fig. 4.24, w is the clearance between the pump rotor and stator; b_L and b_T are the surface length (perimeter) of longitudinal and transversal slips, respectively;

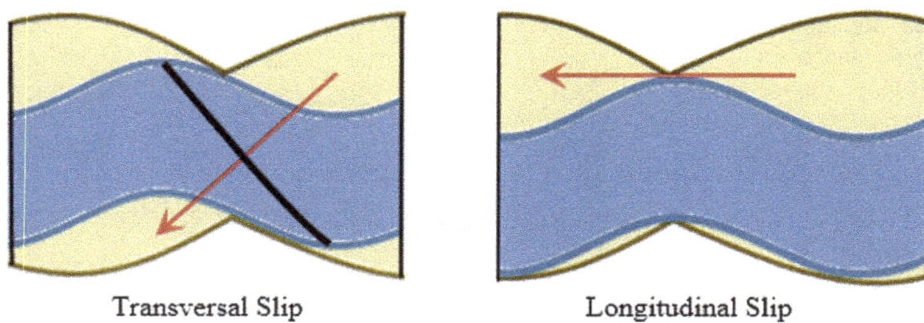

Fig. 4.23 Longitudinal slip and transversal slip [25]

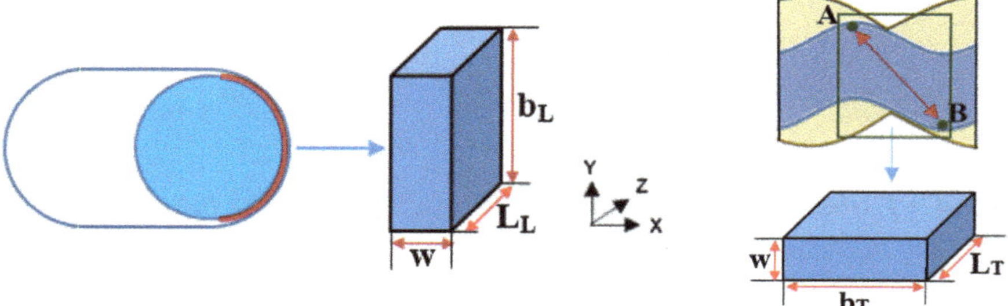

Fig. 4.24 Equivalent geometry for determining the longitudinal slip (left) and transversal slip (right) [9, 10]

L_L and L_T are the channel depths of longitudinal and transversal slip, respectively. To define the channel depth, Paladino et al. [24] and Pessoa et al. [25] used an iterative computational approach to determine L values, which resulted in $L_L = L_T = 1.65$ mm. These values will be used as inputs in all simulation runs in this study. As defined above, b_L is the perimeter of the semicircle with the diameter of d and, hence, it can be expressed as

$$b_L = \pi \times \frac{d}{2} \tag{4.5.38}$$

b_T is the length between points A and B as shown in Fig. 4.24, which can be attained based on the parametric equations of the modified hypocycloid, given by Nguyen et al. [19] as follows:

$$\begin{cases} x_n = r((K-1)\cos\theta + \cos((K-1)\theta)) + \frac{[\cos\theta - \cos((K-1)\theta)]}{\sqrt{2(1-\cos(K\theta))}} \times \frac{d}{2} \\ y_n = r((K-1)\sin\theta - \sin((K-1)\theta)) + \frac{[\sin\theta + \sin((K-1)\theta)]}{\sqrt{2(1-\cos(K\theta))}} \times \frac{d}{2} \\ z_n = \frac{\theta}{2\pi} P_s \end{cases} \tag{4.5.39}$$

For multi-lobe PCPs, b_T can be calculated as:

$$b_T = \int_{\theta=0}^{\theta=\frac{2\pi}{K}} \sqrt{\left(\frac{dx_n}{d\theta}\right)^2 + \left(\frac{dy_n}{d\theta}\right)^2 + \left(\frac{dz_n}{d\theta}\right)^2} d\theta \tag{4.5.40}$$

An analytical solution of Eq. (4.5.40) gives the final equation for calculating the value of b_T as follows:

$$b_T = \int_0^{2\pi} \sqrt{\left[\frac{d(N-2)}{4} - 2e(N-1)\sin\left(\frac{N\theta}{2}\right)\right]^2 + \frac{P_s^2}{4\pi^2}} d\theta \tag{4.5.41}$$

The simplified form of Eq. (4.5.41) was presented by Nguyen et al. [18] and shown in Eq. (4.5.42):

$$b_T = \frac{2\pi}{K}\sqrt{2(K-1)^2 e^2 - \frac{2de(K-2)(K-1)}{\pi} + \frac{d^2(K-2)^2}{16} + \frac{P_s^2}{4\pi^2}} \tag{4.5.42}$$

The discrepancy of the solutions between Eqs. (4.5.41) and (4.5.42) is presented in Fig. (4.25a, b) for single-lobe 2:1 and multi-lobe 8:7, respectively.

with the values of w and b (either b_L or b_T), the rectangular cross-sectional area is converted back to a circular shape by using the definition of hydraulic diameter, D_H. The Reynolds number is then calculated to determine the flow regime, i.e. laminar or turbulent, in the channel in order to calculate the appropriate friction factor, from which the frictional pressure loss is then computed. Pessoa et al. [25] derived the final internal longitudinal and transversal slippage equations as follows:

$$S_L = \frac{b_L w^3 \Delta P_L}{2\mu L_L} \tag{4.5.43}$$

$$S_T = \frac{b_T w^3 \Delta P_T}{2\mu L_T} \tag{4.5.44}$$

where ΔP_L and ΔP_T are the longitudinal and transversal frictional pressure drop.

The total internal slippage of a multi-lobe PCP is calculated as:

$$S_{Total} = (K-1)(S_L + S_T) \tag{4.5.45}$$

The actual pump rate is the difference between theoretical pump rate and total internal fluid slippage:

$$Q_a = Q_{th} - S_{Total} \tag{4.5.46}$$

In summary, for a specific pump geometry, the theoretical pump rate can be determined using Eq. (4.5.33). The actual pump rate can be estimated using

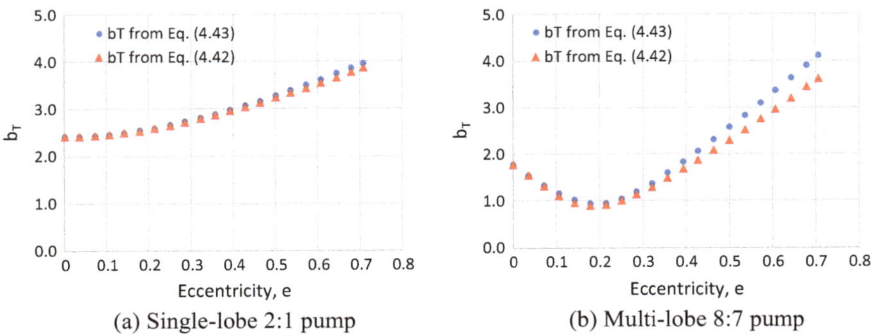

(a) Single-lobe 2:1 pump (b) Multi-lobe 8:7 pump

Fig. 4.25 Comparison between calculated b_T values from Eqs. (4.5.41) to (4.5.42)

Eq. (4.5.46) if the total slippage is known. The procedure to predict the total slippage can be summarized as the following steps:

- Step 1: Calculate the slippage channel geometry for the longitudinal and transversal slips.
- Step 2: Assume values of longitudinal and transversal slippage, S_L and S_T.
- Step 3: Calculate the fluid slip velocities in two components then determine the flow regime.
- Step 4: Calculate the frictional pressure drop in two slip components ΔP_L and ΔP_T.
- Step 5: Calculate the new longitudinal and transversal slippage using Eqs. (4.5.43) and (4.5.44).
- Step 6: Compare the assume slippage values and the calculate slippage value. If the difference is more than the set tolerance, assume new values for S_L and S_T and repeat step 2 to step 5.

4.5.5 Model Validation

The model presented in Sects. 4.5.3 and 4.5.4 is validated first against the experimental data obtained by Gamboa et al. [9, 10], then against the experimental data acquired from the New Mexico Tech Pumping Facility (NMTPF). Gamboa et al. [9, 10] conducted their experiments using a commercial progressing cavity pump with a metallic stator. The geometrical parameters of the tested pump are given in Table 4.3.

The pump has three stages each with a length equals to the stator pitch length. The authors used four different fluid viscosities: 1 cp (water), 42 cp, 134 cp, and 480 cp. Figure 4.26 shows the model validation results against the experimental data conducted by Gamboa et al. [9, 10], for liquid viscosity of 480 cp. The model

Table 4.3 Tested pump geometry [9, 10]

Eccentricity, mm	4.039
Rotor diameter, mm	39.878
Interference, mm	0.185
Stator Pitch, mm	119.99

validation reveals satisfactory results with a relative error in the range of 0.5–7%. The proposed model predicts pump performance quite well with metallic stator PCP due to the consistent clearance between rotor and stator. In addition, the accuracy of the model increases as liquid viscosity increases.

Figure 4.27 demonstrates the validation of the present model against experimental data acquired by NMTPF. The experiments were carried out at pump speed of 300 RPM, and a discharge pressure of 300 psi. The model predicts the experimental data very well with maximum relative error of 4%, especially with high liquid viscosity.

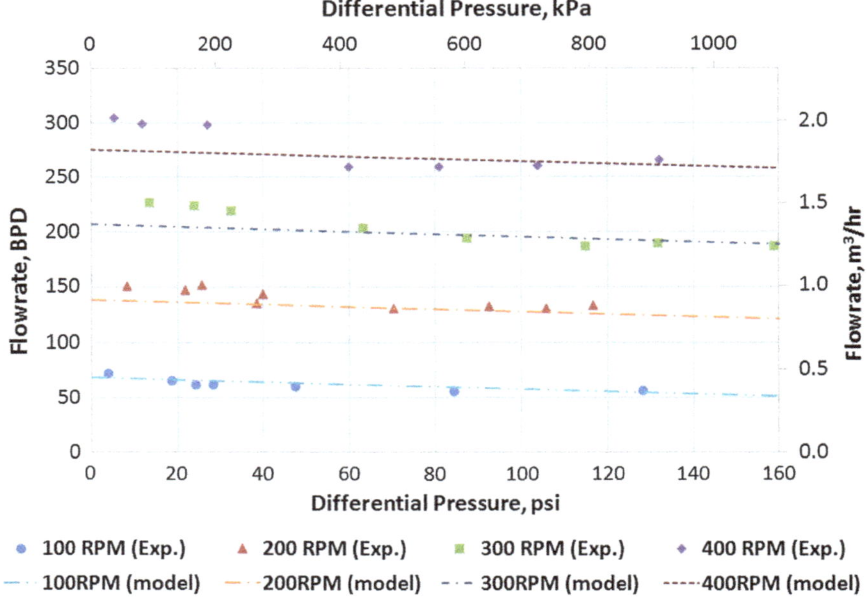

Fig. 4.26 Comparisons between Gamboa et al. [9, 10] experimental data and model predictions for liquid viscosity of 480 cp

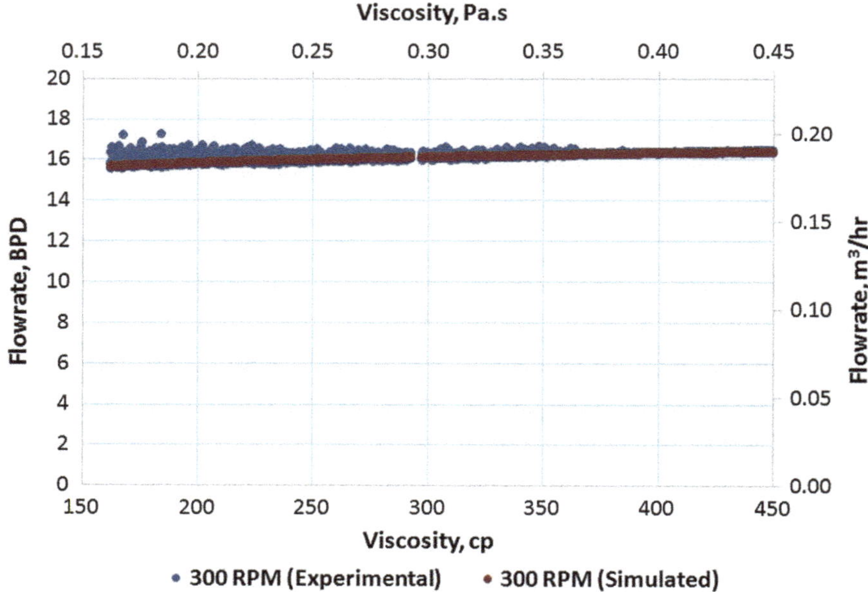

Fig. 4.27 Present model validation against experimental results from NMTPF

4.5.6 Modeling Actual Multi-lobe PCP Performance Using Dimensionless Approach

Al-Safran et al. [1] used dimensionless approach to model the actual multi-lobe PCP performance. In this work, the authors derived three dimensionless groups using mass equation, momentum equation, pump geometry, and appropriate boundary conditions and scaling parameters.

The first dimensionless group: Euler number:

$$Eu = \frac{\Delta P_{pump}}{\rho_f P_s^2 N^2} \tag{4.5.47}$$

The second dimensionless group: Reynolds number:

$$Re = \frac{\rho_f P_s^2 N}{\mu_f} \tag{4.5.48}$$

The third dimensionless group: PCP number:

$$N_{PCP} = \frac{Q_a}{P_s A_f N} \tag{4.5.49}$$

where ΔP_{pump} is the differential pressure at the pump intake and pump discharge; P_s is the stator pitch length; N is the stator speed in RPM, A_f is the cross-sectional flow area defined by Eq. (4.5.27); and ρ_f and μ_f are the fluid density and viscosity.

Using the developed dimensionless groups, a relationship is theorized, relating Euler number to the inverse Reynolds number and PCP number as:

$$Eu = f\left(Re^{-1}, N_{PCP}\right) \tag{4.5.50}$$

Using the developed theorized dimensionless relationship, a non-linear regression model for a steady-state, single-phase, Newtonian fluid flow in 2:1 PCP is developed using Gamboa et al. [9, 10] database. Figure 4.28 shows a 3D plot of the developed dimensionless numbers using Gomboa et al. [9, 10] data, indicating a strong non-linear relationship between Euler number, and inverse Reynolds and PCP dimensionless numbers. A best-fit non-linear empirical correlation, representing the relationship in Fig. 4.28 is generated using TableCurve 3D™ software package given as:

$$ln(Eu) = 5.227 - 3.436 N_{PCP}^3 + \frac{23.51}{ln(Re)} \tag{4.5.51}$$

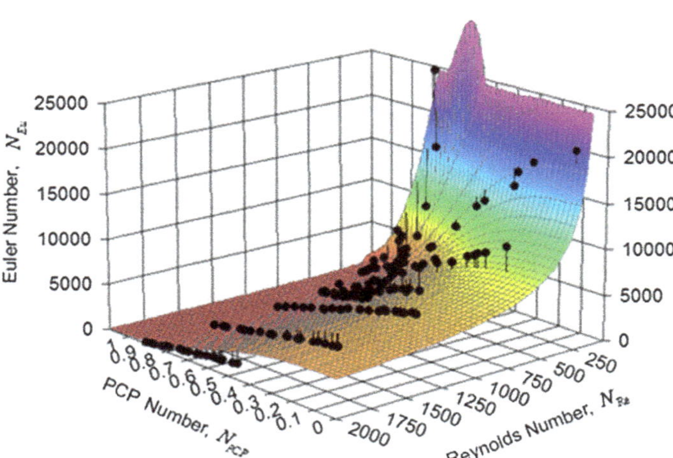

Fig. 4.28 Single-phase proposed model curve fit to Gamboa et al. [9, 10] experimental data

To solve for actual flow rate, Eq. (4.5.51) is rearranged as follows:

$$Q_a = (P_s A_F N)\left[\frac{6.841}{\ln(Re)} + 1.521 - 0.291\ln(Eu)\right]^{1/3} \tag{4.5.52}$$

The authors confirmed that the single-phase model presented in this section is capable of predicting the actual pump rate or pressure drop across PCP for different pump speeds with 85% accuracy.

4.5.7 Extension of This Model to Predict Performance of Positive Displacement Motors

A basic Downhole Mud Motor or Positive Displacement Motor (PDM) is composed of the following parts: sealed bearing assembly, adjustable housing and drive shaft, power section, and dump sub as illustrated in Fig. 4.29. The sealed bearing assembly carries all radial and thrust loading, where the sealing is used to deliver maximum flow to drill-bit. The adjustable housing connects the stator to the bearing assembly housings, and houses the drive shaft assembly. The power section converts hydraulic power from drilling fluid into torque required to rotate the rotor [2]. Dump sub is a by-pass valve installed above the motor, which allows drilling fluid to fill the drill-string when tripping in and to drain the drill-string when tripping out.

When an angle is selected in the adjustable or fixed housing, the bit face is oriented to initiate a build or a drop angle in wellbore. As drill bit reaches the kick-off depth, the motor is tripped out to re-adjust bend angle, then motor is tripped into resume build or drop angle drilling.

Mud motor should be selected properly to match bit torque requirements and flow rates to reach an efficient hole-cleaning, and maximum pump pressure [12]. Furthermore, it is critical to monitor standpipe pressure during downhole mud motor operation [28]. Initially, as motor starts in off-bottom position, standpipe pressure increases until the motor starts rotating, which depends on motor size and geometry, drill-pipe size, drill-collars, and other tools installed in drill-string. If drill-bit is close to the bottom of the hole and the motor is installed near drill bit, standpipe pressure in Pa is given as:

$$P_{standpipe} = P_{wf} - \rho_m g TVD + \Delta P_f^{dp} + \Delta P_{motor} \tag{4.5.53}$$

where P_{wf} is flowing bottom-hole pressure (Pa), ρ_m is drilling fluid density (kg/m^3), TVD is well true vertical depth (m), ΔP_f^{dp} is frictional pressure drop in drill-string (Pa), and ΔP_{motor} is motor pressure drop (Pa). Note that regions, e.g. key-seats may cause higher standpipe pressure to initiate motor rotation if the motor is disoriented.

As the motor rotates and the well is being drilled, the standpipe pressure must be sufficient to overcome the frictional pressure loss inside the drill-string and in the

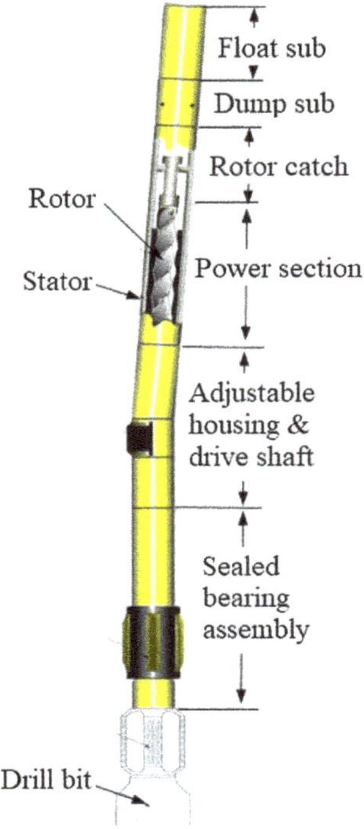

Fig. 4.29 Basic components of a PDM

annuli, the pressure drop across the drill-bit, and the pressure drop inside the motor. The pressure drop inside the motor has two components: the pressure drop to initiate the rotation and the pressure drop to overcome the friction between the bit and the formation rock. Given the Weight On Bit (WOB), the equation to calculate the stand-pipe pressure as the well is being drilled is presented as follows:

$$P_{standpipe} = P_{wf} - \rho_m g TVD + \Delta P_f^{dp} + \Delta P_{motor} \tag{4.5.54}$$

where ΔP_f^{ann} and ΔP_b are the frictional pressure loss in the annulus and the pressure drop at the drill-bit. As drilling commences, the WOB may be increased to reach the desired Rate of Penetration (ROP), maximum torque or maximum motor differential pressure. However, the standpipe pressure should not cause the motor differential pressure to be higher than the motor allowable value. As cuttings are generated into the drilling fluid, some hydraulic power will be lost to lift these cuttings and hence the stand-pipe pressure predicted by Eq. (4.5.54) can vary slightly. If the applied WOB is too high, the motor stall may occur and the bit stops rotating. If the motor stall happens, the stand-pipe pressure increases rapidly; the surface measured torque will be very high causing a possible damage to motor components. Keep pumping drilling fluid through a stalled motor may cause wear to the stator elastomer and damage to the sealed bearing.

Recall Eq. (4.5.33) for calculating the theoretical multi-lobe pump rate:

$$Q_{theo} = \left[2\pi e^2(K-2) + 4de\right](K-1)P_sN \quad (4.5.32)$$

The mechanical horsepower developed by the motor in SI and field units is defined as the product of its torque, T, by its angular velocity, $\frac{2\pi N}{60}$, given respectively as:

$$HP_m(watts) = 2\pi N(RPM) \times T(mN) \quad (4.5.55)$$

$$HP_m(HP) = \frac{T(ft-lbf) \times N(RPM)}{5252} \quad (4.5.56)$$

The hydraulic horsepower for an incompressible fluid in SI and field units are given as:

$$HP_h(watts) = 1.67 \times Q\left(\frac{L}{\min}\right) \times \Delta P(bar) \quad (4.5.57)$$

$$HP_h(HP) = \frac{Q(GPM) \times \Delta P(psi)}{1714} \quad (4.5.58)$$

where Q is flow rate, and ΔP is differential pressure across motor. The overall motor efficiency:

$$\eta = \frac{HP_m}{HP_h} = \frac{T \times N}{3.064 \times Q \times \Delta P} \quad (4.5.59)$$

In Eq. (4.5.59), torque, rotational speed, flow-rate, and pressure drop are in the unit of ft-lbf, RPM, GPM, and psi, respectively. Combining Eqs. (4.5.33) and (4.5.59) gives the torque in ft-lbf as follows:

$$T = 0.01326\left[2\pi e^2(K-2) + 4de\right](K-1)P_s\Delta P\eta \quad (4.5.60)$$

where e, d and P_s are in the unit of inch, and ΔP is in the unit of psi. Equation (6.5.60) indicates that torque of PDM is a function of motor geometry (K, e, and d), pressure drop, and motor efficiency only. Note that torque does not depend on rotor speed or bit speed.

To investigate the effect of motor lobe number on motor torque at different motor differential pressures, the data in Table 4.4 is used. Liquid is pumped through the motor with a constant flow rate of 1363 m^3/D (250 GPM), assuming motor efficiency of 0.75. Equation (4.5.60) is used to calculate motor torque, and the results are illustrated in Figs. 4.30 and 4.31. Note that for a constant stator lobe number and constant differential pressure, torque is independent of the drill-bit speed as given by Eq. (4.5.60). However, as stator lobe number increases, drill-bit speed declines as the differential pressure ramps up, which is shown in the solid-smooth curve in Fig. 4.30. Furthermore, Fig. 4.30 presents the relationship between torque and motor differential pressure for four values of stator lobe number, namely 4 (3:4-lobe motor), 6 (5:6-lobe motor), 10 (9:10-lobe motor), and 12 (11:12-lobe motor). Generally, when increasing the stator lobe number, rotational speed decreases and torque increases for a constant value of differential pressure. As mentioned before, flow rate is proportional to the stator lobe number, leading to an increase in torque as shown in Eq. (4.5.60). However, this increase in torque reduces significantly when the stator lobe number is higher than 10 as shown in Fig. 4.34. The percent increase in torque when the stator lobe number varies from 12 to 14 is approximately 3.6%. Therefore, it is not recommended to design PDM with stator lobe number greater than 12.

In general, as long as the mud pump can provide enough horsepower to PDM, it theoretically can handle high WOB. However, if the applied WOB is too high, differential pressure across power section and motor torque will be too high, causing a damage on the stator elastomer. Depending on the type of the elastomer, there is allowable differential pressure that can be handled by the motor. If differential pressure is higher than the allowable pressure, then the stator will be damaged.

Figure 4.32 shows the theoretical prediction of a complete motor performance, which describes the relationship between torque, horsepower, and drill-bit speed vs. differential pressure across motor for the motor geometry presented in Table 4.4. In this particular simulation, the stator lobe number was kept constant as six (K = 6), as well as the liquid flow rate at 1363 m^3/D (250 GPM), with assumed motor efficiency of 0.75. The differential pressure to initiate motor rotation (to overcome the internal friction between the rotor and the stator) was assumed 689 kPa (100 psi). Using Eqs. (4.5.17) and (4.5.33) to calculate motor eccentricity and theoretical drill-bit speed gave the values of 0.51 cm (0.2 in.) and 214 RPM, respectively. The results also revealed that as differential pressure varies from 689 to 6890 kPa (100–1000 psi), torque linearly increases from 0 to 3641 kPa (0–2685 psi), and horsepower linearly increases from 0 to 110 HP. However, if differential pressure is kept constant at 5512 kPa (800 psi), torque remains at 2913 kPa (2148 psi) regardless to the change in liquid flow rate as presented in Fig. 4.33. Equation (4.5.60) indicates that for a specific motor geometry, when liquid flow rate varies, rotational speed also changes in such a way that torque remains the same for a constant differential pressure. In general, torque is independent of the rotational speed (or liquid flow rate) for a given motor, and operated at a constant differential pressure. Torque

Table 4.4 Simulated PDM geometry

Stator diameter, D_s	6.98 cm	2¾ in.
Stator pitch length, P_s	106.68 cm	42 in.
Semicircle diameter, d	2.28 cm	0.35 in.
Eccentricity, e	Calculated using Eq. (10)	
Stator lobe number, K	Varied	

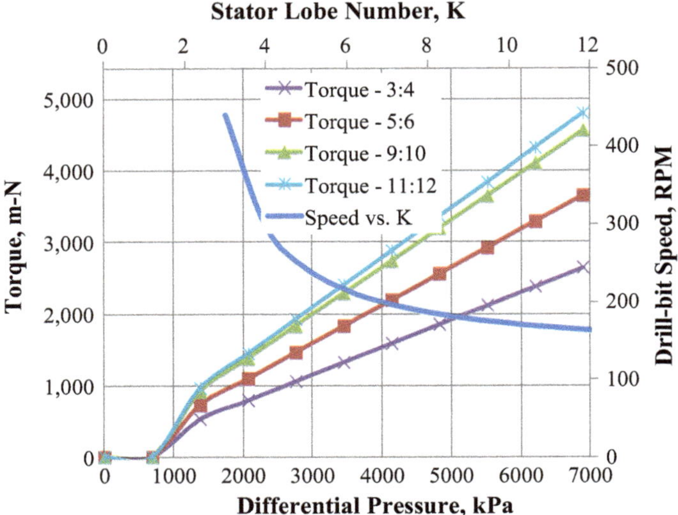

Fig. 4.30 Effect of K on torque (Q = 250 GPM, η = 0.75, D_s = 2 ¾ in., P_s = 42 in., d = 0.35 in.)

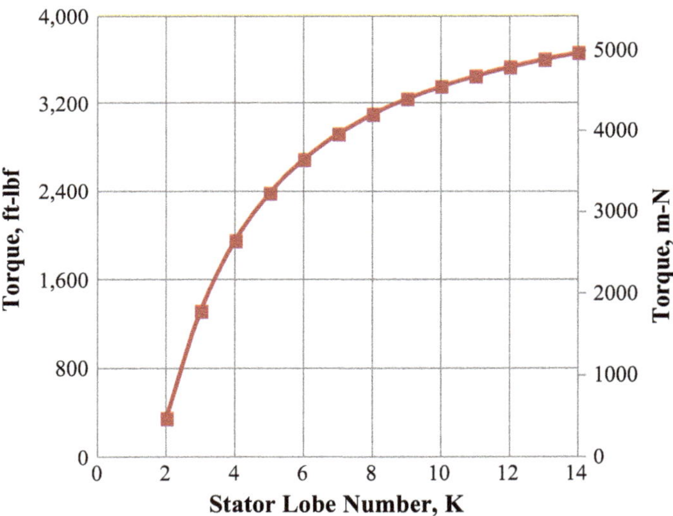

Fig. 4.31 Optimizing torque based K values (Q = 250 GPM, η = 0.75, D_s = 2 ¾ in., P_s = 42 in., d = 0.35 in.)

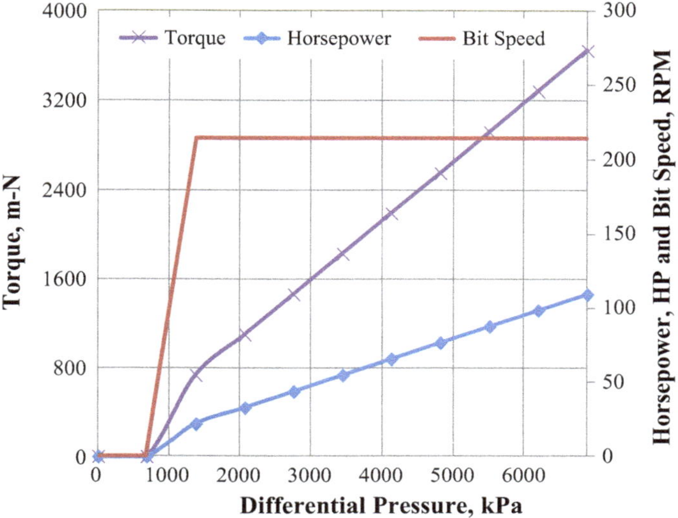

Fig. 4.32 Performance of a PDM (Q = 250 GPM, η = 0.75, D_s = 2 ¾ in., P_s = 42 in., d = 0.35 in. K = 6)

changes when motor geometry (rotor or stator diameter, eccentricity, and stator lobe number) or the differential pressure varies.

Nguyen et al. [20] have drawn several conclusions using the presented model as follows:

- For stator lobe number greater than 5 (K > 5), torque is maximized if the dimensionless motor geometry approaches zero (ϕ = d/e = 0). Practically, the semicircle diameter, d, must be greater than zero to minimize the internal slippage and hence motor eccentricity should be maximized to obtain the highest torque for K > 5.
- For a specific stator lobe number, Eq. (4.5.61) can be used to optimize the ratio between eccentricity and rotor diameter at which the flow rate or torque is maximum.

$$\frac{e}{D_r} = \frac{1}{[4(K-1) - \pi(K-2)]} \tag{4.5.61}$$

- For the stator lobe number greater than four (K > 4), there will be a reduction in the flowing cross-sectional area as the stator lobe number increases. However, flow rate always increases when stator lobe number increases. This is because of the higher total volume of cavities within the rotor and the stator.

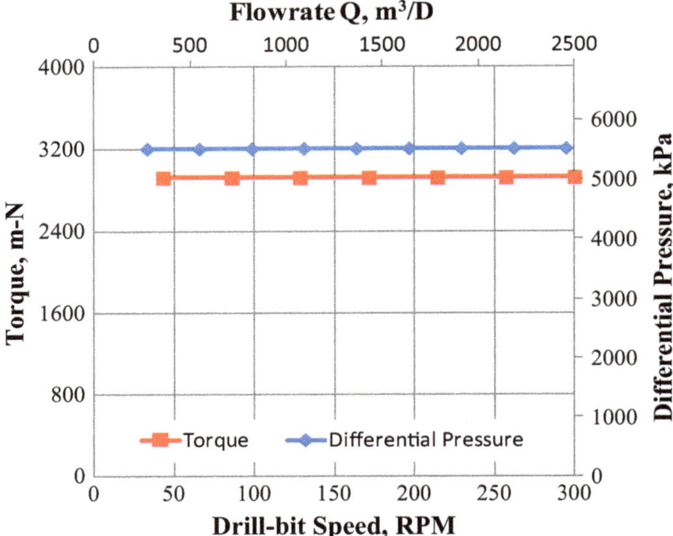

Fig. 4.33 PDM Performance with variable Q and constant ΔP ($\eta = 0.75$, $D_s = 2\ ¾$ in., $P_s = 42$ in., d = 0.35 in.)

- Torque increases as lobe of the PDM increases. However, it is not recommend designing a PDM, which has stator lobe number greater than twelve. The increase in torque is 3.6% or less if the stator lobe number increases from 12 to 14 or higher.

4.6 Example

Example 4.1 Given the radius of a 3-lobe hypocycloid generator circle of 2 in., using the parametric Eqs. (4.5.7) and (4.5.8), plot the 3-lobe hypocycloid. Changing the number of lobes to 6, plot the 6-lobe hypocycloid.

Solution
The parametric equations for a multi-lobe hypocycloid are described in Eqs. (4.5.7) and (4.5.8) as follows:

$$x(\theta) = r(K-1)\cos\theta + r\cos[(K-1)\theta] \tag{4.5.7}$$

$$y(\theta) = r(K-1)\sin\theta - r\sin[(K-1)\theta] \tag{4.5.8}$$

If the number of lobes of stator K = 6 and the radius of the generator circle, r = 2., by changing the value of θ from 0 to 360°, one can easily calculate the values of x and y. The results are plotted and shown in Fig. 4.34.

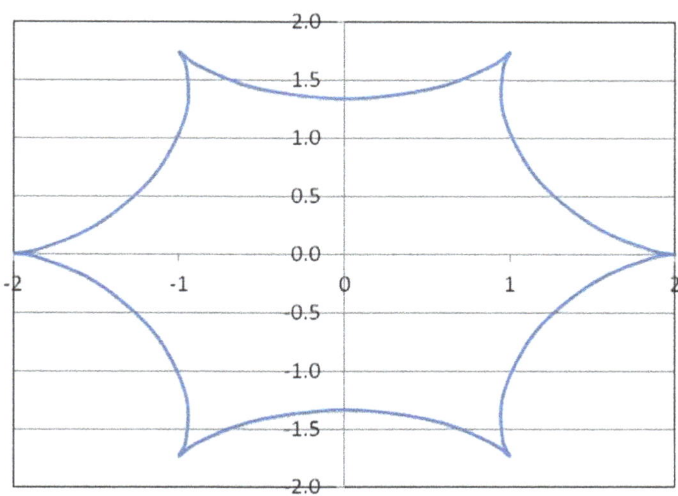

Fig. 4.34 6-lobe original hypocycloid—r = 2 in.

Similar calculation can be done to obtain the 3-lobe original hypocycloid. Readers should practice to gain more confident on the hypocycloid theory.

Example 4.2 Given the radius of a 4-lobe hypocycloid generator circle of 0.5 in. and the semicircle diameter of 2 in. Plot the original 4-lobe hypocycloid and the 4-lobe modified hypocycloid.

Solution

Example 4.1 shows how to obtain an original hypocycloid. To know how to achieve the modified hypocycloid, readers are highly recommended to review Sect. 4.5.3.1. The new components, (x_n, y_n), of the modified hypocycloid are given in Eqs. (4.5.14) and (4.5.15) below where $x_n = x + \Delta x$ and $y_n = y + \Delta y$.

$$x_n = x + \Delta x = r[(K-1)\cos\theta + \cos((K-1)\theta)] + \frac{\cos\theta - \cos((K-1)\theta)}{\sqrt{2[1-\cos(K\theta)]}}\frac{d}{2} \tag{4.5.14}$$

$$y_n = y + \Delta y = r[(K-1)\sin\theta - \sin((K-1)\times\theta)] + \frac{\sin\theta + \sin((K-1)\theta)}{\sqrt{2[1-\cos(K\theta)]}}\frac{d}{2} \tag{4.5.15}$$

With K = 4, r = 0.5 in., and d = 2 in., changing the values of θ from 0 to 360°, one can easily calculate the values of x_n and y_n. The results are shown in Fig. 4.35.

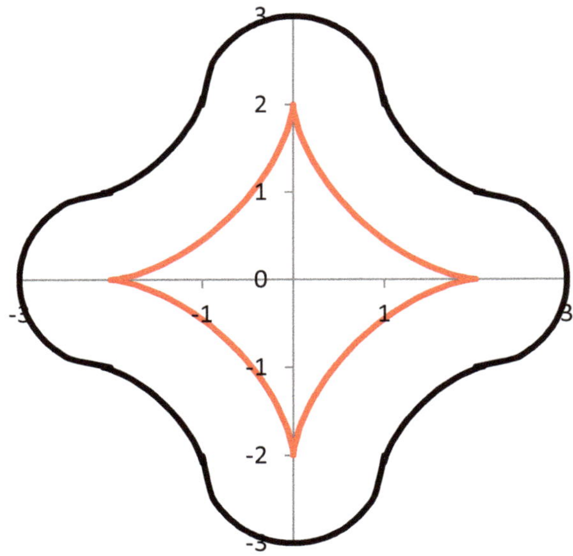

Fig. 4.35 4-lobe original and modified hypocycloids

Example 4.3 A single-lobe PCP, which has the rotor diameter of 2 in. and eccentricity of 0.25 in., stator pitch length of 6 in., is rotating at the speed of 100 RPM. Calculate the semicircle diameter, pump stator diameter and the theoretical pump rate.

Solution

As explained in Sect. 4.5.3.2, the rotor diameter of the single-lobe PCP is the same as the semicircle diameter, d = D_r = 0.5 in. Using Eq. (4.5.22) gives the stator diameter as follows:

$D_s = 4e + D_r = 3$ in.

Using Eq. (4.5.33) for calculating the theoretical pump rate gives:

$$Q_{theo} = 0.00433\left[2\pi e^2(K-2) + 4de\right](K-1)P_sN$$
$$Q_{theo} = 0.00433\left[2\pi \times 0.25^2(2-2) + 4 \times 2 \times 0.25\right](2-1)6 \times 100$$
$$Q_{theo} = 5.19GPM = 178BPD$$

Example 4.4 A PCP, which has the rotor diameter of 2 in., semicircle diameter of 0.5 in., stator pitch length of 6 in., is rotating at the speed of 100 RPM. Calculate the pump eccentricity, pump stator diameter, and the theoretical pump rate.

Solution

Rearranging Eq. (4.5.21) gives an equation to calculate for the pump eccentricity as follows:

$$e = \frac{D_r - d}{2(K - 1)} = \frac{2 - 0.5}{2(3 - 1)} = 0.375\,\text{in.}$$

Pump stator diameter as given in Eq. (4.5.21):

$$D_s = D_r + 2e = 2 + 2 \times 0.375 = 2.75\,\text{in.}$$

Theoretical pump rate can be calculate using Eq. (4.5.33):

$$Q_{theo} = 0.00433\left[2\pi e^2(K - 2) + 4de\right](K - 1)P_s N$$
$$Q_{theo} = 0.00433\left[2\pi \times 0.375^2(3 - 2) + 4 \times 0.5 \times 0.375\right](3 - 1)6 \times 100$$
$$Q_{theo} = 8.49\,GPM = 291\,BPD$$

Example 4.5 A PCP geometry is give as follows: pump stator diameter of 4 in., semicircle diameter of 2 in., stator pitch length of 4 in. Study the impact of rotational speed on the pump rate at K = 2, 4, 6, 8, and 10.

Solution
The following equations will be used for calculating the pump eccentricity (Eq. 4.5.21), and the theoretical pump rate:

$$e = \frac{D_s - d}{2K}$$
$$Q_{theo} = 0.00433\left[2\pi e^2(K - 2) + 4de\right](K - 1)P_s N$$

With K = 2 and N = 10 RPM, the eccentricity and the pump rate are calculated as 0.5 in. and 23.75 BPD. The results are presented in Fig. 4.36.

Example 4.6 Given the following information for a 5:4-lobe PCP: rotor diameter of 2 in., semicircle diameter of 0.5 in., stator pitch length of 12 in., pump speed of 100 RPM, clearance between the rotor and stator is 0.005 in., the channel depths $L_L = L_T = 0.065$ in. The pump is installed at the depth of 6000 ft. The production liquid has a viscosity of 500-cp and a specific gravity of 0.85. The wellhead pressure is maintained at 300 psi. The pump intake pressure is measured and given as 500 psi. Neglecting the frictional pressure drop inside the production tubing and assuming the ΔP_L and ΔP_T are the same and equal to 10 psi. Calculate:

a. The theoretical pump rate.
b. The longitudinal, transversal, and total slippages
c. The actual pump rate and the pump efficiency.

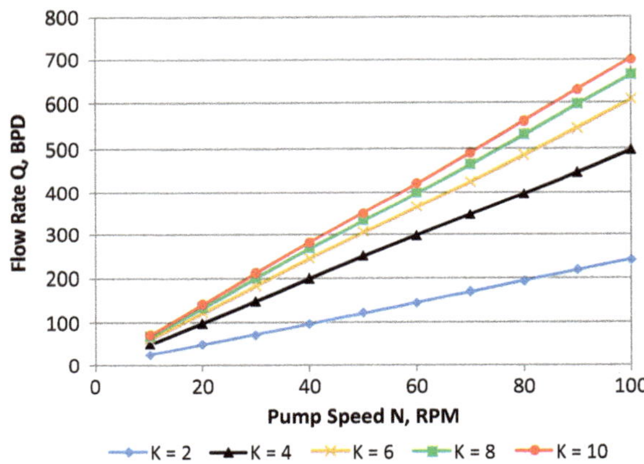

Fig. 4.36 Effect of pump speed and stator lobe number on the pump rate

Solution

a. Calculate the theoretical pump rate

Pump eccentricity

$$e = \frac{D_r - d}{2(K-1)} = \frac{2 - 0.5}{2(3-1)} = 0.1875\,\text{in.}$$

The theoretical pump rate:

$$Q_{theo} = 0.00433\left[2\pi e^2(K-2) + 4de\right](K-1)P_sN$$
$$Q_{theo} = 0.00433\left[2\pi \times 0.1875^2(5-2) + 4 \times 0.5 \times 0.1875\right](5-1)12 \times 100$$
$$Q_{theo} = 21.56\,GPM = 739\,BPD$$

b. Calculate the longitudinal, transversal, and total slippages

Recall the assumed geometries of the longitudinal and transversal slips as shown in Fig. 4.24.

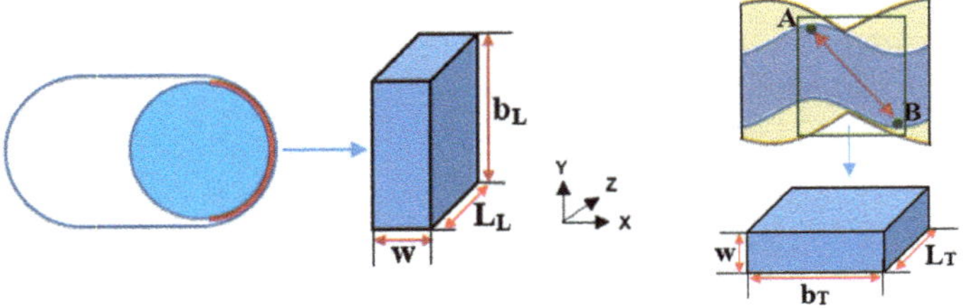

For simplicity, we will use SI unit for this calculation.

e = 0.1875 in. = 0.00476 m
d = 0.5 in. = 0.0127 m
w = 0.005 in. = 0.000127 m
$L_L = L_T$ = 0.065 in. = 0.001651 m
μ = 500 cp = 0.5 PaS
$\Delta P_L = \Delta P_T$ = 10 psi = 68,950 Pa

Applying Eqs. (4.5.38) and (4.5.42) to calculate the surface lengths of the longitudinal and transversal slips, b_L and b_T respectively.

$$b_L = \pi \times \frac{d}{2} = \pi \times \frac{0.0127}{2} = 0.785\,\text{in.} = 0.02\,\text{m}$$

$$b_T = \frac{2\pi}{K}\sqrt{2(K-1)^2 e^2 - \frac{2de(K-2)(K-1)}{\pi} + \frac{d^2(K-2)^2}{16} + \frac{P_s^2}{4\pi^2}}$$

$$b_T = 2.574\,\text{in.} = 0.065\,\text{m}$$

Applying Eqs. (4.5.43) and (4.5.44) for calculating the slippage in longitudinal and transversal directions:

$$S_L = \frac{b_L w^3 \Delta P_L}{2\mu L_L} = \frac{0.02 \times 0.000127^3 \times 68,950}{2 \times 0.5 \times 0.00165} = 1.7066 \times 10^{-6}\frac{\text{m}^3}{\text{s}} = 0.93\,BPD$$

$$S_T = \frac{b_T w^3 \Delta P_T}{2\mu L_T} = \frac{0.065 \times 0.000127^3 \times 68,950}{2 \times 0.5 \times 0.00165} = 5.5938 \times 10^{-6}\frac{\text{m}^3}{\text{s}} = 1.02\,BPD$$

Total slippage from Eq. (4.5.45):

$$S_{Total} = (K-1)(S_L + S_T) = 0.00541\frac{\text{m}^3}{\text{s}} = 4.06\,BPD$$

c. The actual pump rate and the pump efficiency

Actual pump rate:

$$Q_{Actual} = Q_{theo} - S_{Total} = 739.45 - 4.06 = 735.39\,BPD$$

Pump efficiency

$$\eta_{pump} = \frac{Q_{Actual}}{Q_{theo}}100 = \frac{735.39}{739.45}100 = 99.45\%$$

Please note that the actual pump rate is directly related to slippage. Slippage strongly depends on the pump clearance and the pressure difference across the pump. Under downhole conditions, the pump clearance is a function of fluid temperature, fluid pressure, and fluid properties. In other words, to have a better prediction of a PCP's performance, the correlation to predict the clearance under downhole conditions is needed. This is one of the ongoing research directions for improving the pump prediction.

Example 4.7 Using inputs from Example 4.6, calculate the actual pump rate using the dimensionless approach presented in Sect. 4.5.6.

Solution
Stator pitch length: Ps = 12 in. = 0.3048 m
Rotor speed: N = 100 RPM = 2 × π × 100 = 628.3 rad/min
Cross-sectional flowing area:

$$A_F = 2\pi e^2(K-2) + 4de = 1.26\,\text{in.}^2 = 0.000812\,\text{m}^2$$

Fluid density and viscosity:

$$\rho_f = 0.85 \times 8.3 = 7.05\,ppg = 850\frac{\text{kg}}{\text{m}^3}$$

μ = 500 cp = 0.5 PaS
Pump discharge and pump intake pressure:

$$P_d = P_{wh} + 0.052 \times \rho_f \times TVD = 300 + 0.052 \times 0.85 \times 8.3 \times 6{,}000 = 2,501\,\text{psi}$$

$$P_i = 500\,\text{psi}$$

Differential pressure across the pump:

$$\Delta P_{pump} = P_d - P_i = 2,001\,\text{psi} = 13,797,998\,\text{Pa}$$

Apply Eqs. (4.5.47), (4.5.48) to calculate the Euler and Reynolds dimensionless groups:

$$Eu = \frac{\Delta P_{pump}}{\rho_f P_s^2 N^2} = \frac{13,797,998}{850 \times 0.3048^2 \times 628.3^2} = 0.4426$$

$$Re = \frac{\rho_f P_s^2 N}{\mu_f} = \frac{850 \times 0.3048^2 628.3}{0.5} = 99,234$$

The actual pump rate can be estimated using Eq. (4.5.52):

$$Q_a = (P_s A_F N)\left[\frac{6.841}{\ln(Re)} + 1.521 - 0.291\ln(Eu)\right]^{1/3}$$

$$Q_a = (0.3048 \times 0.000812 \times 100)\left[\frac{6.841}{\ln(99,234)} + 1.521 - 0.291\ln(0.4426)\right]^{1/3}$$

$$Q_a = 0.033\,\frac{\text{m}^3}{\text{s}} = 17,888\,BPD$$

This example has revealed that the dimensionless approach is not applicable for high discharge pressure. This is understandable because this approach is based on the experimental data collected at the pump discharge pressure less than 500 psi.

References

1. Al-Safran E, Aql A, Nguyen T (2017) Analysis and prediction of fluid flow behavior in progressing cavity pumps. J Fluids Eng 139(121):102–111
2. Ba S, Pushkarev M, Anton K, Lijun S, Ling Ling Y (2016) Positive displacement motor modeling: skyrocketing the way we design, select, and operate mud motors. In: 183298-MS SPE conference paper
3. Baldenko D, Baldenko F, Gnoevykh A (2005) Single-screw hydraulic machines: volume 1 and 2. IRTs Gazprom
4. Beauquin J, Boireau C, Lemay L, Seince L (2005) Development status of a metal progressing cavity pump for heavy oil and hot production wells. In: SPE international thermal operations and heavy oil symposium, Alberta, Canada, November 2005
5. Becerra O, Mena ME (2007) Integrated analysis for PCP systems. In: SPE 107899 presented at the SPE Latin American and Caribbean petroleum engineering conference, Buenos Aires, Argentina, April 2007
6. Briggs PJ, Baron PR, Fulleylove RJ et al (1988) Development of heavy-oil reservoirs. J Pet Technol 40(2):206–214. SPE-15748-PA
7. Cholet H (1997) Progressing cavity pumps. Editions Technip, Paris
8. Cougar Drilling Solutions (2012) Motor operations handbook. Version 5.0
9. Gamboa J, Aurelio O, Sorelys E (2003) New approach for modeling progressive cavity pumps performance. In: SPE 84137 presented at the SPE annual technical conference and exhibition. Denver, Colorado, October 2003
10. Gamboa J, Olivet A, Iglesias J, Gonzalez P (2003) Understanding the performance of a progressive cavity pump with a metallic stator. In: Presented at the 20th international pump users symposium program. Houston, Texas, March 2003
11. Li J, Tudor R, Ginzburg L, Robello G, Xu H, Grigor C (1998) Evaluation and prediction of the performance of positive displacement motor. In: SPE international conference on horizontal well technology, Alberta, Canada, November 1998
12. Makohl F, Jurgens R (1986) Evolution and differences of directional and high-performance downhole motors. In: Presented at the IADC/SPE drilling conference held in Dallas, TX, February 1986—IADC/SPE 14742
13. Martinez AR (1987) The orinoco oil belt. J Petrol Geol 10:125, Venezuela
14. Meyer RF, Mitchell RW (1987) A perspective on heavy and extra heavy oil, natural bitumen, and shale oil. In: Paper presented at the 1987 twelfth world petroleum congresses, Houston
15. Mitchell R, Miska S (2011) Fundamentals of drilling engineering. Soc Petrol Eng 12. SPE Textbook Series

16. Moineau J (1932) "Pompe" Patent US No. 1 892 217, 27 December 1932
17. Nguyen TC, Tu H, Al-Safran E, Saasen A (2016) Simulation of single-phase liquid flow in progressing cavity pump. J Petrol Sci Eng 147:617–623
18. Nguyen K, Nguyen T, Al-Safran E (2019) Experimental and theoretical study on slippage effect of pcp performance. MLF59—Presented at the Middle East Artificial Lift Forum, Oman 2019
19. Nguyen TC, Al-Safran E, Saasen A, Nes, OM (2014) Modeling the design and performance of progressing cavity pump using 3-D vector approach. J Petrol Sci Eng 122:180–186
20. Nguyen T, Al-Safran E, Nguyen V (2018) Theoretical modeling of positive displacement motors performance. J Pet Sci Eng 166:188–197
21. Noble E, Dunn L (2011) Pressure distribution in progressing-cavity pumps: test results and implications for performance and run life. SPE 153944 Submitted to SPE for a reprint volume
22. Noonan S (2008) The progressing cavity pump operating envelope: you cannot expand what you don't understand. In: SPE international thermal operations and heavy oil symposium, Calgary, Canada, October 2008
23. Noonan S, Langer D, Klaczek W, Yip C (2013) Technical challenges and learnings from a high temperature metallic progressing cavity pump test. In SPE progressing cavity pumps conference, Calgary, Canada, August 2013
24. Paladino E, Lima J, Almeida R, Assmann B (2008) Computational modeling of the three-dimensional flow in a metallic stator progressing cavity pump. In: SPE 114110 presented at the SPE progressing cavity pump conference held in Houston, Texas, April 2008
25. Pessoa P, Paladino E, De Lima J (2009) A simplified model for the flow in a progressive cavity pump. In: Presented at the COBEM conference, Gramado, Brazil, November 2009
26. PetroWiki PEH (2015) Progressing cavity pumping systems. Modified on June 2015
27. Rassenfoss S (2013) New uses keep emerging for a deceptively simple pump. J Petrol Technol 65(10)
28. Roberts B, Mohr C (1972) Down-hole motors for improved drilling. J Petrol Technol (JPT) 3343:1484–1490
29. Robles J (2001) Another look to multilobe progressive cavity pump. In: SPE progressing cavity pump workshop, Puerto La Cruz, January 2001
30. Samuel R, Miska S (2003) Performance of positive displacement motor (PDM) operating on air. J Energy Res Technol 125:119–125
31. Samuel R, Miska S, Volk L (1997) Analytical study of the performance of positive displacement motor (PDM): modeling for incompressible fluid. In: Presented at the fifth latin American and caribbean petroleum conference and exhibition held in Rio de Janeiro, Brazil, September 1997—SPE 39026
32. Vetter G, Wirth W (1995) Understand progressing cavity pumps characteristics and avoid abrasive wear. In: Presented at the 12th international pump users symposium program. Houston, Texas

Chapter 5
Sucker Rod Pump

5.1 Fundamentals of Sucker Rod Pump

5.1.1 Introduction and Main Principles of Sucker Rod Pump

Sucker rod pump, rod pump, reciprocating pump or pump jack is different names of an oldest and most widely used artificial lift method for oil wells. There is a high percentage of oil wells in North America and other areas in the world that are vertical and produce less than 10 barrels of oil per day. These wells are called stripper wells and are commonly lifted with sucker rod pumps.

Rod pump uses rods to connect a downhole pump and a surface driving unit. The downhole pump is a positive displacement plunger pump type. The surface driving unit converts the rotational motion of a motor to the reciprocating motion via a mechanical linkage system. Figure 5.1 shows a basic rod pump system. The rotation from the prime mover will be transmitted to the gear reducer to reduce the speed of the prime mover. The counterbalance (sometimes called equalizer), crank, pitman arm, walking beam, and horsehead as shown in Fig. 5.1a are to convert the rotation motion to the reciprocating motion. This reciprocating motion is transferred to the polished rods via a rod string coming out from the horsehead. The polished rods are connected to the downhole rod string which drives the downhole pump plunger.

As the plunger is moving upward (upstroke), the traveling valve shown in Fig. 5.1b moves downward due to the liquid column above it and the standing valve starts to open allowing formation fluids to enter the working barrel. In other words, during the upstroke period, the liquid volume in the working barrel increases and the pressure in the working barrel decreases. The formation fluids are lifted up in the annulus between the rod string and the production tubing. When the plunger is moving downward (downstroke), the standing valve closes due to the piston effect and the traveling valve will open due to an increase in pressure of the fluid inside the working barrel. As the plunger is moving downward further, the fluids

T. Nguyen, *Artificial Lift Methods*, Petroleum Engineering,
https://doi.org/10.1007/978-3-030-40720-9_5

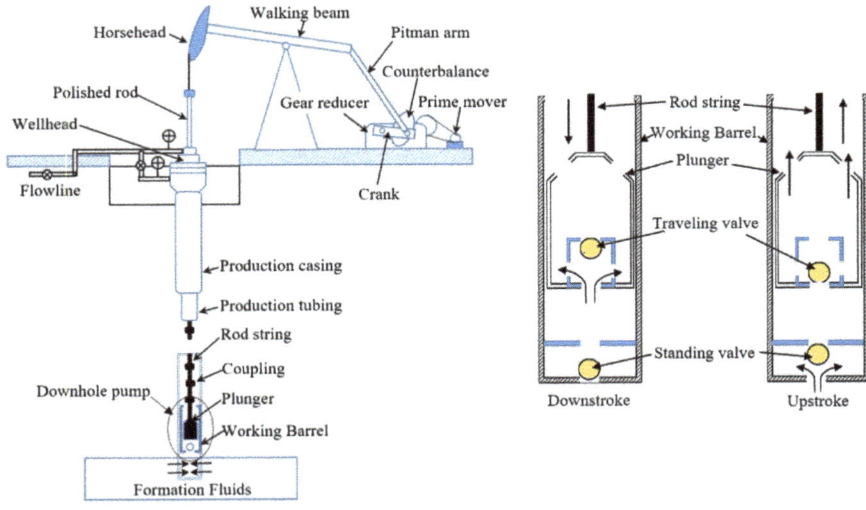

(a) Main components of a rod pump system (b) Positive displacement plunger pump

Fig. 5.1 A basic rod pump system

inside the working barrel between the traveling and standing valves pass the traveling valve and are stored in the upper part of the working barrel. This cycle keeps repeating as long as the rod pump system is operated and hence the surface production is intermittent; not continuous.

Readers should note that the main principle of how rod pump systems work to produce hydrocarbon is similar to that of gas lift, Electrical Submersible Pump (ESP), and Progressive Cavity Pump (PCP). All of these lift methods help wells to improve hydrocarbon production by reducing the flowing bottomhole pressure (reducing back pressure on the reservoir at the perforation). In more specific, gas lift system decreases the flowing bottomhole pressure by injecting gas into the fluids to reduce the hydrostatic pressure of the fluid column. ESP system converts kinetic energy to hydraulic energy to reduce the bottomhole pressure by lifting the fluid up. PCP and Rod pump systems reduce the bottomhole pressure by lifting the fluid up continuously and periodically, respectively using the pump differential pressure, ΔP_{pump}. Figure 5.2 shows the main principle of how rod pump helps to reduce the bottomhole pressure and to and hence to produce more liquid production.

Advantages of sucker rod pump can be listed as follows:

- Because rod pumping works similar to positive displacement pump, it can reduce the bottomhole pressure to a very low level and hence deplete the reservoir better.
- Rod pump systems are relatively simple in design, operation, and maintenance. Therefore, they are quite understood by the industry in terms of design and operation.

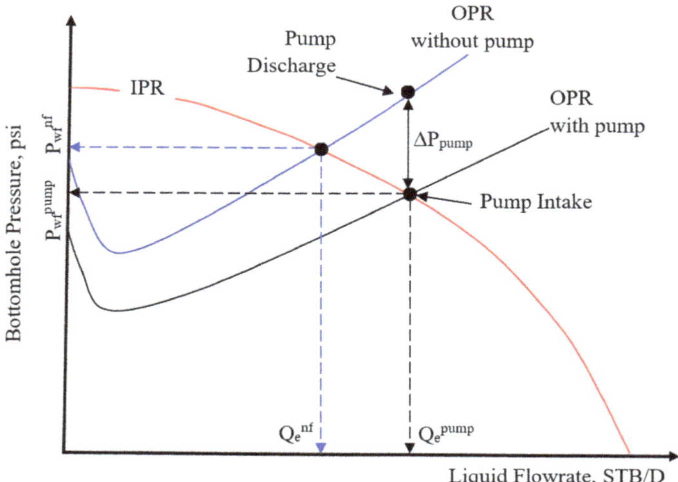

Fig. 5.2 Main working principle of a rod pump system

- Because rod pump systems are the most common lift methods, replacement parts are normally widely available, compatible and interchangeable.
- The operating cost is quite competitive in comparing to other lift methods.

Disadvantages of sucker rod pump are as follows:

- Wells with medium or high Dog Leg Severity (DLS) (normally greater than 5 °/100ft) are not suitable for rod pump systems. The mechanical friction between rod string and production tubing in high DLS wells may cause rod/coupling failures [7], [9], and [11]. In addition, the system requires more surface power to operate due to this friction.
- Rod pump systems are only applicable for wells with shallow and medium depths. The pump depth is normally limited by the strength of the rod string and coupling materials.
- Like any positive displacement pumps, rod pump systems do not operate well with the presence of gas and solids.
- Due to its high surface space requirement, rod pump system is mainly applicable for onshore wells.

5.1.2 API Rod Pump Classification

5.1.2.1 API Rod Pump Unit Classification

API [2], [3], and [4] classifies rod pump units into three main types: conventional unit (class I lever unit), air balanced unit, and Mark II unit based on the position of the fulcrum and the design of the counterbalance.

Conventional rod pump unit (or the class I lever unit): is the oldest and most common rod pump units used in the field because it is relatively simple in operation, low maintenance requirements, and adaptable to a wide range of field applications. This unit has the fulcrum (Samson post bearing) located at the middle of the walking beam as shown in Fig. 5.3.

Air balanced rod pump unit (or class III lever unit): This unit has the fulcrum located in the rear of the walking beam. The counterbalance in the conventional pump unit is replaced by a piston and air cylinder to counterbalance the well load by adjusting air pressure using a pressure switch in the cylinder as presented in Fig. 5.4.

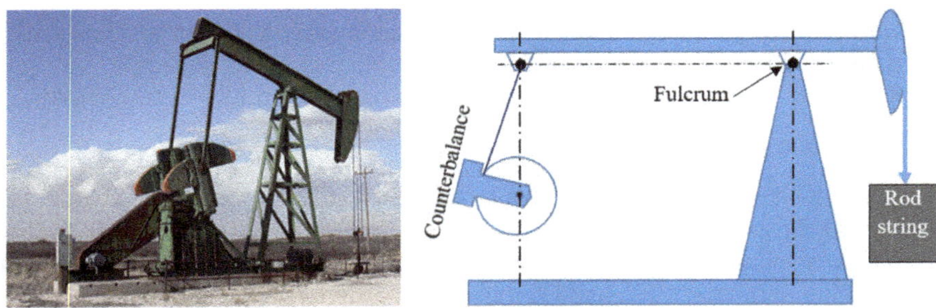

Fig. 5.3 Conventional rod pump unit or class I lever unit. Photo from: https://www.flickr.com/photos/stuartwildlife/2414763009

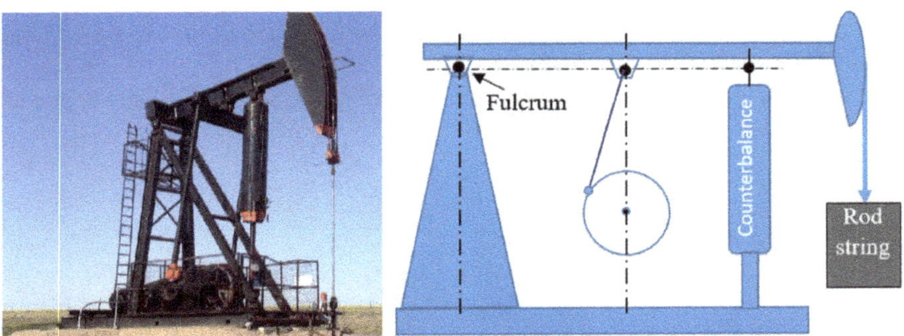

Fig. 5.4 Air balanced unit or class III lever unit. Photo from: http://curvetube.com/Lufkin_Beam_pumping_unit/aMX3ord9GB0.video

The most important benefit of this unit is that the piston and air cylinder allow users to control the counterbalance more accurately than using the counterweights used in the conventional unit. Therefore, the air balanced unit is more energy efficient than the conventional pumping unit. In addition, the air balanced unit is preferable for wells that need longer pump strokes.

Mark II rod pump unit: This unit has the fulcrum located in the rear of the walking beam similar to the air balanced unit. In addition, the Mark-II unit has the cross yoke bearing located very close to the horsehead as shown in Fig. 5.5. The crank has an angular offset to produce an out-of-phase condition between the torque exerted by the well load and the torque exerted by the counterbalance weights. This unique design reduces torque peaks that are commonly encountered with conventional pump units.

A typical rod pump designation consists of four parts:

- The first part is a letter either A, B, C, or M. A is for air balanced, B is for beam balanced, C is for conventional, and M is for mark II unitorque.
- The second part is normally a number following by a letter. The number representing the peak torque rating in thousands of inch-pounds. The letter represents single or double reduction gear reducer.
- The third part is also a number representing the polished rod load rating in hundreds of pounds.
- The last part is a number designating the stroke length in inches.

Example: a rod pump unit with a designation of **C-140D-117-64** means that this is a conventional pump unit; 140,000 in. lbf of peak torque and double reduction gear reducer; polished rod load rating of 11,700 lbf; and stroke length of 64 in.

5.1.2.2 Downhole Plunger Pump Classification

API [3] classifies downhole plunger pumps into two main categories, namely tubing pump and insert pump (or rod pump) as shown in Fig. 5.6a and b, respectively.

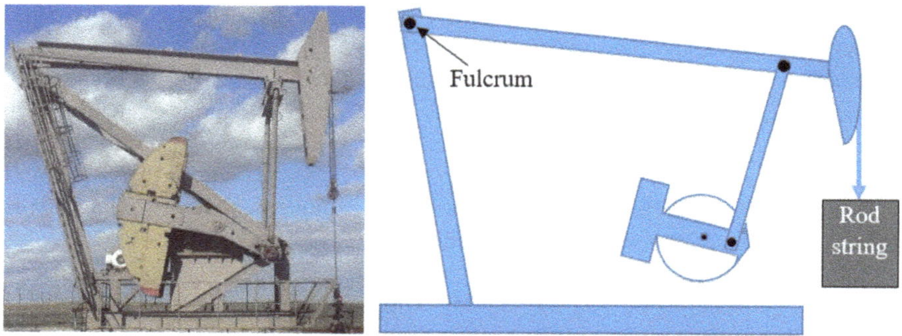

Fig. 5.5 Mark II rod pump units. Photo from: https://en.wikipedia.org/wiki/Lufkin_Industries

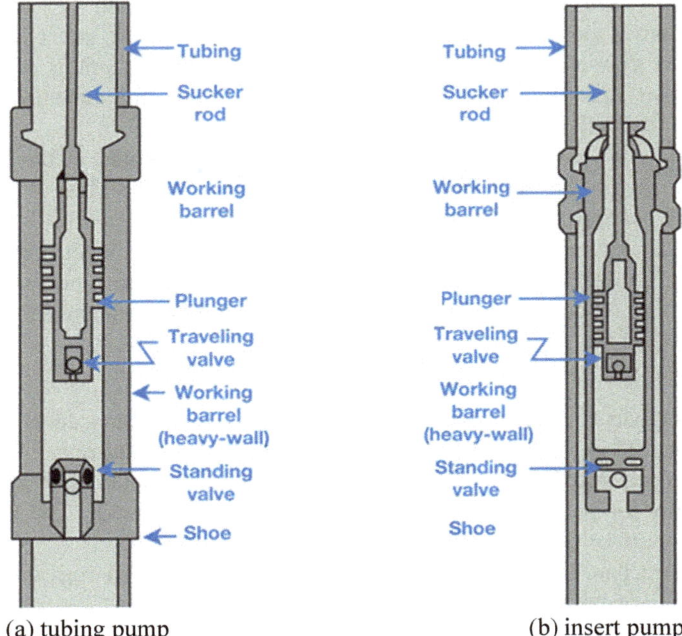

Fig. 5.6 Downhole plunger pump classification—stationary working barrel

Please note that these common types have stationary working barrel and moving plunger.

For tubing pumps, as the name states, the working barrel is an integral part of the tubing. This design allows bigger plunger diameters to be used inside the working barrel in comparison with plungers used in rod pumps. Therefore, tubing pumps normally offer more liquid production rate compared to insert pumps when the tubing diameter and the surface stroke length are the same. In addition, tubing pumps are stronger and more durable than insert pumps during the pump operations. The only drawback is that the whole tubing string must be pulled to repair, service, or replace the plunger pump.

Insert pumps, on the other hands, has the working barrel inserted and secured inside the tubing and run on sucker rods rather than on tubing. Therefore, the plunger diameter is smaller when compared to that of in tubing pump. This will limit the amount of fluid produced per stroke length. The benefit of the insert pump is that the working barrel and the plunger can be removed easily using the rods for repair or service without the need of pulling the entire tubing.

Sometimes, rod pumps are designed with moving working barrel and stationary plunger. The principle of how the moving working barrel pump works is similar to that of the stationary working barrel pump. The upstroke and downstroke during the operation of moving working barrel pumps are shown in Fig. 5.7a, b.

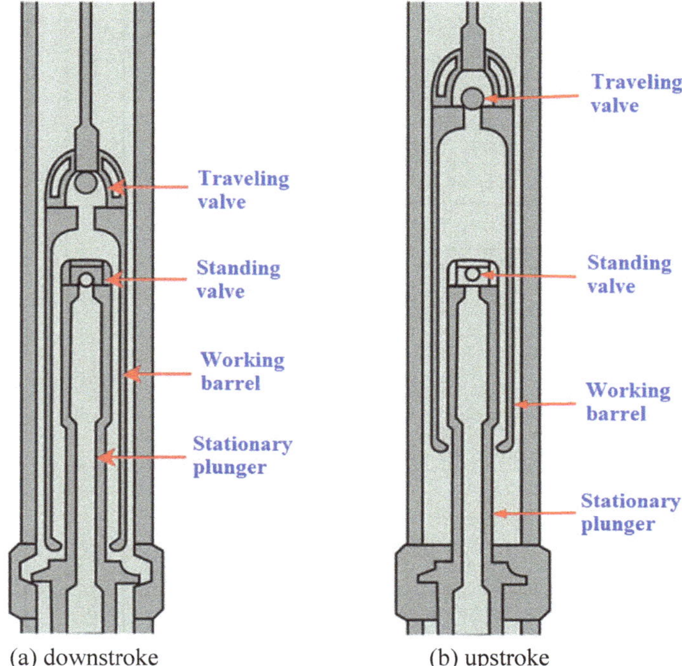

Fig. 5.7 Downhole plunger pump classification—moving working barrel

API specification 11AX has developed standard designations for a complete downhole plunger pump. The designation consists of 12 characters, alphanumeric code as XX-XXX-X-X-X-X-X-X-X and explained in Fig. 5.8.

5.1.3 Rod String

Rod string consists of sucker rods joined together by threaded couplings as shown in Fig. 5.9. Rod string is used to transmit the reciprocating motion from the surface driving unit to the downhole plunger pump. The sucker rods are standardized by API in either 25 or 30 ft with outside diameter of 1/2, 5/8, 3/4, and 1-1/8 in. The API threaded couplings are about 4 in. in length. To meet the exact length of a rod string in a well, shorter length rods called pony rods are used. API specification 11B provides dimensions and material standards for sucker rods and couplings.

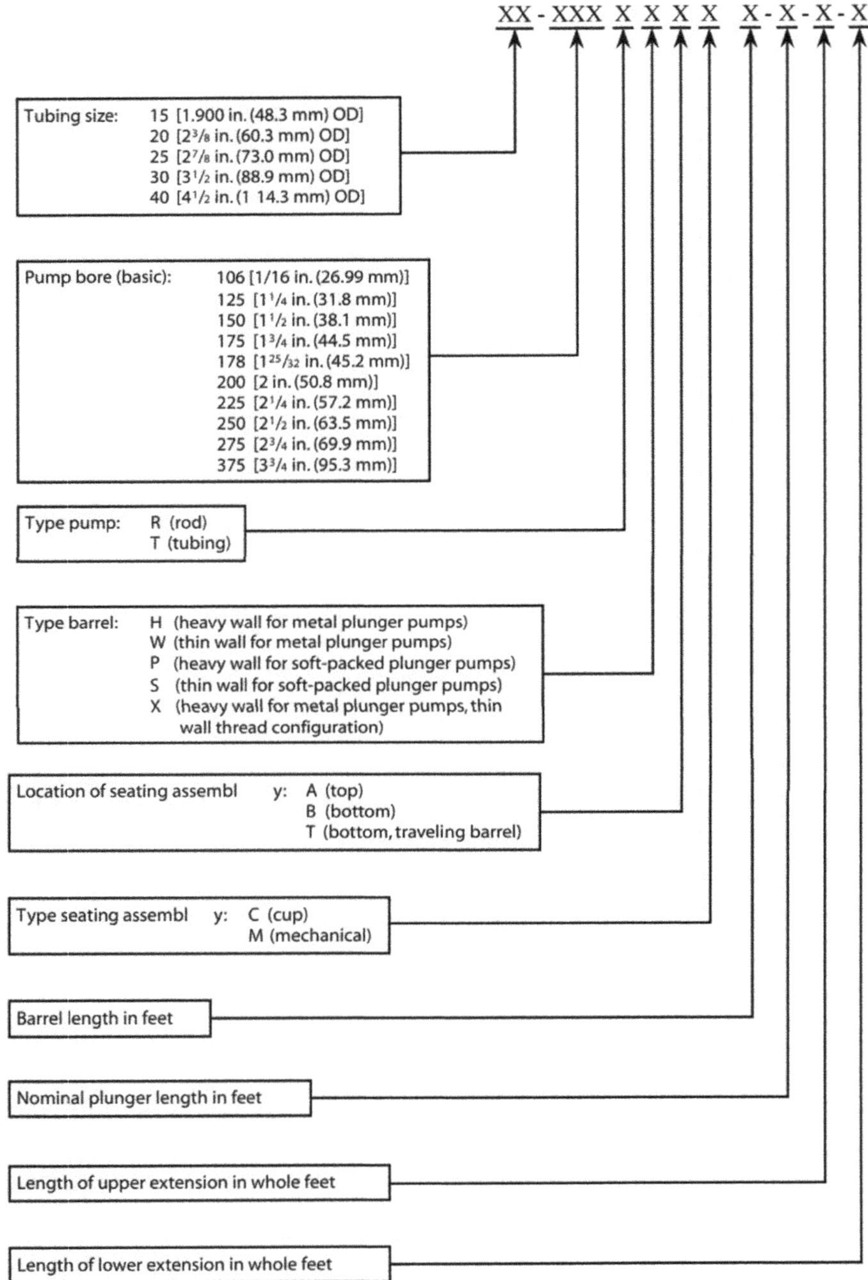

Fig. 5.8 Designation of downhole plunger rod pump

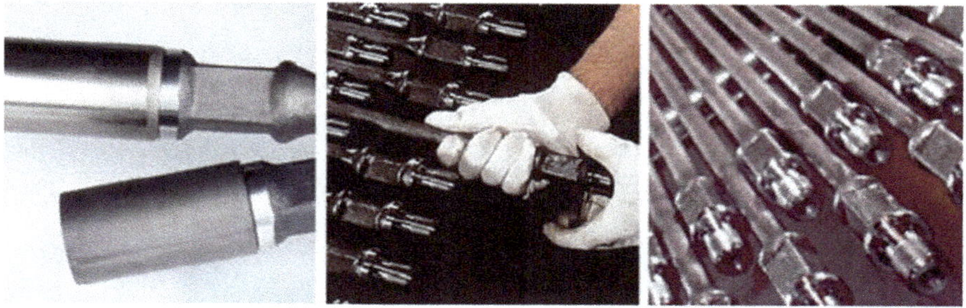

Fig. 5.9 API sucker rods and threaded couplings

5.2 Pumping Motion

Understand pumping motion is important because the rotational and reciprocating motions control polished rod velocities and accelerations. The rod velocity and rod acceleration will have a great impact on the rod string load, the durability of rod, and the downhole pump displacement. As mentioned, the downhole plunger pump unit is operated due to the reciprocating motion of the rod string and the rotational motion of the prime mover. The complete motion of a rod pump system can be understood by visualizing two basic theoretical motions: the Simple Harmonic Motion (SHM) and the Crank and Pitman Motion (CPM).

5.2.1 Simple Harmonic Motion (SHM)

SHM is a type of periodic motion where the restoring force is directly proportional to the displacement. Vertical motion of an object installed at the end of a spring is an example of the SHM as shown in Fig. 5.10a. The restoring force, F, is proportional to the displacement X by a constant k; or $F = -kX$. Another common SHM is the rotation of an object, P, around its center, O, at a constant angular velocity, ω, as shown in Fig. 5.10b. Let P′ is the projection of P to the y-axis. As the object, P, travels from B to C (first quadrant), P′ will accelerate and move vertically from point B to point O. P′ reaches its maximum velocity at the center point when P arrives to point C. The magnitude of the P′ velocity depends on the displacement angle θ. As P′ travels from C to D, the P′ velocity decelerates and reaches to zero when P arrives to point D. This completes a first half motion. The velocity performance of P′ in the second half motion is the same as that of in the first half motion.

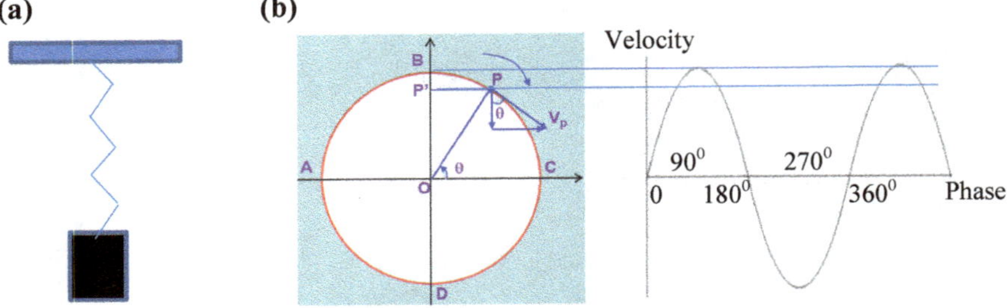

Fig. 5.10 Examples of simple harmonic motion

5.2.2 *Crank and Pitman Motion (CPM)*

The second motion that we are going to consider is the CPM. This motion is very similar to the SHM and applied to rod pump system. The main difference between these two motions is that the vertical motion of point P′ in the SHM shown in Fig. 5.10b is generated using a pitman as shown in Fig. 5.11. The upper pitman connection (V) called the cross yoke or equalizer, is assumed to move vertically up and down a straight line. In reality, the connection (V) does not move in a perfect vertical direction even though the crank is rotated with a constant angular velocity. This is due to the angularity of the pitman-crank mechanism. The smaller the angle between crank and pitman when the crank is horizontal, the greater the divergence of the motion of the equalizer (V) from that of the SHM by point P′.

Note that when crank pin P moves around the circle from A to B to C, the equalizer (V) moves vertically from a′ to b′ and back to c′. This distance is greater than the distance from c′ to d′ and back to a′ when (P) moves from C to D and back to A. This is because the connection (V) travel is a function of the vertical components of the crank OP and the pitman PV as shown in Fig. 5.11.

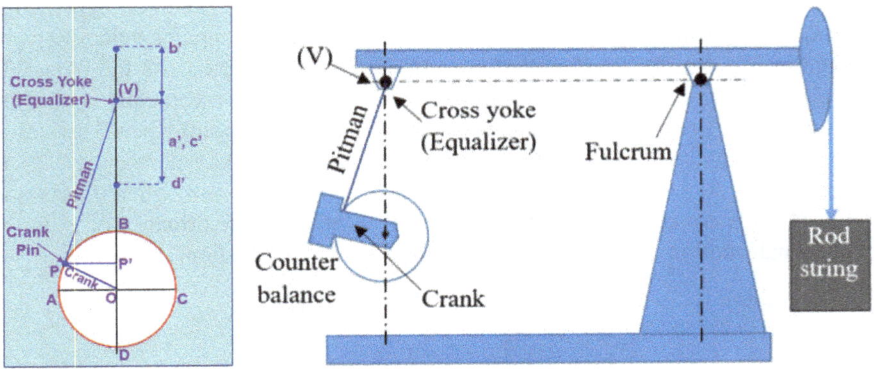

Fig. 5.11 Crank and pitman motion

5.3 Basic Rod Pump Design

The basic rod pump design consists of the calculations of the downhole pump rate, surface pump rate, surface torque, minimum and maximum polished rod loads.

5.3.1 Theoretical and Actual Pump Displacement (Downhole Pump Rate)

The key behind the selection of downhole pump is that for a specific pump depth and specific liquid production rate, there is an optimal size of the working barrel and plunger at which the stroke length and the polished rod load will be optimal.

For a constant desired liquid rate, if the plunger is too large (high plunger cross-sectional area), there will be high loads imposed upon the equipment. In addition, when the plunger diameter is large, the stroke length will be smaller to produce the same amount of liquid resulting in inefficient operations. On the other hand, if the plunger is too small, pumping speeds become too high and the increased acceleration effects can result in increased peak loads on the equipment.

The total theoretical pump displacement can be determined as follows:

$$PD = A_p\left(\text{in.}^2\right) \times S_p\left(\frac{\text{in.}}{stroke}\right) \times N\left(\frac{stroke}{\text{min}}\right) \times \frac{1440\,\text{min/D}}{9702\,\frac{\text{in.}^3}{\text{bbl}}} \tag{5.3.1}$$

$$PD = 0.1484 \times A_p \times S_p \times N \tag{5.3.2}$$

where PD is the pump displacement in barrel per day (BPD); A_p is the cross-sectional area of the pump plunger in.^{2}; S_p is the effective plunger stroke in inch; and N is the pump speed in stroke/min.

From Eq. (5.3.2), one can introduce a constant K to represent for a given pump with a given plunger diameter.

$$K = 0.1484 \times A_p \tag{5.3.3}$$

Combining Eqs. (5.3.2) and (5.3.3) gives:

$$PD = K \times S_p \times N \tag{5.3.4}$$

Common values of K are given in Table 5.1.

The actual production rate at the surface, Q, may be less than the total theoretical pump displacement because of the volumetric efficiency of the pump.

Table 5.1 Common values of K calculated using Eq. (5.3.3)

Plunger dia. (In.)	Area of plunger (A_p, in.^{2})	Constant (K)	Plunger dia. (In.)	Area of plunger (A_p, in.^{2})	Constant (K)
5/8	0.307	0.046	1 ¾	2.405	0.357
3/4	0.442	0.066	1 25/32	2.488	0.370
15/16	0.690	0.102	2	3.142	0.466
1	0.785	0.117	2 ¼	3.976	0.590
1 1/16	0.886	0.132	2 ½	4.909	0.728
1 1/8	0.994	0.148	2 ¾	5.940	0.881
1 ¼	1.227	0.182	3 ¾	11.045	1.640
1 ½	1.767	0.262	4 ¾	17.721	2.630

$$Q = E_{\upsilon} \times PD \tag{5.3.5}$$

The volumetric efficiency depends on the effective plunger stroke, mechanical opening and closing of the traveling and standing valves, presence of gas and solid. In practice, this efficiency varies from 70–80%. The displacement calculations presented in this section are applicable for conventional and unconventional pumping units.

5.3.2 *Calculation of Polished Rod Loads*

Surface loads on the polished rods change from minimum value (downstroke) to maximum value (upstroke) as the rod string completes a cycle. The surface polished rod loads can be estimated using a simple force analysis acting on the rod string. Assuming a uniform rod string is reciprocating in a straight inclined tubing and assuming there are no centralizers used along the rod string. Let consider a small rod string element or Free Body Diagram (FBD), which has a unit length of dL. Forces acting on the FBD consist of tension at the lower end of the FBD, F, tension at the upper end of the FBD, F + dF, weight of the FBD in fluid or buoyant weight, $w_b dL$, contact force, $w_c dL$, and mechanical friction force, $\mu_m w_c dL$. All the forces acting on the FBD are shown in Fig. 5.12.

In the y-direction under steady state motion:

$$w_c dL - w_b dL \sin \varphi = 0 \tag{5.3.6}$$

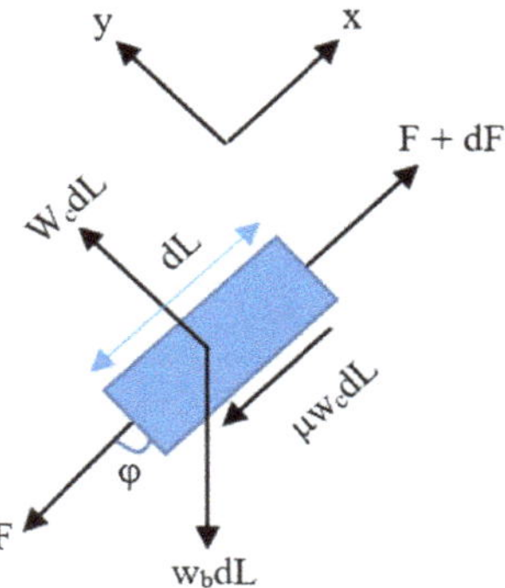

Fig. 5.12 Forces acting on the FBD

In the x-direction under steady state motion:

$$(F + dF) - F - w_b dL \sin \varphi - \mu_m w_c dL = 0 \tag{5.3.7}$$

Combining Eqs. (5.3.6) and (5.3.7) gives:

$$dF = w_b \cos \varphi dL \pm \mu_m w_b \sin \varphi dL \tag{5.3.8}$$

Integrating Eq. (5.3.8) gives an equation to calculate for the Polished Rod Load, PRL, at the surface:

$$PRL = w_b \cos \varphi L \pm \mu_m w_b \sin \varphi L \tag{5.3.9}$$

where μ_m is the mechanical friction factor and L is the total length of the rod string. Please note that Eq. (5.3.9) neglects the friction due to the fluid flow in the annulus between the rod string and the production tubing. The positive and negative signs are used for upstroke and downstroke, respectively. The weight of rod string in fluid is calculated as follows:

$$w_b = w_r \left(1 - \frac{\rho_f}{\rho_s}\right) = w_r \times BF = w_r (1 - 0.127 \times SG) \tag{5.3.10}$$

where w_b and w_r are the weight per unit length of the rod string in fluid and in air; ρ_f and ρ_s are the densities of fluid and steel used to make the rods, respectively; BF is the buoyant factor; and SG = ρ_f/ρ_{water} is the fluid specific gravity. The density of steel is approximated as 7874 kg/m^3 = 492 lbm/ft^3. For a perfect vertical well, there is no mechanical friction and hence Eq. (5.3.9) becomes:

$$PRL = w_b L \tag{5.3.11}$$

Equation (5.3.11) tells us that if there is no contact between the rod string the production tubing and neglecting the friction due to the fluid flow inside the annulus, the polished PRL is the same as the total weight of the rod string in the fluid. In reality, rod string is reciprocating following the crank and pitman motion where the PRL values change from minimum load to maximum load and back to minimum load. In other words, there must be an additional force due to the acceleration, F_a, adding to Eq. (5.3.9) or Eq. (5.3.11) to take care of the inertia effects. By definition F_a is calculated as follows:

$$F_a = ma = m\frac{dv}{dt} \tag{5.3.12}$$

If S is the polished rod stroke length in inch and N is the pump speed in stroke per minute, Mills introduced an equation to calculate for the acceleration factor as follows:

$$\alpha = \frac{SN^2}{70{,}500} \tag{5.3.13}$$

where S is the stroke length in inches/stroke and N is the pump speed in stroke per minute.

The peak polished rod load, PPRL, for the conventional pumping units can now be calculated as follows:

$$PPRL = PRL + w_r L\alpha = w_b L + w_r L\alpha \tag{5.3.14}$$

During the upstroke, there will be an additional force due to the fluid column inside the tubing acting downward on the donut area formed by the plunger and the rod cross-sectional areas as shown in Fig. 5.13. Please note that the pressure at the bottom of the plunger, F_{bottom}, during the upstroke is much smaller than that of at the top of the plunger. The final equation for calculating the peak polished rod load is given:

$$PPRL = w_b L + w_r L\alpha + F_o \tag{5.3.15}$$

$$F_o = 0.433 \times L \times SG \times \left(A_p - A_r\right) \tag{5.3.16}$$

Because A_r is much smaller than A_p, API suggests to disregard A_r. Equation (5.3.16) becomes:

$$F_o = 0.433 \times L \times SG \times A_p \tag{5.3.17}$$

where A_p is the cross-sectional area of the plunger in in.^{2}. Please note that 0.433 is the pressure gradient of water in psi/ft. 0.433 × L × SG is the pressure of oil at the pump depth.

During the downstroke, there is a fluid communication at the top and the bottom of the plunger. The forces acting on the top and bottom of the plunger due to the fluid hydrostatic pressure are very similar and in opposite directions. Therefore they will cancel out from each other as shown in Fig. 5.14.

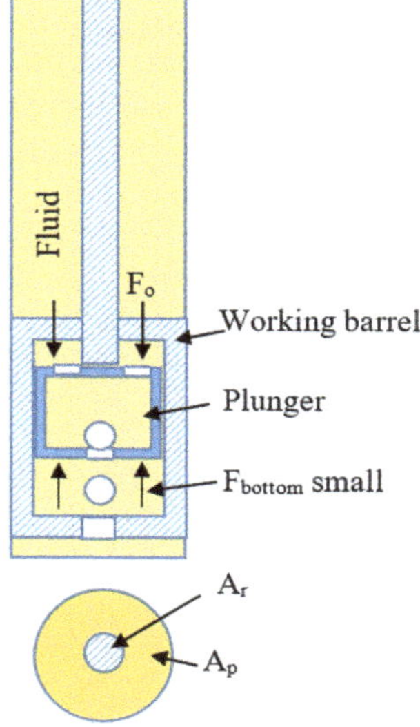

Fig. 5.13 Fluid pressure acting on the plunger during the upstroke

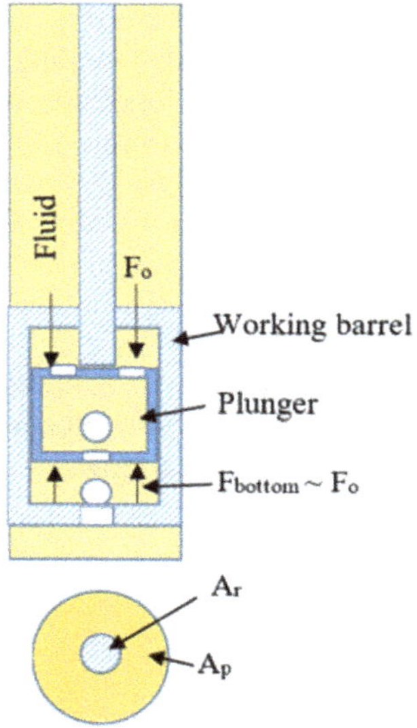

Fig. 5.14 Fluid pressure acting on the plunger during the downstroke

The final equation for calculating the minimum polished rod load, MPRL, for the conventional rod pumping units is calculated as follows:

$$MPRL = w_b L - w_r L \alpha \tag{5.3.18}$$

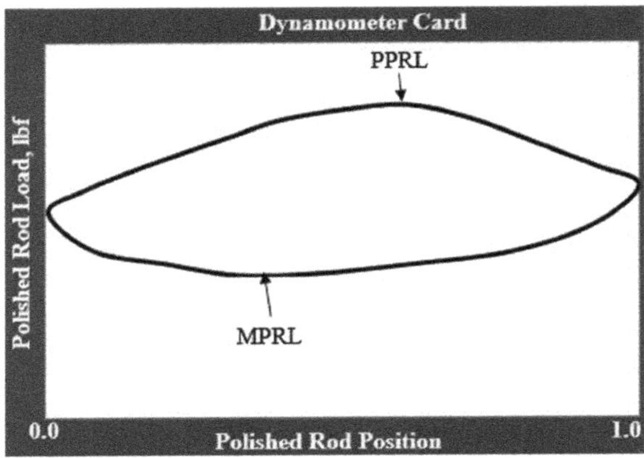

Fig. 5.15 A typical surface and downhole loads from a dynamometer card

Combining Eqs. (5.3.18) and (5.3.10) gives:

$$MPRL = w_r L(1 - \alpha - 0.127 \times SG) \tag{5.3.19}$$

A typical surface and downhole loads change when the rod string completes a cycle is presented in Fig. 5.15.

For air balance pumping units, the industry accepts that these units generate only 70% of the acceleration compared to the conventional units. The PPRL and MPRL are modified as follows:

$$PPRL = w_b L + 0.7 w_r L\alpha + F_o \tag{5.3.20}$$

$$MPRL = w_b L - 1.3 w_r L\alpha \tag{5.3.21}$$

For Mark II pumping units, the industry accepts that these units generate only 60% of the acceleration compared to the conventional units. The PPRL and MPRL are modified as follows:

$$PPRL = w_b L + 0.6 w_r L\alpha + F_o \tag{5.3.22}$$

$$MPRL = w_b L - 1.4 w_r L\alpha \tag{5.3.23}$$

5.3.3 Calculations of Counterbalance

Counterbalance is a weight that balances another load. In other words, counterbalance is a force used to offset an opposing force. Assuming there is no counterbalance used in rod pump systems. During the upstroke, the prime mover must provide enough torque (or energy) to pull the entire rod string upward. During the downstroke, the weight of the rod string is normally more than enough for the rod string to move downward by itself. Therefore, the prime mover again has to provide high torque to hold this downward movement of the rod string constant. According to energy perspective, the system is being operated in a very inefficient manner. To store the energy lost during the downstroke and use it during the upstroke, a counterbalance is used in a rod pump system. By storing this lost energy, the use of a counter balance can (1) reduce the size of the prime mover and the gear reducer; (2) reduce the peak polished rod load; (3) prolong the life of the surface equipment, especially the prime mover.

To achieve an optimal weight of the counterbalance can be a challenging task. The actual or optimal counterbalance depends on the geometry of a specific pump unit (position of the counterbalance on the walking crank, position of the fulcrum, length of the crank), the stroke length. However, based on field experience, the industry accepts that the ideal counterbalance can be approximated by taking the average of the PPRL and the MPRL. Combining Eqs. (5.3.15) and (5.3.18) gives an estimation of the ideal counterbalance:

$$CB = \frac{PPRL + MPRL}{2} = w_b L + 0.5F_o \tag{5.3.24}$$

In other words, the approximate ideal counterbalance is equal to the buoyant weight of the rod (rod of the rod in the fluid) plus half the weight of the fluid at the pump depth. If the counterbalance is designed following Eq. (5.3.24) then the actual amount of force the prime mover needs to overcome during the upstroke and downstroke, respectively, can be estimated as follows:

$$PPRL - CB = (w_b L + w_r L\alpha + F_o) - (w_b L + 0.5F_o) = w_r L\alpha + 0.5F_o \tag{5.3.25}$$

$$CB - MPRL = (w_b L + 0.5F_o) - (w_b L - w_r L\alpha) = w_r L\alpha + 0.5F_o \tag{5.3.26}$$

From Eqs. (5.3.25) and (5.3.26), one can recognize that the work requirements of the prime mover will be approximately equal if the counterbalance is designed using this principle. The work requirements of the prime mover during the upstroke and downstroke are considerably less than half the work required for an uncounterbalanced system. In addition, the design with counterbalanced system also helps to stabilize the rod pump system, especially the surface equipment.

The counterbalance calculations presented in this section are applicable for conventional and unconventional pumping units.

5.3.4 Surface Torque Calculation

In general, torque is defined as a force, F, acting at the end of a lever arm, multiplied by the length of the arm, tending to produce rotation around the axis of

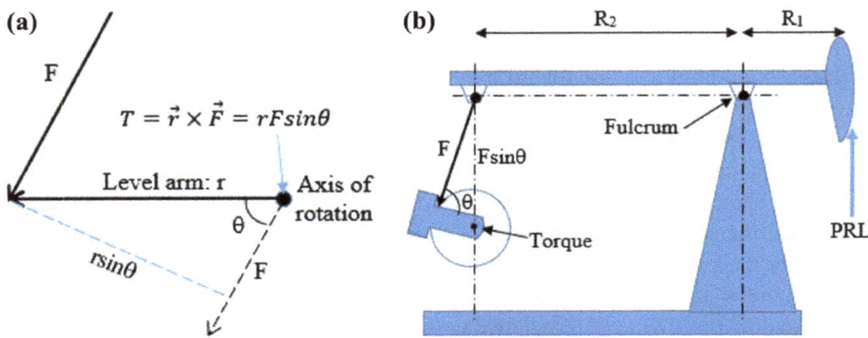

Fig. 5.16 **a** Demonstration of torque, **b** torque in a rod pump unit

rotation following the cross product definition as shown in Fig. 5.16a. The level arm in a conventional rod pump unit is the crank; F represents for the pitman. Torque at the shaft of the gear reducer as shown in Fig. 5.16b is calculated as follows:

$$T = \vec{r} \times \vec{F} = rF\sin\theta \tag{5.3.27}$$

Applying the definition of moment to the pump system during the upstroke as shown in Fig. 5.16b gives:

$$(F\sin\theta)R_2 = (PPRL - CB) \times R_1 \tag{5.3.28}$$

$$F = \frac{PPRL - CB}{\sin\theta}\frac{R_1}{R_2} \tag{5.3.29}$$

A typical recorded torque at surface is presented in Fig. 5.17.

The maximum torque or peak torque, PT, generally occurs twice during each revolution of the crank (one complete cycle of the polished rods) and near the middle of the stroke (S/2). Assuming the θ is 90° and $R_1 = R_2$ in Fig. 5.16b, combining Eqs. (5.3.27) and (5.3.29) gives an equation for calculating the PT during the upstroke as follows:

$$PT_u = (PPRL - CB)\frac{S}{2} \tag{5.3.30}$$

Similarly, one can drive an equation for calculating the PT during the downstroke using the same assumptions above.

$$PT_d = (CB - MPRL)\frac{S}{2} \tag{5.3.31}$$

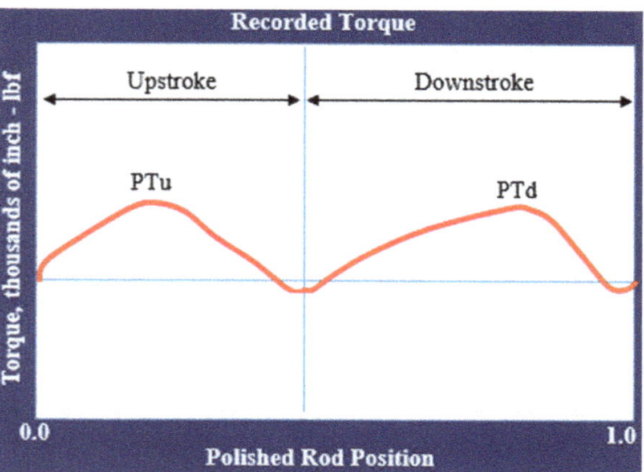

Fig. 5.17 Typical recorded torque at surface

Normally, the PT during the upstroke is slightly higher than the PT during the downstroke. When designing and selecting the gear reducer, the PT during the upstroke should be used for safety reason. Assuming the mechanical friction loss due to friction in the structural bearing is 7%, Eq. (5.3.30) becomes:

$$PT = \frac{(PPRL - CB)MTF}{0.93} \tag{5.3.32a}$$

where MTF is the maximum torque factor and depends on the stroke length. The values of the MTF are given in Table 5.2a.

Equation (5.3.32a) is applicable only for the conventional unit. For the unconventional unit such as air balanced and Mark II units, the peak torque is calculated as follow:

$$PT = \frac{0.5(PPRL \times MTF_u - MPRL \times MTF_d)}{0.93} \tag{5.3.32b}$$

where the values of MTF_u and MTF_d are the maximum torque factor for upstroke and downstroke and given in Table 5.2b.

Table 5.2a Values of the MTF

Stroke (in.)	MTF	Stroke (in.)	MTF
16	8.5	64	34
24	13	74	39
30	16	86	45
36	19	100	52
42	22	120	63
48	26	144	75
54	29	168	87

Table 5.2b Approximation of maximum torque factor for unconventional units

Stroke (in.)	MTF_u	MTF_d
64	29	37
74	34	43
86	39	51
100	47	57
120	55	71
144	66	88
168	79	102

5.3.5 *Calculation of Nameplate Motor Horsepower*

There are two common types of prime mover used in the field: internal combustion engines and electric motors. The internal combustion engines are common in places where electric is not available. However, if electric is available, electric motors are preferable due to its cost effective, less noise, and more reliable.

The total nameplate motor horsepower can be calculated as follows:

$$HP_{np} = \frac{PRHP \times CLF}{\eta} \tag{5.3.33}$$

where PRHP is the polished rod horsepower per stroke; CLF is the cyclic load factor; and η is the efficiency of the pumping system.

The CLF is the additional power needed to handle the cyclic pumping load. CLF values for different pump units are given by API and presented in Table 5.3.

The PRHP is determined as

$$PRHP = \frac{PT \times N}{63{,}025} \tag{5.3.34}$$

where PT is the peak torque in in-lbf and N is the pump speed in SPM.

Table 5.3 CLF values for different pumping units

Unit type	Prime mover type	CLF value
Conventional and air balanced	NEMA "D" electric motors and slow speed engines	1.375
	NEMA "C" electric motors and multi-cylinder engines	1.897
Mark II	NEMA "D" electric motors and slow speed engines	1.10
	NEMA "C" electric motors and multi-cylinder engines	1.517

5.4 API Recommended Design Procedure

Because of the nature of pump design, different pump units have different geometries and different operating conditions. Therefore, API tried to standardize the design of the pumping units by introducing a new design method based on dimensionless groups. API adopted the experimental work conducted at the Midwest Research Institute (1964) on sucker rod pumping and expanded to the API design procedure described in API RP 11L [1]. The work at the Midwest Research Institute consisted of numerous analog computer simulations over a wide range of pump operating conditions. The results were correlated based on seven dimensionless variables: (1) dimensionless pump speed; (2) dimensionless fluid rod stretch; (3) dimensionless upstroke rod stretch; (4) dimensionless downstroke rod stretch; (5) dimensionless plunger stroke;

Dimensionless pump speed variable may have two forms: N/N_o or N/N'_o. N/N_o is the pump speed in strokes per minute (N), divided by the natural frequency of the string (N_o). N/N'_o is equal to $(N/N_o)/F_c$, where F_c is the frequency factor of the rod string. F_c is equal to 1.0 for non-tapered rod strings, and is greater than 1.0 for tapered rod strings. If a rod string is non-tapered, then $N/N_o = N/N'_o$. The values of F_c are presented in Table 5.4.

The natural frequency of the string, N_o, is calculated as follows:

$$N_o = \frac{V_{SS}}{4L} \tag{5.4.1}$$

where V_{SS} is the speed of sound in steel rods (approximately 980,000 ft/min); L is the rod length in ft. The dimensionless pump speed becomes:

$$\frac{N}{N_o} = \frac{NL}{245{,}000} \tag{5.4.2}$$

$$\frac{N}{N'_o} = \frac{NL}{245{,}000 F_c} \tag{5.4.3}$$

Dimensionless fluid rod stretch, $\frac{F_o}{Sk_r}$: is defined as the rod stretch caused by the static fluid load, F_o, and given as a fraction of the polished rod stroke length, Sk_r. Fo is the static fluid load imposed on the rod string at pump depth; S is the polished rod stroke length in inches; and k_r is the spring constant of the rod string calculated using Eq. (5.4.9). Sk_r represents the load in pound required to stretch the rod string the length of the polished rod stroke. If $\frac{F_o}{Sk_r} = 1$ then the rod string will stretch an amount equal to the length of the polished rod.

Table 5.4 Frequency factor, Fc, for different rod numbers

Rod no.	Plunger diameter (in.)	Rod weight (lb/ft)	Frequency factor	Elastic constant (in/lb ft)	Rod no.	Plunger diameter (in.)	Rod weight (lb/ft)	Frequency factor	Elastic constant (in/lb ft)
44	All	0.726	1.000	1.99E−06	65	1.06	1.307	1.098	1.14E−06
54	1.06	0.908	1.138	1.69E−06	65	1.25	1.321	1.104	1.13E−06
54	1.25	0.929	1.140	1.63E−06	65	1.50	1.343	1.110	1.11E−06
54	1.50	0.957	1.137	1.58E−06	65	1.75	1.369	1.114	1.09E−06
54	1.75	0.99	1.122	1.53E−06	65	2.00	1.394	1.114	1.07E−06
54	2.00	1.027	1.095	1.46E−06	65	2.25	1.426	1.110	1.05E−06
54	2.25	1.067	1.061	1.39E−06	65	2.50	1.46	1.099	1.02E−06
54	2.50	1.108	1.023	1.32E−06	65	2.75	1.497	1.082	9.90E−07
55	All	1.135	1.000	1.27E−06	65	3.25	1.574	1.037	9.30E−07
64	1.06	1.164	1.229	1.38E−06	66	All	1.634	1.000	8.83E−07
64	1.25	1.211	1.215	1.32E−06	75	1.06	1.566	1.191	9.97E−07
64	1.50	1.275	1.184	1.23E−06	75	1.25	1.604	1.193	9.73E−07
64	1.75	1.341	1.145	1.14E−06	75	1.50	1.664	1.189	9.35E−07
					75	1.75	1.732	1.174	8.92E−07
					75	2.00	1.803	1.151	8.47E−07
					75	2.25	1.875	1.121	8.01E−07

<u>*Dimensionless upstroke rod stretch*</u>, $\frac{F_1}{Sk_r}$: is defined as the rod stretch caused by the total loads due to the fluid and the acceleration force during the upstroke, F_1, and a fraction of the polished rod stroke length, Sk_r.

Recall Eq. (5.3.15): $PPRL = w_b L + w_r L\alpha + F_o$

$$F_1 = w_r L\alpha + F_o \tag{5.4.4}$$

$$PPRL = w_b L + \frac{F_1}{Sk_r} Sk_r \tag{5.4.5}$$

The value of $\frac{F_1}{Sk_r}$ can be obtained from Fig. 5.18.

Nguyen et al. [10] digitized Fig. 5.18 and derived the following equation:

$$\begin{aligned}\frac{F_1}{Sk_r} = & -187.78\left(\frac{N}{N_o}\right)^6 + 341.15\left(\frac{N}{N_o}\right)^5 - 210.78\left(\frac{N}{N_o}\right)^4 + 55.62\left(\frac{N}{N_o}\right)^3 - 5.58\left(\frac{N}{N_o}\right)^2 \\ & + 0.75\left(\frac{N}{N_o}\right) + 1.252434\left(\frac{F_o}{Sk_r}\right)^4 - 0.39\left(\frac{F_o}{Sk_r}\right)^3 - 0.91\left(\frac{F_o}{Sk_r}\right)^2 + 1.23\left(\frac{F_o}{Sk_r}\right) - 0.14\end{aligned} \tag{5.4.6}$$

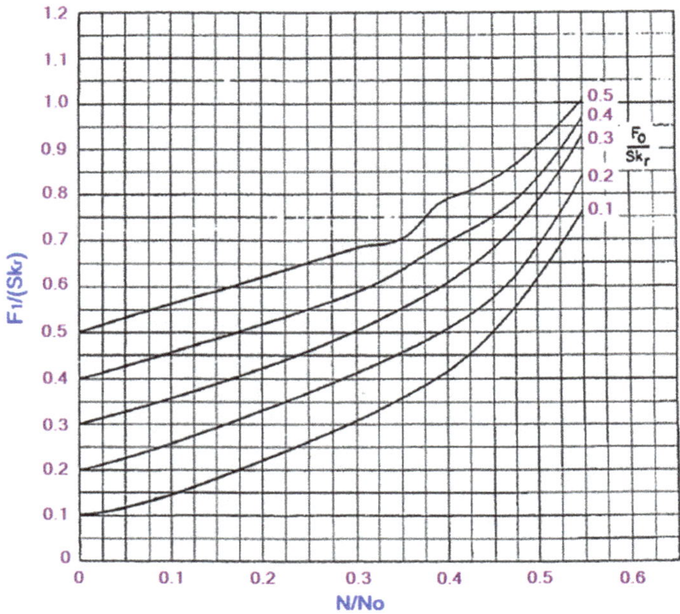

Fig. 5.18 Determination of the dimensionless upstroke rod stretch (API RP 11L)

Dimensionless downstroke rod stretch, $\frac{F_2}{Sk_r}$: is defined as the rod stretch caused by the acceleration force during the downstroke, F_2, and a fraction of the polished rod stroke length, Sk_r.

Recall Eq. (5.3.18): $MPRL = w_b L - w_r L \alpha$

$$F_2 = w_r L \alpha \tag{5.4.7}$$

$$MPRL = w_b L - \frac{F_2}{Sk_r} Sk_r \tag{5.4.8}$$

The value of $\frac{F_2}{Sk_r}$ can be obtained from Fig. 5.19.

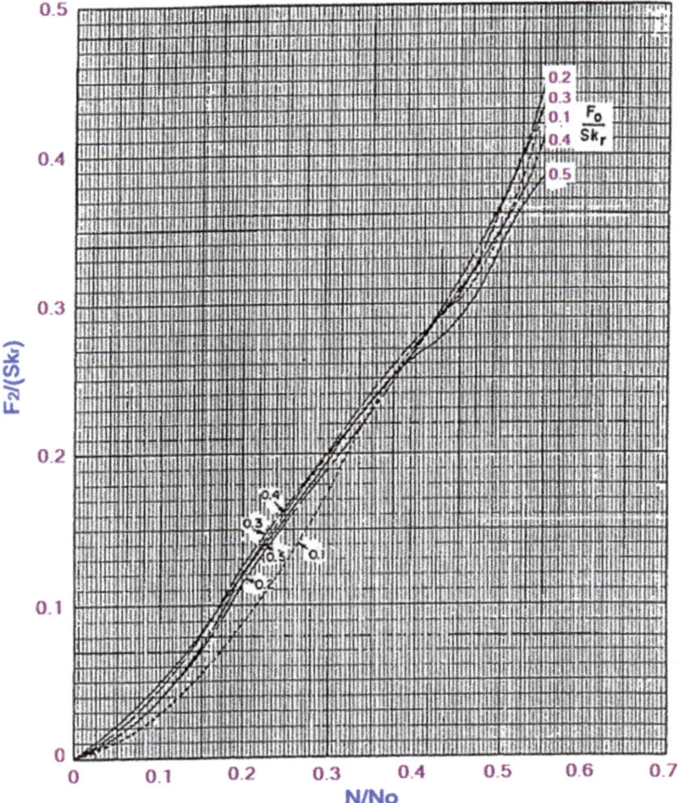

Fig. 5.19 Determination of the dimensionless downstroke rod stretch (API RP 11L)

Nguyen et al. [10] digitized Fig. 5.19 and derived the following equation:

$$\frac{F_2}{Sk_r} = -21.21\left(\frac{N}{N_o}\right)^6 + 86.58\left(\frac{N}{N_o}\right)^5 - 82.55\left(\frac{N}{N_o}\right)^4 + 29.57\left(\frac{N}{N_o}\right)^3 - 3.49\left(\frac{N}{N_o}\right)^2 + .63318\left(\frac{N}{N_o}\right) - 30.56\left(\frac{F_o}{Sk_r}\right)^4 + 37.86\left(\frac{F_o}{Sk_r}\right)^3 - 16.73\left(\frac{F_o}{Sk_r}\right)^2 + 3.11\left(\frac{F_o}{Sk_r}\right) - 0.21 \quad (5.4.9)$$

Dimensionless plunger stroke, S_p/S: is defined as the ratio between the plunger stroke length, S_p, and the polished stroke length, S. If the production tubing is anchored the ratio $S_p/S = 1$. To take into account the effect of the tubing stretch on the plunger stroke length, the following equation is given:

$$S_p = \left(\frac{S_p}{S}S\right) - \left(F_o\frac{1}{k_t}\right) \quad (5.4.10)$$

F_o can be calculated using Eq. (5.3.17). k_t is the spring constant of the production tubing and depends on the elastic constant. Please note that k_r can also be calculated using Eq. (5.4.11). The elastic constant rod string, E_{tr}, is given in Table 5.4. Readers can get the elastic constant values for production tubing, E_{tt}, in the tubing catalog.

$$k_t = \frac{1}{E_{tt}L} \quad (5.4.11)$$

The value of $\frac{S_p}{S}$ can be obtained from Fig. 5.20.

Nguyen et al. [10] digitized Fig. 5.20 and derived the following equation:

$$\frac{S_p}{S} = -288.46\left(\frac{N}{N'_o}\right)^6 + 464.43\left(\frac{N}{N'_o}\right)^5 - 285.94\left(\frac{N}{N'_o}\right)^4 + 88.95\left(\frac{N}{N'_o}\right)^3 - 13.19\left(\frac{N}{N'_o}\right)^2 + 1.12\left(\frac{N}{N'_o}\right) + 5.55\left(\frac{F_o}{Sk_r}\right)^4 - 7.10\left(\frac{F_o}{Sk_r}\right)^3 + 3.33\left(\frac{F_o}{Sk_r}\right)^2 - 1.54\left(\frac{F_o}{Sk_r}\right) + 0.99 \quad (5.4.11)$$

Dimensionless torque, $\frac{2T}{S^2k_r}$: Rearrange Eq. (5.3.32a) gives:

$$PT = \frac{2T}{S^2k_r} \times Sk_r \times T_a\frac{S}{2} \quad (5.4.12)$$

where Ta is an adjustment for peak torque and determined using Fig. 5.21. Let consider one example to learn how to use Fig. 5.21. Given the weight of rod in fluid, $w_bL/Skr = 0.6$; $N/N'_o = 0.2$; $F_o/Sk_r = 0.19$. From Fig. 5.21, we can arrive the percent of 3.0%.

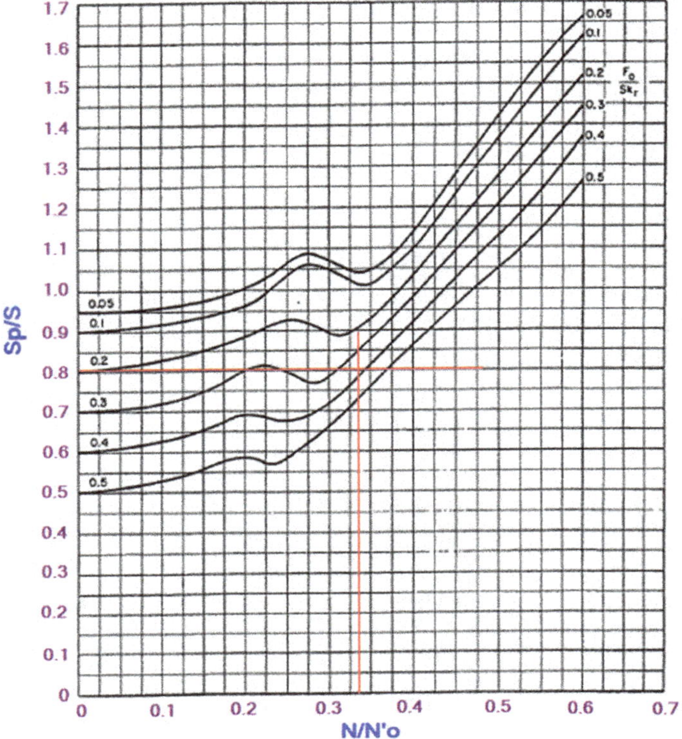

Fig. 5.20 Determination of dimensionless plunger stroke (API RP 11L)

$\frac{\frac{w_bL}{Sk_r}-0.3}{0.1} = \frac{0.6-0.3}{0.1} = 3$. The total adjustment is $\frac{\frac{w_bL}{Sk_r}-0.3}{0.1} \times 3\% = 9\%$. Therefore, Ta = 1.00 + 0.09 = 1.09.

Similarly, from Eq. (5.3.34), one can derive the following equation to calculate for the polished rod horsepower:

$$PRHP = 2.53 \times 10^{-6} \frac{F_3}{Sk_r} \times Sk_r \times S \times N \tag{5.4.13}$$

The values of ($2T/S^2k_r$) and (F_3/Sk_r) can be obtained from Figs. 5.22 and 5.23.

Nguyen et al. [10] digitized Figs. 5.22 and 5.23 and derived the following equations:

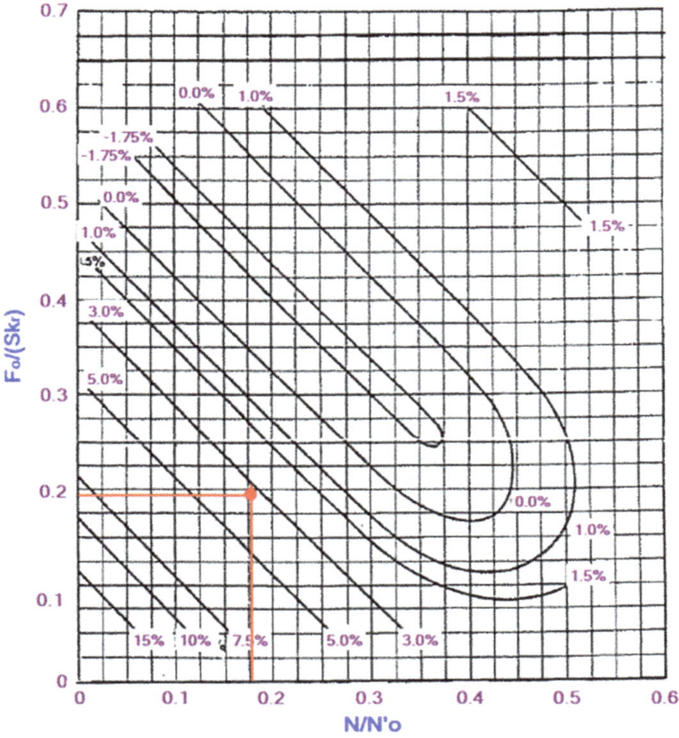

Fig. 5.21 Determination of the peak torque adjustment, T_a. Multiply the % indicated on the curve by $\frac{\frac{w_b L}{Sk_r}-0.3}{0.1}$ (API RP 11L)

$$\frac{2T}{S^2k_r} = -74.99\left(\frac{N}{N_o'}\right)^6 + 132.86\left(\frac{N}{N_o'}\right)^5 - 87.85\left(\frac{N}{N_o'}\right)^4 + 27.17\left(\frac{N}{N_o'}\right)^3 - 3.52\left(\frac{N}{N_o'}\right)^2 + 0.66\left(\frac{N}{N_o'}\right) + 5.71\left(\frac{F_o}{Sk_r}\right)^4 - 5.63\left(\frac{F_o}{Sk_r}\right)^3 + 1.03\left(\frac{F_o}{Sk_r}\right)^2 + 0.57\left(\frac{F_o}{Sk_r}\right) - 0.04 \tag{5.4.14}$$

$$If\ \frac{F_o}{Sk_r} < 0.3$$

$$\frac{F_3}{Sk_r} = -1.71\left(\frac{N}{N_o}\right)^6 - 12.87\left(\frac{N}{N_o}\right)^5 + 18.30\left(\frac{N}{N_o}\right)^4 - 6.66\left(\frac{N}{N_o}\right)^3 + 1.49\left(\frac{N}{N_o}\right)^2 + 0.07\left(\frac{N}{N_o}\right) + 11.87\left(\frac{F_o}{Sk_r}\right)^4 - 20.42\left(\frac{F_o}{Sk_r}\right)^3 + 8.68\left(\frac{F_o}{Sk_r}\right)^2 - 0.59\left(\frac{F_o}{Sk_r}\right) + 0.08 \tag{5.4.15}$$

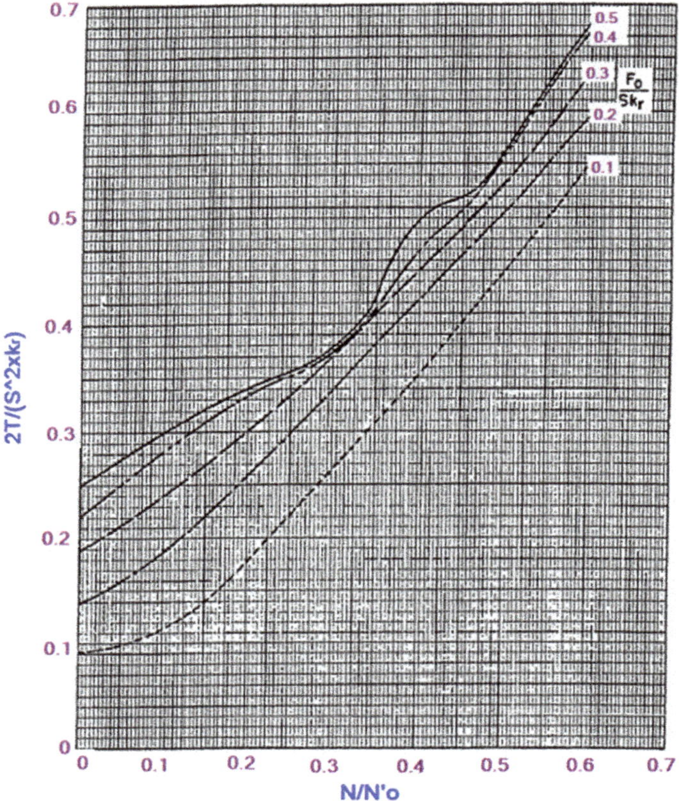

Fig. 5.22 Determination of dimensionless torque (API RP 11L)

$$If\ \frac{F_o}{Sk_r} > 0.3$$

$$\frac{F_3}{Sk_r} = -4.56\left(\frac{N}{N_o}\right)^6 - 15.57\left(\frac{N}{N_o}\right)^5 + 11.95\left(\frac{N}{N_o}\right)^4 + 3.11\left(\frac{N}{N_o}\right)^3 - 1.20\left(\frac{N}{N_o}\right)^2 + 0.30\left(\frac{N}{N_o}\right) + 13.81\left(\frac{F_o}{Sk_r}\right)^4 - 20.11\left(\frac{F_o}{Sk_r}\right)^3 + 8.65\left(\frac{F_o}{Sk_r}\right)^2 - 0.40\left(\frac{Fo}{Sk_r}\right) - 0.06 \tag{5.4.16}$$

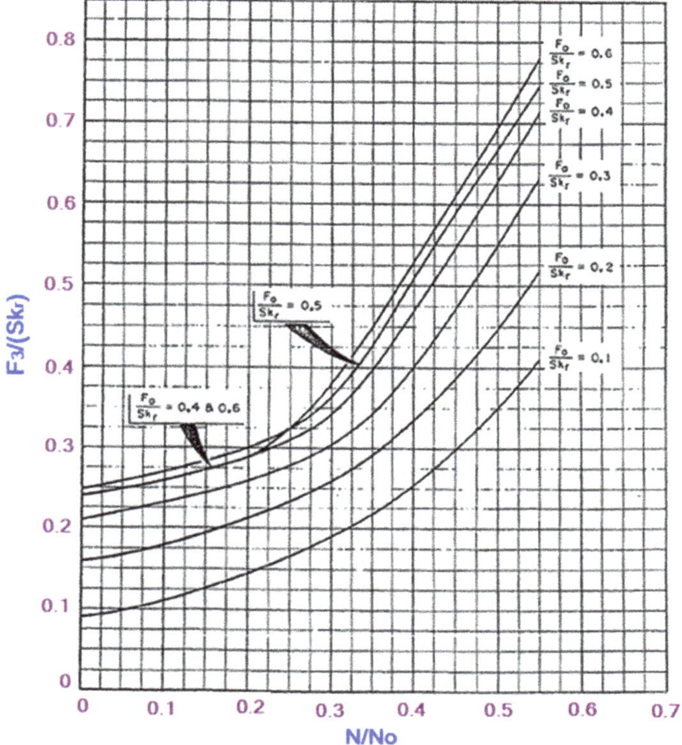

Fig. 5.23 Determination of (F_3/Sk_r) (API RP 11L)

5.5 Viscous Effect on Rod Pump Performance

During the upstroke, the rod string moves upward at a speed of u_r and the liquid travels in the annulus formed by the rod string and the production tubing at an average velocity of $\bar{u}_f$. The fluid is normally moves faster than the speed of the rod and hence there exist two friction forces due to the moving fluid: one friction force is at the inner wall of the production tubing and the other force is at the outer wall of the rod string as shown in Fig. 5.24.

For light oils with API of 30 °API or above, the fluid viscosity of these fluids is normally quite low. The friction forces can practically negligible. However, if the fluid viscosity is high (API of 20 or smaller), the friction forces may significantly impact the polished rod loads and the torque.

Applying the momentum equation for an incompressible fluid flowing in one direction, isothermal, and under steady state condition give the relationship between the frictional pressure force gradient and the wall shear stress in cylindrical coordinates as expressed in Eq. (5.5.1):

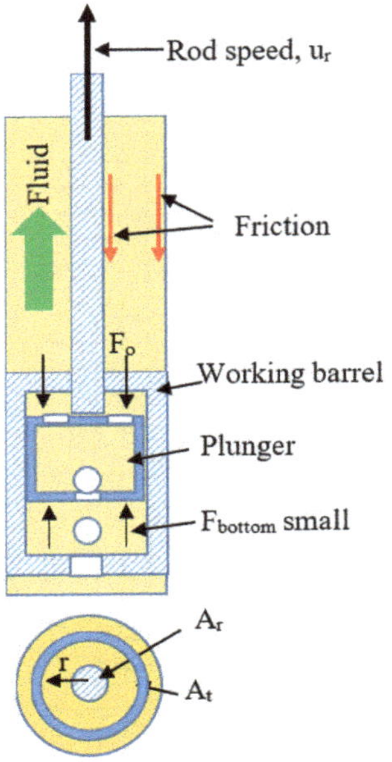

Fig. 5.24 Fluid flow in the annulus during the upstroke

$$\tau_w = \frac{r}{2}\frac{dP}{dL}\bigg|_f + \frac{c_1}{r} \tag{5.5.1}$$

where τ_w is the wall shear stress; $\frac{dP}{dL}\big|_f$ is the frictional pressure drop gradient; r is the radius of the donut of the free body diagram as shown in Fig. 5.24 and r changes from R_r and R_t; and c_1 is the integral constant. The subscripts r and t are for rod and tubing. Combining Eq. (5.5.1) with the definition of a Newtonian fluid which has viscosity of μ_f gives:

$$\tau_w = -\mu_f\frac{du}{dr} = \frac{r}{2}\frac{dP}{dL}\bigg|_f + \frac{c_1}{r} \tag{5.5.2}$$

Integrate Eq. (5.5.2) gives the general form of the velocity profile in an annulus as follows:

$$u = -\frac{r^2}{4\mu_f}\frac{dP}{dL}\bigg|_f - \frac{c_1}{\mu_f}\ln r + c_2 \tag{5.5.3}$$

To obtain the integral constants c_1 and c_2, the two boundary conditions will be applied: (1) @ $r = R_r$, $u = u_r$ and @ $r = R_t$, $u = 0$. The final equation to describe the velocity profile in an annulus is expressed as follows:

$$u = \frac{1}{4\mu_f} \frac{dP}{dL}\bigg|_f \left(\left(R_t^2 - r^2\right) - \left(R_t^2 - R_r^2\right) \frac{\ln\left(\frac{R_t}{r}\right)}{\ln\left(\frac{R_t}{R_r}\right)} \right) + u_r \tag{5.5.3}$$

Note that u_r is the speed of the rod string.

Flowrate can be expressed in terms of average fluid velocity as

$$Q = \int_{R_r}^{R_t} u \times 2\pi r dr = \bar{u}_f \times \pi\left(R_t^2 - R_r^2\right) \tag{5.5.4}$$

Combining Eqs. (5.5.3) and (5.5.4) gives the final equation to calculate for the frictional pressure drop gradient of Newtonian fluids flowing in an annulus:

$$\frac{dP}{dL}\bigg|_f = \frac{8\mu_f\left(\bar{u}_f - u_r\right)}{\left[R_t^2 + R_r^2 - \frac{R_t^2 - R_r^2}{\ln\frac{R_t}{R_r}}\right]} \tag{5.5.5}$$

Equation (5.5.5) was presented by Nguyen et al. [10]. The friction force calculated from Eq. (5.5.5) is added into the PPRL and MPRL to take into account the effect of fluid viscosity. It can be seen that the frictional pressure loss in the annulus is a function of the relative velocity between the average fluid velocity and the rod speed. For the upstroke, the friction force causes an increase in the PPRL. For downstroke, the average fluid velocity in the annulus can be assumed to be zero and hence the friction force causes a decrease in the MPRL. Readers should be careful with the unit when using Eq. (5.5.5). If the PPRL and MPRL are in the unit of pressure such as psi, the results from Eq. (5.5.5) is simply multiplied with the rod length and added into the PPRL and MPRL. If the PRL and the MPRL are in the unit of force such as lbf or kN, the results from Eq. (5.5.5) must be multiplied with the rod length and the cross-sectional area of the annulus first to obtain the force unit before adding it into the PPRL and MPRL.

Nguyen et al. [10] derived Eq. (5.5.5) and did the sensitivity analysis to study the effect of heavy oil on rod pump performance. They have drawn the following conclusions:

Heavy oil has a major effect on rod loads and torque. However, the results reveal that the effect on the torque is much higher than is seen on the rod loads due to the stress that is put on the gear reducer (>50% vs. ~25%). For conventional units, the results show that:

- 10% increase on PPRL when the fluid viscosity increases from 0–4000 cp
- 25% decrease on MPRL when the fluid viscosity increases from 0–4000 cp
- 35% increase on the PT Upstroke when the fluid viscosity increases from 0–4000 cp

- 40% decrease on the PT Downstroke when the fluid viscosity increases from 0–4000 cp
- 55% increase on PT when the fluid viscosity is from 0–6000 cp.

5.6 Common Sucker-Rod Pump Failures

5.6.1 Rod String Failures and Design

Rod strings are composed of sucker rods, pony rods, and couplings. There are different ways to classify for rod string failures such as stress failures, mechanical failures, corrosion fatigue failures, and manufacturing defects.

Stress failures: associate with tensile and fatigue failures. The tensile failures happen when the applied load is higher than the tensile strength of the rod. Tensile failures normally occur when (1) pulling too much on the rod string such as during an attempt to unseat a stuck pump. This will cause the string to failure at some points along the string, especially on the smaller diameter size rods. (2) pushing too much on the rod string such as during running the pump into high dog-leg severity holes. The rod string may be buckled and failed. The fatigue failures happen in a period of time. It begins with small cracks and grow under the action of cyclic stresses. The maximum stresses associated with these failures are normally smaller than the yield strength of the rods in the final treated conditions. During the operating conditions, couplings and rods are constantly in contact with the production tubing and hence erosion may occur at these contacts. As the erosion progresses, the cross-sectional areas get smaller and pressure at these locations is higher. After a period of time the fatigue cracks may occur and propagate perpendicular to the principle stress.

Mechanical failures: account for a large percentage of the total number of all rod string failures. These failures range from bent rod failures, surface damage failures, and connection failures.

Corrosion fatigue failures: Corrosion-fatigue failures happen when rods encounter with produced fluids with the presence of salt water, H_2S, or CO_2.

Manufacturing defects: Failures due to a manufacturing defect seldom occur. The common manufacturing defects are mill defects, forging defects, and processing defects.

The main constituent of all sucker rods is iron, which is about 90% in the composition of most of the rods. Pure iron is weak and hence other additives such as carbon, manganese, nickel, copper, boron, etc. must be added to the composition to make the rods stronger, to resist abrasive environments, reduce oxidation and corrosion. Common API rods and their properties are shown in Table 5.5.

Sucker rod failures can be mitigated and prevented if correct design and selection process are applied. Calculation of load distribution along the rod string should be the first step in designing a rod string.

Table 5.5 Common API rod and their properties

API grade	Load applications	Environment	Min. tensile strength (psi)	Max. tensile strength (psi)
C	Light to medium	Non-corrosive	90,000	115,000
D	Heavy	Non-corrosive	115,000	140,000
K	Light to medium	Corrosive	N/A	N/A
KD63	Heavy	Corrosive	N/A	N/A

Let's review the general equation for calculating the PPRL during the upstroke in an inclined wellbore presented in Sect. 5.3.2:

$$PPRL = w_b L \cos\varphi + \mu_m w_b L \sin\varphi + w_r L\alpha \cos\varphi + F_o \cos\varphi + F_f \cos\varphi \quad (5.6.1)$$

For vertical wells, $\varphi = 0°$ and hence Eq. (5.6.1) becomes:

$$PPRL = w_b L + w_r L\alpha + F_o + F_f \quad (5.6.2)$$

The first, second, third, and fourth term on the right hand side of Eq. (5.6.2) represent for the buoyant weight of the rod string, force due to acceleration, force due to the fluid column acting on the plunger, and the friction force, respectively.

When selecting sucker rods, one needs to make sure that the minimum tensile strength of the selected rods must be greater than the PPRL calculated using Eq. (5.6.2). Another practical approach is to select sucker rods based on the maximum allowable rod stress, S_A, instead of based on the minimum tensile strength. In other words, designers must make sure that the S_A of the selected rods is greater than the PPRL.

To calculate for the S_A, the modified Goodman equation can be used.

$$S_A = (0.25T + 0.5625 \times MPRL)SF \quad (5.6.3)$$

where S_A is the maximum allowable rod stress in psi; T is the minimum tensile strength in psi; and SF is the service factor. Different SF values in different environments are given in Table 5.6 [5]

Table 5.6 Service factor in different environments

Service	API grade "C" rods	API grade "D" rods
Non-corrosive	1.00	1.00
Salt water	0.65	0.90
Hydrogen sulfide	0.50	0.70

Based on Eq. (5.6.2), one can recognize that the load at the bottom of the rod is minimum and the load at the surface polished rod is maximum. If the well depth is high (more than 3,500 ft), it is more economical to design the rod string with different lengths and different sizes. The smallest rods should be placed at the bottom of the string and the largest rods should be placed at the top of the string. Therefore, it is recommended to use tapered rod string for wells with depths greater than 3,500 ft. API recommends designing a tapered sucker rod string by assign to each of the graduated sections of the string its maximum stress. From this point up, a larger size rod is used. With this design, the maximum allowable stress is placed on the top rod of the smallest size.

At the final step in designing a tapered rod string, the maximum anticipated stress must be checked to certify it does not exceed the safe allowable working stress. The maximum stress at the top of the entire rod string will be the peak polished rod load divided by the cross-sectional area of the top section of rods.

The % of each size in tapered rod string design is given by Brown [5] and presented in Tables 5.7.

5.6.2 Pump Barrel Failures or Improper Operations

A downhole rod pump consists of a working barrel, a plunger, a standing valve and a traveling valve as shown in Fig. 5.1b. The presence of gas and solid entrained in the fluid may cause these components to work improperly.

Solid interference: With high solid concentration entering the pump, the ports of the standing and traveling valves may be eroded causing a leak and a reduction in pump efficiency.

Gas interference: is also a common cause of low pump efficiency. As the plunger starts moving downward during the downstroke, the standing valve does not close right away. At the meantime, the traveling valve opens slowly without the rapid load change experience in fluid pound as shown in Fig. 5.25a. This is because when the plunger moves downward, the compressed gas reaches the pressure needed to open the traveling valve before the traveling valve reaches the fluid. The pump efficiency can be very low [12]. To avoid or reduce the gas interference and to improve the pump efficiency, a downhole gas separator should be installed below the downhole pump.

Fluid pound: is also a common cause of low pump efficiency when the downhole pump rate is higher than the liquid rate supplied from the formation. This will cause a void space with low pressure gas between the standing and traveling valves.

As the plunger starts moving downward during the downstroke, the standing valve does not close right away. At the meantime, the traveling valve still remains closed until it hits the fluid and opens as shown in Fig. 5.25b. At this moment, the weight on the rod string can suddenly drop thousands of pounds in a fraction of a second and cause an extreme stress on the rod string. To avoid this problem, one may consider to go with smaller downhole pump and operate with smaller stroke length.

Table 5.7 Sucker rod and % of each size [5]

Rod no.	Plunger diameter (in.)	Rod weight (lb/ft)	Frequency factor (Fc)	Elastic constant E (in./lb ft)	Tapered rod string, % of each size					
					1 1/8 in.	1.00 in.	7/8 in.	3/4 in.	5/8 in.	1/2 in.
44	All	0.726	1.000	1.99E−06	0.0	0.0	0.0	0.0	0.0	100.0
54	1.06	0.908	1.138	1.69E−06	0.0	0.0	0.0	0.0	44.6	55.4
54	1.25	0.929	1.140	1.63E−06	0.0	0.0	0.0	0.0	49.5	50.5
54	1.50	0.957	1.137	1.58E−06	0.0	0.0	0.0	0.0	56.4	43.6
54	1.75	0.99	1.122	1.53E−06	0.0	0.0	0.0	0.0	64.6	35.4
54	2.00	1.027	1.095	1.46E−06	0.0	0.0	0.0	0.0	73.7	26.3
54	2.25	1.067	1.061	1.39E−06	0.0	0.0	0.0	0.0	83.4	16.6
54	2.50	1.108	1.023	1.32E−06	0.0	0.0	0.0	0.0	93.5	6.5
55	All	1.135	1.000	1.27E−06	0.0	0.0	0.0	0.0	100.0	0.0
64	1.06	1.164	1.229	1.38E−06	0.0	0.0	0.0	33.3	33.1	33.6
64	1.25	1.211	1.215	1.32E−06	0.0	0.0	0.0	37.2	35.9	26.9
64	1.50	1.275	1.184	1.23E−06	0.0	0.0	0.0	42.3	40.4	17.3
64	1.75	1.341	1.145	1.14E−06	0.0	0.0	0.0	47.4	45.2	7.4
65	1.06	1.307	1.098	1.14E−06	0.0	0.0	0.0	34.4	65.6	0.0
65	1.25	1.321	1.104	1.13E−06	0.0	0.0	0.0	37.3	62.7	0.0
65	1.50	1.343	1.110	1.11E−06	0.0	0.0	0.0	41.8	58.2	0.0
65	1.75	1.369	1.114	1.09E−06	0.0	0.0	0.0	46.9	53.1	0.0
65	2.00	1.394	1.114	1.07E−06	0.0	0.0	0.0	52.0	48.0	0.0
65	2.25	1.426	1.110	1.05E−06	0.0	0.0	0.0	58.4	41.6	0.0
65	2.50	1.46	1.099	1.02E−06	0.0	0.0	0.0	65.2	34.8	0.0
65	2.75	1.497	1.082	9.90E−07	0.0	0.0	0.0	72.5	27.5	0.0
65	3.25	1.574	1.037	9.30E−07	0.0	0.0	0.0	88.1	11.9	0.0
66	All	1.634	1.000	8.83E−07	0.0	0.0	0.0	100.0	0.0	0.0
75	1.06	1.566	1.191	9.97E−07	0.0	0.0	27.0	27.4	45.6	0.0
75	1.25	1.604	1.193	9.73E−07	0.0	0.0	29.4	29.8	40.8	0.0
75	1.50	1.664	1.189	9.35E−07	0.0	0.0	33.3	33.3	33.4	0.0
75	1.75	1.732	1.174	8.92E−07	0.0	0.0	37.8	37.0	25.2	0.0
75	2.00	1.803	1.151	8.47E−07	0.0	0.0	42.4	41.3	16.3	0.0
75	2.25	1.875	1.121	8.01E−07	0.0	0.0	46.9	45.8	7.3	0.0

5.7 Dynamometer

Dynamometer is a tool for measuring polished rod load, peak load, peak torque, and horsepower as the surface polished rod complete a cycle [6]. In other words, a dynamometer measures the load forces acting on a rod string during a complete pumping cycle and records the forces on a chart or computer display. This display is

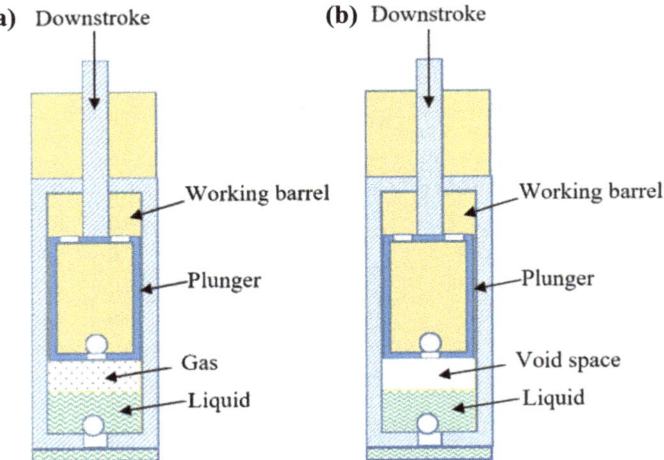

Fig. 5.25 **a** Gas interference, **b** fluid pound

often called a *dynamometer card*. The card records changes in the rod load versus rod displacement, or changes in the rod load versus pumping time as shown in Fig. 5.26. The dynamometer can be an effective tool to do the trouble-shooting to determine pump problems before making a decision of doing a workover [8]. The correct interpretations of the dynamometer card data can help to evaluate downhole problems such as gas lock, liquid pound, improper closing or opening of the standing and traveling valves.

A theoretical dynamometer card is shown in the Fig. 5.27a for non-elastic rods. At point (a) the traveling valve closes at the beginning of the upstroke and the rod string takes up the whole load instantaneously, represented from (a) to (b). From (b) to (c) the load remains constant and equals to the PPRL until the top of the upstroke is reached. Then the traveling valve opens, the standing valve closes and the load is removed from the rods from (c) to (d). The downstroke begins and the load value is the same as the MPRL. For elastic rods, the surface polished rod stroke length is different from the downhole plunger stroke length. This will cause a delay in closing/opening in the standing and traveling valves as shown in Fig. 5.27b. The theoretical load difference between bc and ad can be estimated as:

$$bc - ad = PPRL - MPRL = 2w_r L\alpha + F_o \tag{5.7.1}$$

Equation (5.7.1) tells us that the theoretical upstroke polished rod load is higher than the theoretical downstroke polished rod load an amount which equals to twice the force due to acceleration plus the force due to liquid acting on the top of the plunger.

The theoretical card is seldom encountered in reality because of many factors such as the improper opening/closing of the standing and traveling valves, rod and

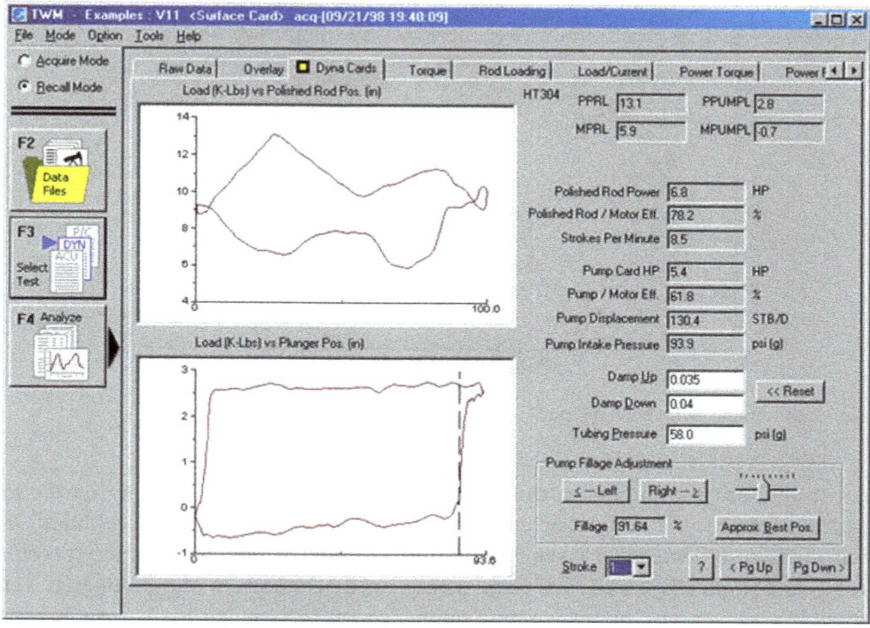

Fig. 5.26 Example of a dynamometer card (http://echometer.com/Products/WellAnalyzer/DynamometerTest/tabid/112/Default.aspx)

walking beam acceleration, etc. Figures 5.28, 5.29a–e shows a typical pumping cycle on a dynamometer card.

At point 1 in Fig. 5.28, the plunger is at the lowest position and ready for the upstroke cycle. The crank is on the highest vertical position as shown in Fig. 5.29a. At this moment, the standing valve is close and the traveling valve is open.

When the polished rod travels from point 1 to point 2, the crank and the plunger are accelerating and hence the PRL is increasing. At point 2, the traveling valve is

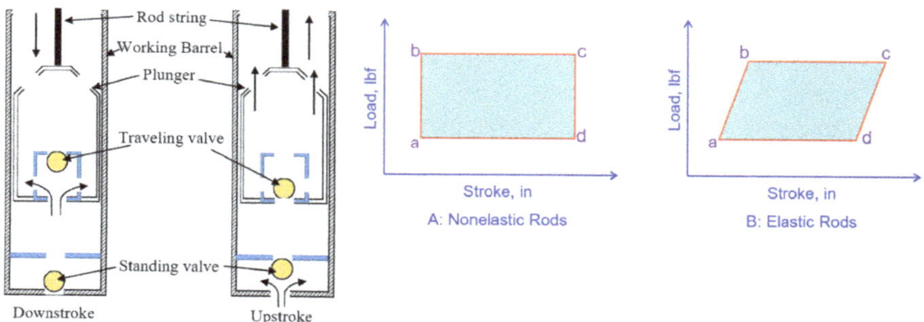

Fig. 5.27 Theoretical dynamometer card

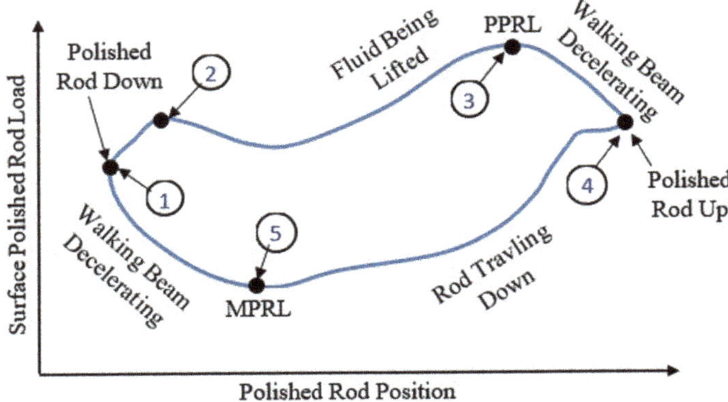

Fig. 5.28 Actual dynamometer card

not fully close yet and hence the forces due to the fluid column acting on the top and the bottom of the plunger are very similar and canceled out. In addition, the acceleration is decreasing causing a slight reduction in the PRL. As the traveling valve is completely close, the force acting on the bottom of the plunger is very small in compared to the force acting on the top of the plunger, F_o. F_o now is being added to the PRL causing it to increase. During this period, the fluid is being lifted and the polished rod is moving toward point 3 as shown in Fig. 5.28. Figure 5.29b, c. At point 3, the PRL reaches a maximum value.

From point 3 to point 4 in Fig. 5.28, the crank and the walking beam are decelerating causing a reduction in the PRL. During this period, the traveling valve is still close and the standing valve is still open.

From point 4 in Fig. 5.28, the plunger begin the downstroke cycle. The crank and the plunger are accelerating in the downward direction causing the PRL to reducer further. The traveling valve is opening and the standing valve is closing. During this period, the forces acting on the top and the bottom of the plunger are very similar and canceled out. The PRL will reach the minimum value at point 5.

5.8 Example

Example 5.1 A sucker-rod pump well with a 1.5 in. plunger is operating at a speed of 20 SPM. The effective plunger stroke is 55 in. The liquid production (SG = 0.85) at the surface is 210 BPD. Calculate the total theoretical pump displacement and the volumetric efficiency of the pump.

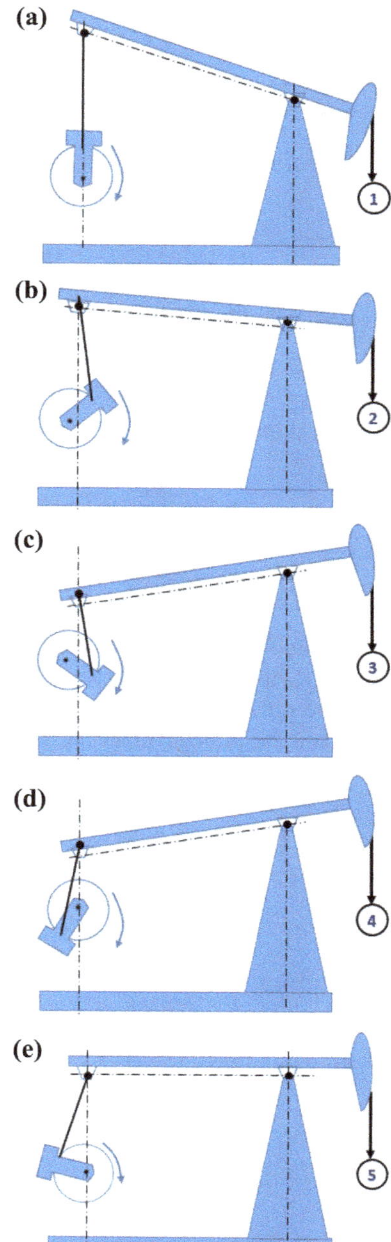

Fig. 5.29 **a** Demonstration of rod position, **b** demonstration of rod position, **c** demonstration of rod position, **d** demonstration of rod position, **e** demonstration of rod position

Solution

$$D_p = 1.5\,\text{in.} \quad N = 20\,\text{SPM} \quad S_p = 55\,\text{in.} \quad Q = 210\,\text{BPD}$$

The cross-sectional area of the plunger:

$$A_p = \frac{\pi D_p^2}{4} = \frac{\pi \times 1.5^2}{4} = 1.767\,\text{in.}^2$$

The total theoretical pump displacement:

$$PD = 0.1484 \times A_p \times S_p \times N = 0.1484 \times 1.767 \times 55 \times 20 = 288\,\text{BPD}$$

The pump efficiency

$$E_v = \frac{Q}{PD} = \frac{210}{288}100 = 73\%$$

Users can also use the constant K = $0.1484A_p$ given in Table 5.1 to obtain the same total theoretical pump displacement.

Example 5.2 A rod string consists of sucker rods with diameters D_r = 1.5 in. and a plunger with a diameter D_p = 2.5 in. and is reciprocating in a formation fluid with a specific gravity SG = 0.8 as shown in Fig. 5.30. The weight per foot of the sucker rods and the plunger in air are 1.343 lb/ft and 1.460 lb/ft, respectively. The length

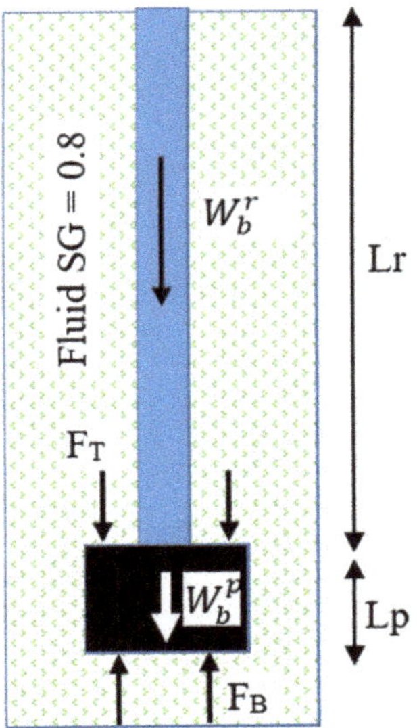

Fig. 5.30 Diagram for Example 5.2

of the sucker rods and the plunger are 4,000 ft and 0.5 ft, respectively. The rod speed and stroke length are 30 SPM and 50 in., respectively. Calculate

a. The surface rod load under static conditions
b. The surface rod load under the upstroke (neglect fluid friction)
c. The surface rod load under the downstroke (neglect fluid friction).

Please note that this example is to explain the concept of how to calculate the surface load. There is NO subsurface pump in this example.

Solution

a. Calculate the surface rod load under static conditions:

Using water density of 62 lb/ft^3 and steel density of 492 lb/ft^3, one can derive the buoyant weight of the rods (weight of the rods in fluid) as follows:

$$w_b = w_r\left(1 - \frac{\rho_f}{\rho_s}\right) = w_r\left(1 - \frac{62 \times SG}{492}\right) = w_r(1 - 0.127 \times SG)$$

Buoyant weight per foot of the sucker rods and the plunger:

$$w_b^r = w_r(1 - 0.127 \times SG) = 1.343(1 - 0.127 \times 0.8) = 1.206\,\text{lb/ft}$$
$$w_b^p = w_p(1 - 0.127 \times SG) = 1.46(1 - 0.127 \times 0.8) = 1.312\,\text{lb/ft}$$

Buoyant weight of the sucker rods and the plunger:

$$W_b^r = w_b^r \times L_r = 1.206 \times 4{,}000 = 4{,}826\,\text{lbf}$$
$$W_b^p = w_b^p \times L_p = 1.312 \times 0.5 = 0.66\,\text{lbf}$$

Cross-sectional areas of the sucker rods and the plunger:

$$A_r = \frac{\pi D_r^2}{4} = \frac{\pi \times 1.5^2}{4} = 1.767\,\text{in.}^2$$
$$A_p = \frac{\pi D_p^2}{4} = \frac{\pi \times 2.5^2}{4} = 4.908\,\text{in.}^2$$

The forces acting upon the top and the bottom of the plunger due to the fluid hydrostatic pressure:

$$F_T = 0.052 \times \rho_f \times L_r(A_p - A_r) = 0.052 \times 8.3 \times 0.8 \times 4{,}000(4.908 - 1.767) = 4{,}339\,\text{lbf}$$
$$F_b = 0.052 \times \rho_f \times (L_r + L_p)A_p = 0.052 \times 8.3 \times 0.8 \times 4000.5 \times 4.908 = 6{,}780\,\text{lbf}$$

Note that the fluid hydrostatic pressure in psi: $P = 0.052\rho_f TVD$ where ρ is the fluid density in pound per gallon (ppg) and TVD is the true vertical depth in ft. $\rho_f = \rho_{water} \times SG = 8.3 \times SG$.

The surface rod load under static conditions:

$$SRL = W_b^r + W_b^p + F_T - F_b = 4{,}826 + 0.66 + 4{,}339 - 6{,}780 = 2{,}385\,\text{lbf}$$

The stress on the rod at the surface:

$$\sigma_s = \frac{SRL}{A_r} = \frac{2{,}385}{1.767} = 1{,}349\frac{\text{lbf}}{\text{in.}^2} = 1{,}349\,\text{psi}$$

If A_p is used instead of $(A_p - A_r)$ to calculate for the force acting on the top of the plunger, F_T, there will be a significant error, about 50%, in the calculation.

$$F_T = 0.052 \times \rho_f \times L_r \times A_p = 0.052 \times 8.3 \times 0.8 \times 4{,}000 \times 4.908 = 6{,}779\,\text{lbf}$$
$$SRL = W_b^r + W_b^p + F_T - F_b = 4{,}826 + 0.66 + 6{,}779 - 6{,}780 = 4{,}826\,\text{lbf}$$

b. The surface rod load under the upstroke (neglect fluid friction)

The Mill's acceleration factor:

$$\alpha = \frac{SN^2}{70{,}500} = \frac{50 \times 30^2}{70{,}500} = 0.638$$

The surface rod load during the upstroke:

$$SRL_u = SRL + w_r L\alpha = 2{,}385 + 1.343 \times 4{,}000 \times 0.638 = 5{,}814\,\text{lbf}$$

c. The surface rod load under the downstroke (neglect fluid friction)

$$SRL_d = SRL - w_r L\alpha = 2{,}385 - 1.343 \times 4{,}000 \times 0.638 = -1{,}043\,\text{lbf}$$

The negative sign is telling us there is force acting upward on the rod at the surface.

Example 5.3 A uniform rod string with a diameter of 1.5 in. is reciprocating in a formation fluid with a specific gravity SG = 0.8 as shown in Fig. 5.31. The weight per foot in air of the rod string is 1.343 lb/ft. The wellbore has a Kick of Point of KOP = 3,000 ft and the True Vertical Depth of TVD = 6,000 ft. The Dog Leg Severity of the inclined section is DLS = 5 °/100 ft. The rod speed and stroke length are 30 SPM and 50 in., respectively. Calculate The loads at the KOP and at the surface during the:

a. upstroke (neglect fluid friction)
b. downstroke (neglect fluid friction)

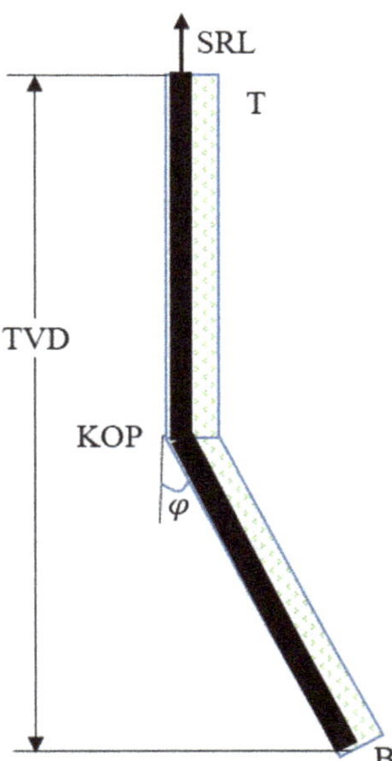

Fig. 5.31 Diagram for Example 5.2

The mechanical friction between the rod string and the production tubing is 0.45.

Solution

a. Calculate the loads at the KOP and at the surface during the upstroke

The Mill's acceleration factor:

$$\alpha = \frac{SN^2}{70{,}500} = \frac{50 \times 30^2}{70{,}500} = 0.638$$

Because the DLS = 5 °/100 ft is constant, the angle $\varphi = 5°$.
The length of the inclined section:

$$L_{KOP-B} = \frac{TVD - KOP}{\cos \varphi} = \frac{6{,}000 - 3{,}000}{\cos 5} = 3{,}012\,\text{ft}$$

The buoyant weight per foot of the rod string (weight in the fluid):

$$w_b = w_r(1 - 0.127 \times SG) = 1.343(1 - 0.127 \times 0.8) = 1.2\,\text{lb/ft}$$

The buoyant weight of the rod string from the KOP to the bottom (inclined section)

$$W_b^{inclined} = w_b \times L_{KOP-B} = 3{,}633\,\text{lbf}$$

Neglecting the friction due to the fluid flow inside the annulus, the rod load at the KOP can be calculated as follows:

$$RL_{KOP} = W_b^{inclined}\cos\varphi + \mu_m W_b^{inclined}\sin\varphi + w_r L_{KOP-B}\alpha\cos\varphi$$
$$RL_{KOP} = 3{,}633 \times \cos 5 + 0.25 \times 3{,}633 \times \sin 5 + 1.343 \times 3{,}004 \times 0.638 \times \cos 5$$
$$RL_{KOP} = 6{,}073\,\text{lbf}$$

If the mechanical friction is neglected $(\mu_m W_b^{inclined}\sin\varphi = 0$ then RL_{KOP} = 5,930 lbf.

The rod load at the surface:

$$RL_{surf} = RL_{KOP} + w_b \times KOP + w_r L_{KOP}\alpha\cos$$
$$RL_{surf} = 6{,}073 + 1.2 \times 3{,}000 + 1.343 \times 3{,}000 \times 0.638 \times \cos 5 = 12{,}254\,\text{lbf}$$

b. Calculate the loads at the KOP and at the surface during the downstroke

The rod load at the KOP:

$$RL_{KOP} = W_b^{inclined}\cos\varphi - \mu_m W_b^{inclined}\sin\varphi - w_r L_{KOP-B}\alpha\cos\varphi$$
$$RL_{KOP} = 3{,}633 \times \cos 5 - 0.25 \times 3{,}633 \times \sin 5 - 1.343 \times 3{,}004 \times 0.638 \times \cos 5$$
$$RL_{KOP} = 1{,}167\,\text{lbf}$$

The rod load at the surface:

$$RL_{surf} = RL_{KOP} + w_b \times KOP - w_r L_{KOP}\alpha\cos$$
$$RL_{surf} = 6{,}073 + 1.2 \times 3{,}000 = 2{,}224\,\text{lbf}$$

Example 5.4 A pump with a 1.5 in. plunger is to be set at 6,000 ft using a three-way taper. Sucker rods are available in 25 ft lengths. Determine the length of each section of the tapered rod string if rod number 75 is used.

Solution

From Table 5.7; for API rod number 75 and plunger diameter of 2.00 in., the three rod diameters are recommended as follows:

$$\begin{aligned} R_1 &= 42.4\% \text{ of } 7/8 \text{ in. rods} \quad \text{Then} \quad L_1 = 6000(0.328) = 2{,}544 \text{ ft} \\ R_2 &= 41.3\% \text{ of } 3/4 \text{ in. rods} \quad \text{Then} \quad L_2 = 6000(0.332) = 2{,}478 \text{ ft} \\ R_3 &= 16.3\% \text{ of } 5/8 \text{ in. rods} \quad \text{Then} \quad L_3 = 6000(0.339) = 978 \text{ ft} \end{aligned}$$

Number of rods for each rod diameter:

$$\begin{aligned} N_1 &= 102 \quad \text{Then} \quad L_1 = 2{,}550 \text{ ft} \\ N_2 &= 99 \quad \text{Then} \quad L_2 = 2{,}475 \text{ ft} \\ N_3 &= 39 \quad \text{Then} \quad L_3 = 975 \text{ ft} \end{aligned}$$

Total length: L_{total} = 2,550 + 2,475 + 975 = 6,000 ft.

Example 5.5 Given the following inputs for a sucker rod pump unit:

Pumping depth	= 5,900 ft
Desired fluid production	= 150 B/D
Volumetric efficiency	= 0.8
Stroke length	= 64 in.
Pumping speed	= 16.5 SPM
Plunger diameter	= 1 1/4 in.
Rod number	= API No.7 6
Fluid specific gravity	= 1.0

Estimate the following parameters for conventional, air balanced and Mark II rod pump unit

a. The peak and minimum polished rod loads.
b. The counterbalance required.
c. The peak torque
d. The nameplate horsepower of the NEMA “D” electric prime mover and low speed engines.

Solution

a. The PPRL and MPRL for conventional, air balanced and Mark II units

From Table 5.7, for API rod number 76 and plunger diameter of 1.25 in., the weight per foot rod the rod in air is 1.814 lbf/ft.

Cross-sectional area of the Plunger

$$A_p = \frac{\pi D_p^2}{4} = \frac{\pi 1.25^2}{4} = 1.227 \text{ in.}^2$$

The total buoyant weight of rod (total weight of rod in the fluid):

$$W_b = w_r L(1 - 0.127 \times SG) = 1.814 \times 5{,}900(1 - 0.127 \times 1) = 9{,}343\,\text{lbf}$$

The Mill's acceleration factor:

$$\alpha = \frac{SN^2}{70{,}500} = \frac{64 \times 16.5^2}{70{,}500} = 0.247$$

The force due to the acceleration

$$F_a = w_r L\alpha = 1.814 \times 5{,}900 \times 0.247 = 2{,}645\,\text{lbf}$$

The force due to fluid hydrostatic pressure acting on the plunger

$$F_o = 0.433 \times L \times SG \times A_p = 0.433 \times 5{,}900 \times 1 \times 1.227 = 3{,}135\,\text{lbf}$$

The peak polished rod load for the conventional, air balanced, and Mark II unit, respectively:

$$PPRL = W_b + F_a + F_o = 9{,}343 + 2{,}645 + 3{,}135 = 15{,}124\,\text{lbf}$$
$$PPRL = W + 0.7F_a + F_o = 14{,}330\,\text{lbf}$$
$$PPRL = W + 0.6F_a + F_o = 14{,}065\,\text{lbf}$$

The minimum polished rod load for the conventional, air balanced, and Mark II unit, respectively:

$$MPRL = W_b - F_a = 9{,}343 - 2{,}645 = 6{,}698\,\text{lbf}$$
$$MPRL = W - 1.3F_a = 5{,}905\,\text{lbf}$$
$$MPRL = W - 1.4F_a = 5{,}640\,\text{lbf}$$

The counterbalance for conventional, air balanced and Mark II units
The counterbalance required is calculated as follows:

$$CB = \frac{PPRL + MPRL}{2}$$

$$\text{CB} = 10{,}911\,\text{lbf} \quad \text{For conventional unit}$$
$$\text{CB} = 10{,}117 \quad \text{For air balanced unit}$$
$$\text{CB} = 9{,}853 \quad \text{For Mark II unit}$$

b. The peak torque for conventional, air balanced and Mark II units

The peak torque for conventional unit can be estimated using Eq. (5.3.32a)

$$PT = \frac{(PPRL - CB)MTF}{0.93}$$

From Table 5.2, the value of MTF = 34 for the stroke length of 64 in.

$$PT = \frac{(15{,}124 - 10911)34}{0.93} = 154{,}012\,\text{in. lbf}$$

Peak torque for air balanced and Mark II can be calculated using Eq. (5.3.32b). The values of MTF_u and MTF_d are 29 and 37, respectively given in Table 5.2b.

$$PT = \frac{0.5(PPRL \times MTF_u - MPRL \times MTF_d)}{0.93}$$

$$PT = 105{,}967\,\text{in. lbf} \quad \text{For air balanced}$$
$$PT = 107{,}104\,\text{in. lbf} \quad \text{For Mark II}$$

c. The nameplate horsepower of the NEMA "D" electric prime mover and low speed engines

The polished rod horsepower is calculated using Eq. (5.3.34)

$$PRHP = \frac{PT \times N}{63{,}025}$$

$$\text{PRHP} = 40.3\,\text{HP} \quad \text{For conventional unit}$$
$$\text{PRHP} = 27.7\,\text{HP} \quad \text{For air balanced unit}$$
$$\text{PRHP} = 28.0\,\text{HP} \quad \text{For Mark II unit}$$

The nameplate horsepower

$$HP_{np} = \frac{PRHP \times CLF}{\eta}$$

From Table 5.3: the cyclic load factor, CLF, for conventional, air balanced, and Mark II units using NEMA "D" motor and slow speed engines is: CLF = 1.375, 1.375, 1.10. The pump efficiency is $\eta = 0.8$.

$$\text{HP}_{\text{np}} = 69.3\,\text{HP} \quad \text{For conventional unit}$$
$$\text{HP}_{\text{np}} = 47.7\,\text{HP} \quad \text{For air balanced unit}$$
$$\text{HP}_{\text{np}} = 38.5\,\text{HP} \quad \text{For Mark II unit}$$

The results are summarized in Table 5.8.

Table 5.8 Summary the results of Example 5.5

Pump unit	PPRL (lbf)	MPRL (lbf)	CB (lbf)	PT (in.-lbf)	PRHP (HP)	HPnp (HP)
Conventional	15123.6	6698.2	10910.9	154012.1	40.3	69.3
Air balanced	14330.1	5904.7	10117.4	105966.7	27.7	47.7
Mark II	14065.5	5640.2	9852.9	107104.4	28.0	38.6

Example 5.6 Consider a well that is to be pumped at a rate of 100 B/D. The well conditions are:

- Pump depth = 5000 ft
- Anchored Tubing diameter = 1.9 in.
- Plunger diameter = 1.06 in.
- For preliminary design, pump speed, N = 16 SPM
- Polished rod stroke length, S = 64 in.
- Specific gravity of the pumped fluid = 0.88.
- A tapered sucker rod string (API 75), consisting of 1350 ft of 7/8 in. rods, 1375 ft of ¾ in. rods, and 2275 ft of 5/8 in. rods, is calculated to be sufficient for this design. These grade D rods have a maximum allowable rod stress of 34,000 psi.

Solution

From Table 5.7, for API rod number 75 and the plunger diameter of 1.06 in., the WPF of the rod in air is w_b = 1.566 lbf/ft.

The cross-sectional area of the plunger:

$$A_p = \frac{\pi D_p^2}{4} = \frac{\pi \times 1.06^2}{4} = 0.882\,\text{in.}^2$$

The theoretical pump displacement

$$PD = 0.1484 \times A_p \times S \times N = 0.1484 \times 0.882 \times 64 \times 16 = 134\,BPD$$

a. <u>*Calculation of actual pump rate*</u>

The dimensionless pump speed becomes:

$$\frac{N}{N_o} = \frac{NL}{245{,}000} = \frac{16 \times 5{,}000}{245{,}000} = 0.326$$

For the selected tapered rod string, the frequency factor from Table 5.7: Fc = 1.191

$$\frac{N}{N_o'} = \frac{NL}{245{,}000F_c} = 0.274$$

The force due to fluid hydrostatic pressure acting on the plunger

$$F_o = 0.433 \times L \times SG \times A_p = 0.433 \times 5{,}000 \times 0.88 \times 0.882 = 1{,}681\,\text{lbf}$$

From Table 5.4, the elastic constant, E_r = 0.997E−06 in./lbf ft. The spring constant k_r:

$$k_r = \frac{1}{E_r L} = \frac{1}{(0.997 \times 10^{-6})5{,}000} = 200.6\,\frac{\text{lbf}}{\text{in.}}$$

The dimensionless fluid rod stretch:

$$\frac{F_o}{Sk_r} = \frac{1{,}681}{64 \times 200.6} = 0.13$$

Using Eq. (5.4.11) or Fig. 5.20 gives the dimensionless plunger stroke:

$$\frac{S_p}{S} = 0.97$$

This means if the tubing is anchored, the bottomhole plunger stroke length is equal to 97% of the polished rod stroke length at the surface. Therefore, the bottomhole stroke length will be 62 in.

The actual pump displacement:

$$Q = 0.1484 \times A_p \times S \times N = 0.1484 \times 0.882 \times 62.7 \times 16 = 130\,\text{BPD}$$

This more than the expected rate rate of 100 BPD.

b. <u>*Calculation of PPRL, MPRL and CB*</u>

Using Eq. (5.4.6) or Fig. 5.18 gives the dimensionless upstroke rod stretch:

$$\frac{F_1}{Sk_r} = 0.37$$

The PPRL can be calculated using Eq. (5.4.5)

$$PPRL = w_b L + \frac{F_1}{Sk_r} Sk_r = 1.566 \times 5{,}000 + 0.37 \times 64 \times 200.6 = 12{,}422\,\text{lbf}$$

Using Eq. (5.4.9) for Fig. 5.19 gives the dimensionless downstroke rod stretch:

$$\frac{F_2}{Sk_r} = 0.21$$

The MPRL can be calculated using Eq. (5.4.8)

$$MPRL = w_b L - \frac{F_2}{Sk_r} Sk_r = 1.566 \times 5{,}000 + 0.21 \times 64 \times 200.6 = 5{,}223\,\text{lbf}$$

The ideal counterbalance

$$CB = \frac{PPRL + MPRL}{2} = \frac{12{,}422 + 5{,}223}{2} = 8{,}823$$

c. *Calculation of PT and HP_{np}*

Using Eq. (5.4.14) for Fig. 5.22 give the dimensionless torque:

$$\frac{2T}{S^2 k_r} = 0.32$$

With $\frac{N}{N_o'} = 0.274$ and $\frac{F_o}{Sk_r} = 0.13$, from Fig. 5.21 gives the percentage correction of 2.5%.

$$\frac{\frac{w_b L}{Sk_r} - 0.3}{0.1} = \frac{\frac{6{,}948}{62 \times 200.6} - 0.3}{0.1} = 3.3$$

$$\text{Ta} = 1.0 + 2.5\% \times 3.3 = 1.08$$

The peak torque can be obtained using Eq. (5.4.12)

$$PT = \frac{2T}{S^2 k_r} \times Sk_r \times T_a \frac{S}{2} = 0.32 \times 62 \times 200 \times 1.08 \times \frac{62}{2} = 133{,}000\,\text{in. lbf}$$

Using Eq. (5.4.15) or Fig. 5.23 gives:

$$\frac{F_3}{Sk_r} = 0.225$$

The polished rod horsepower:

$$PRHP = 2.53 \times 10^{-6} \frac{F_3}{Sk_r} \times Sk_r \times S \times N = 7.48\,HP$$

The nameplate horsepower for the NEMA Design D electric motor:

$$HP_{np} = \frac{PRHP \times CLF}{\eta} = \frac{7.48 \times 1.375}{0.63} = 16.3\,HP$$

References

1. API RP 11L (1988) Recommended practice for design calculation for sucker rod pumping system, 4th edn., June
2. API Specification 11B (1980) Specification for sucker rods, 18th edn., January
3. API Specification 11AX (2015) Subsurface sucker rod pump assemblies, components, and fittings, 13th edn. American Petroleum Association
4. Baker Hughes—A GE Company (2018) Lufkin beam pumping units. Conventional and reverse Makr. Installation and operations manual
5. Brown K (1980) The technology of artificial lift methods, vol 2a. PennWell Publishing Company, Tulsa, OK
6. Fagg L (1949) Dynamometer charts and well weighing. Presented at the Petroleum Branch Meeting in San Antonio, TX
7. Heinze L, Ge Z, Rahman M (1999) Sucker-rod pumping failures in the Permian Basin, SPE 56661. Presented at the ATCE in Houston, TX, October
8. Marsh H, Watts E (1938) Practical dynamometer tests. Presented at Eighth Mid-year Meeting, Wichita, May
9. Moises G, Andrade S, Carcia A (2010) Sucker-rod pumping failures diagnostic system, SPE 134975. Presented at the ATCE in Florence, Italy, September
10. Nguyen TC, Bhargava U, Al-Safran E (2019) Effect of viscosity on rod pump performance, MLF67. Middle East Artificial Lift Forum (MEALF), Oman
11. Norris (2014) A special report from Norris—Sucker rod failure analysis
12. Williams B, Fischer H (2010) Gas locking and gas interference solutions for sucker rod pumps. In: 6th annual sucker rod pumping workshop. Dallas, TX, September

Chapter 6
Plunger Lift

6.1 Fundamentals of Plunger Lift

6.1.1 Introduction and Main Principles of Plunger Lift

The primary challenge for gas well production is the removal of liquids from the wellbore. Early in the life of a gas well, reservoir pressure is high and the well usually has continuous, steady production. The gas velocity inside the production tubing is normally high enough to carry all the liquid upward to the surface in the form of mist entrained in the gas stream as shown in Fig. 6.1. Over time, well production causes the reservoir pressure to decline, which causes a reduction in gas velocity. If the gas velocity is not high enough, mist flow can no longer be sustained. The liquids will begin to coalesce on the inner wall of the production tubing to form a film. This flow pattern is called annular flow, as shown in Fig. 6.1. As gas velocity inside the production tubing continues to decrease, liquid droplets coalesce in the center of the tubing, and the annular film becomes thicker. As the gas velocity decreases further, the gravitational effect on the liquid starts to dominate the drag force between the gas and liquid. This causes the liquids on the tubing walls to stall and fall against the direction of gas flow [13]. Under these conditions, known as critical lift slug flow occurs, as shown in Fig. 6.1 At this point, liquids begin to accumulate in the wellbore and significantly hinder gas production. The well begins a cyclic process of unloading liquids commonly referred to heading or slugging. The liquid slugs will get bigger with time and apply more back pressure on the formation. If the reservoir pressure is not high enough to lift the liquid slug, all the liquid will accumulate at the bottom of the well, resulting in bubble flow as shown in Fig. 6.1. Often, this means the end of the well's life unless an artificial lift method is applied to remove the liquid from the wellbore [3].

Plunger lift is one of the artificial lift methods commonly used for vertical liquid-producing gas wells to remove liquids from the wellbore and maintain gas production [4]. It is typically applied at the time that slugging begins to occur in the

T. Nguyen, *Artificial Lift Methods*, Petroleum Engineering,
https://doi.org/10.1007/978-3-030-40720-9_6

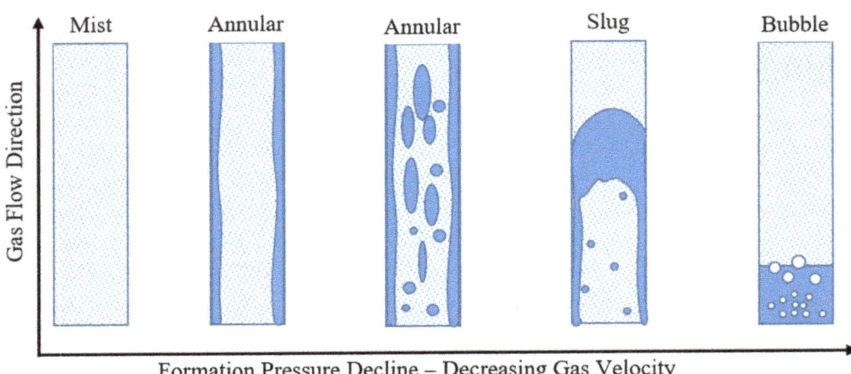

Fig. 6.1 Potential flow patterns existing in the production tubing during the life of a gas well

production tubing. Often, the well will be configured to flow intermittently for a few weeks using a controller and valve. The well is shut into build up pressure, and then a control valve is opened to lift large slugs of fluid and flow the well. This is inefficient because there is a significant amount of liquid fallback as the slug is lifted. Gas quickly breaks through the slugs, and the liquid falling back creates new slugs as the mixture travels up the production tubing. A large amount of the pressure energy is wasted due to this churning effect. The plunger lift method is used to prevent liquid fallback and lift the fluid more efficiently [17]. A plunger is used to serve as a mechanical interface between the gas and the liquid slug. This solid interface helps to prevent gas from breaking through the slug and hence decrease the liquid fallback. The slug is brought up all at once, and the well can then flow for a period of time with minimal bottomhole pressure.

A typical plunger lift system is shown in Fig. 6.2.

In general, plunger lift systems are best applicable for high liquid-producing gas wells or high gas-liquid-ratio oil wells where a combination of low reservoir pressure and/or permeability are causing liquid slugging and intermittent production [2]. Plunger lift is used to efficiently remove liquid slugs, reduce back pressure on the formation, and prolong the life of the well. Note that plunger lift is an intermittent lift method that relies completely on energy from the reservoir. Every plunger lift cycle, the well is shut for a period of time to build up pressure before production is resumed.

A complete plunger cycle can be divided into four stages namely: (1) buildup, (2) upstroke, (3) blow-down for gas well or after-flow for oil well, and (4) downstroke as shown in Fig. 6.3.

Buildup stage: The control valve is closed and the well is completely shut in. The plunger is seated on top of the bumper spring as shown in Fig. 6.3a. Produced gas is stored and builds pressure inside the casing annulus (between the production casing and the production tubing) and hence the surface casing pressure is increasing. Note that the plunger, by design, does not create a perfect seal because

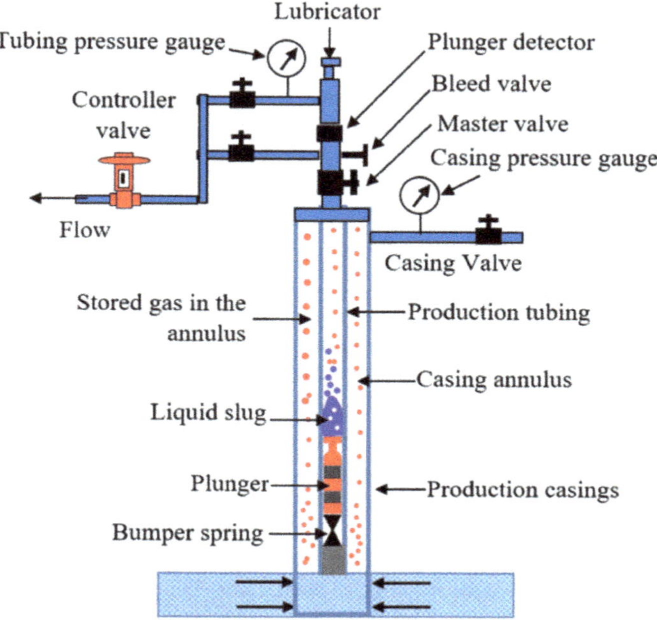

Fig. 6.2 Schematic of a typical plunger lift system

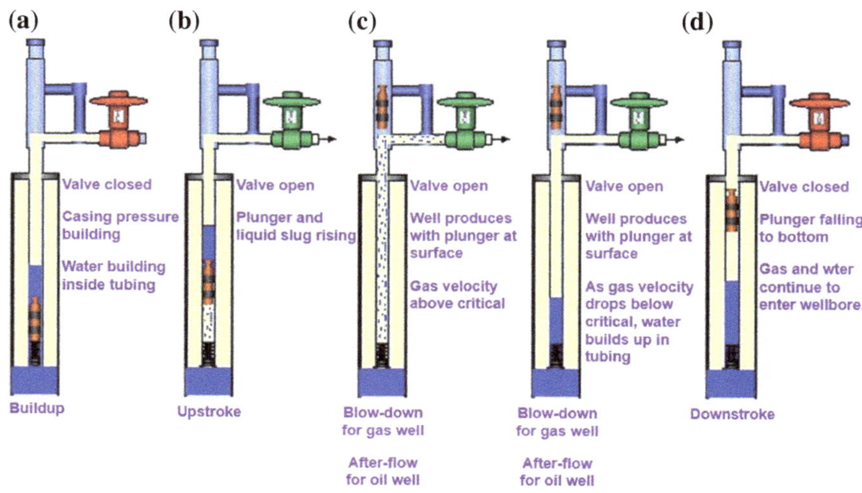

Fig. 6.3 Four stages during a complete plunger cycle. **a** Buildup stage, **b** upstroke stage, **c** blow-down stage, **d** downstroke stage

of the need to minimize mechanical friction with the tubing. In other words, there is always pressure communication across the plunger. Therefore, as gas and liquid flow into the well during this stage, gas and liquid build up inside the tubing. The surface tubing pressure increases as shown in Fig. 6.4. The difference between the tubing and casing pressure is due to the hydrostatic pressure of the liquid column in the production tubing.

Upstroke stage: When the casing pressure reaches a desired value, the control/ sales valve is opened, which causes a rapid reduction in tubing pressure. This will cause a pressure difference across the plunger which drives the plunger to accelerate and travel upward as shown in Fig. 6.3b. The liquid slug will rise at more or less the same speed as the plunger. The efficiency of a plunger lift system is dependent on the amount of liquid falling back through the small clearance between the plunger and the tubing, and often by how much gas slips upward past the plunger [5]. More on this later. As the plunger moves upward, gas in the casing annulus expands and enters the production tubing, which can be seen at the surface as a slow reduction in casing pressure. The upstroke stage ends when the plunger reaches the surface. During this stage, the bottomhole pressure decreases allowing more formation fluids to enter the well.

Blow-down stage: The plunger arrives in the lubricator above the wellhead and the slug of liquid travels down the flowline to a separator. The plunger remains suspended in the lubricator between the dual flow lines by a small pressure differential. The control valve remains open as shown in Fig. 6.3c. Because the weight of the liquid slug is no longer impeding flow, gas from the casing annulus rapidly enters the production tubing, and the surface casing pressure rapidly drops. The lowest

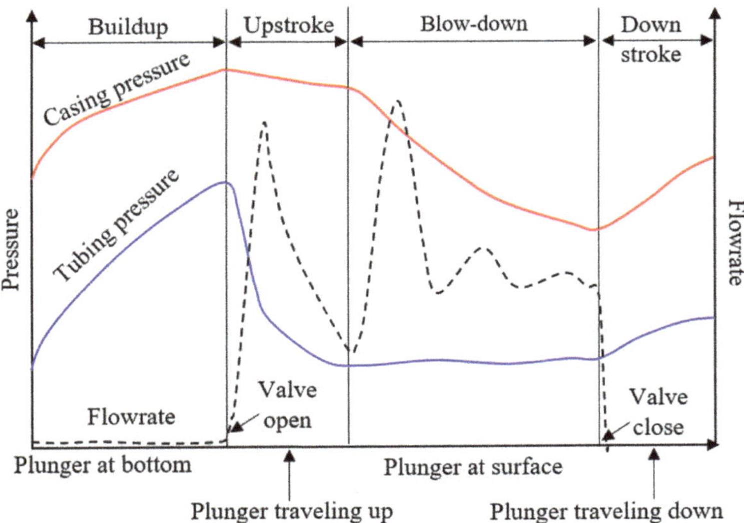

Fig. 6.4 Pressures and flowrate response during a plunger cycle

bottomhole pressure occurs during the end of this stage, and thus the maximum inflow occurs. Tubing pressure drops to just above the line pressure of the gathering system, and then remains there. As the gas velocity drops below the critical lift point, liquids begin to accumulate in the production tubing. The gas rate drops, and as more liquids enter the wellbore, the casing surface pressure starts to flatten.

Downstroke stage: The control valve closes, stopping the flow of gas, and the plunger begins to fall as shown in Fig. 6.3d. During this stage, tubing and casing pressures start to increase as shown in Fig. 6.4. The buildup stage begins when the plunger lands on the bumper spring, completing the cycle. This can be detected as a 1–3 psi bump in the surface tubing pressure on low liquid rate gas wells equipped with high resolution SCADA capability. For practical applications, it is accepted that most commercial plungers fall at a velocity of 1,000 ft/min in the gas phase and 170 ft/min in the liquid phase. Of course, these values are just a guideline for quick calculations. The actual falling velocity of a plunger depends on its weight, its design, wellbore deviation, and the clearance between the tubing and the plunger. Wear is also a factor. As the plunger wears out, the fall velocity increases, and the sealing efficiency on the upstroke decreases.

A properly optimized plunger lift system uses the available reservoir energy as efficiently as possible to lift the plunger and remove fluids from the wellbore [2]. This allows the system to operate at the lowest possible average casing pressure and maximizes well drawdown. This also maximizes the number of cycles the plunger can make per day. The buildup stage should only be long enough to build up enough energy to lift the plunger. Any longer will result in wasted energy by lifting the plunger excessively fast, and will cause the average bottomhole pressure to be higher. Average plunger lift velocity should be in the range of 600–1200 ft/min to avoid stall out. Excessively high speeds (2000–3000 ft/min) can cause damage to the plunger and surface equipment.

In general, a relatively strong well will build up enough pressure before the plunger falls to bottom. In this case, the buildup stage should be very short, just long enough to ensure the plunger is on bottom before starting the upstroke. Plunger speed can then be controlled by adjusting the amount of blowdown. As the well weakens, the blowdown is shortened to maintain the lift velocity. Eventually the well is operated with minimum buildup and minimum blowdown. As soon as the plunger reaches bottom, the control valve opens, and as soon as the plunger arrives, it shuts again. At this stage of the well's life, the system has the most cycles per day. As the well depletes further, the buildup must be lengthened to maintain lift velocity. In this stage, there should be little or no blowdown time. If the plunger system is operated with significant buildup time, and significant blowdown time at the same time, then there is room to optimize the system and increase uplift.

Advantages of using plunger lift are as follows [4]:

- It is the most cost effective lift method for liquid-producing gas wells. The equipment and installation costs are relatively small compared to gas lift or subsurface pumps, and the operating and maintenance costs are low.

- Plunger lift is quite reliable. The main wear item is the plunger. It can be easily inspected for wear and damage, and replaced without disrupting production. The main causes of damage are operating with excessive rise velocity, or operating with little or no fluid, both of which can be avoided with proper control settings.
- The external power required is minimal. The main energy source to drive the plunger is reservoir pressure. A small amount of power is required for the control panel, control valve, and any SCADA system devices. Solar panels and batteries are enough to provide power for this system. Instrument gas is typically used to operate the control valve, but all-electric options are available.
- Can operate at very low liquid rates, as low as 0.25 bbl/day. These rates are impractical for other lift methods.
- Plunger lift is tolerant of paraffin, which often forms in the tubing midway up the wellbore. The plunger acts as a scraper to keep it from building up and causing a blockage.

Limitations of using plunger lift are as follows [4]:

- Modeling, optimizing and troubleshooting can be a challenge [10]. Plunger lift is dynamically complex, involving compressible gas expansion, gas and fluid accelerations, and different multiphase flow patterns existing in the tubing below and above the plunger [15].
- Typically limited to liquid rates less than 20 bbl/day, but higher rates are possible depending on reservoir pressure and inflow. If the reservoir pressure is not high enough to lift the plunger and the liquid slugs upward, plunger lift is not applicable.
- Plungers cannot run in horizontal sections of wells. Maximum inclination is about 60°, which is typically the limit for setting bottomhole equipment via slickline. Plungers are typically used in vertical wells. They can tolerate some wellbore deviation, but efficiency decreases with inclination.
- Plunger lift is not suitable when continuous production is required, such as in horizontal wells where intermittent flow causes liquid to fall back into the horizontal. However, bypass plungers and gas-assisted plunger lift may be useful in such circumstances.
- This method is not suitable for wells which have high sand production problems. However, very small amounts of sand can be tolerated by using bar stock or brush type plungers, which are often employed for a short time after well workovers to help clean up the well.

6.1.2 Surface and Subsurface Plunger Lift Equipment

6.1.2.1 Surface Equipment: Lubricator

The lubricator is installed directly on the top of the tree, above the master valve. The primary function of the lubricator is to catch the plunger and to absorb the kinetic energy of the plunger when it arrives the surface. The lubricator consists of a shock spring, a bumper plate, a catcher, a sensor for detecting the arrival of the plunger, and a removable cap to allow for inspection of the plunger as shown in Fig. 6.5.

The main function of the shock spring is to absorb the impact of the plunger, which preserves the integrity of the lubricator and reduces the risk of damaging the plunger. Excessive arrival speed can result in mushrooming of the plunger fishneck, or fracturing of the plunger. This usually requires slickline intervention to retrieve broken pieces of the plunger from the tubing. The spring may be a steel coil, or it may be a polymer design. Both designs should be inspected regularly and replaced as needed. The catcher is manually engaged to catch the plunger in the lubricator to facilitate inspection and replacement. Once caught, the master valve and flow lines can then be closed, the lubricator bled-down, and then the cap can be removed. The flow ports tie the lubricator/catcher assembly into the flowline piping. Dual flow ports are preferred over a single flow port because they help to handle the arrival of the liquid slug, and they are better at providing the pressure differential required to suspend the plunger during blowdown. However, even with dual flow ports, the plunger will partially block the flow path. The manual catcher can be replaced with an automatic catcher and motor valve. This can be used to ensure the plunger is suspended out of the flow path and doesn't cause a restriction to flow. This can be beneficial on wells that have very long blowdown times.

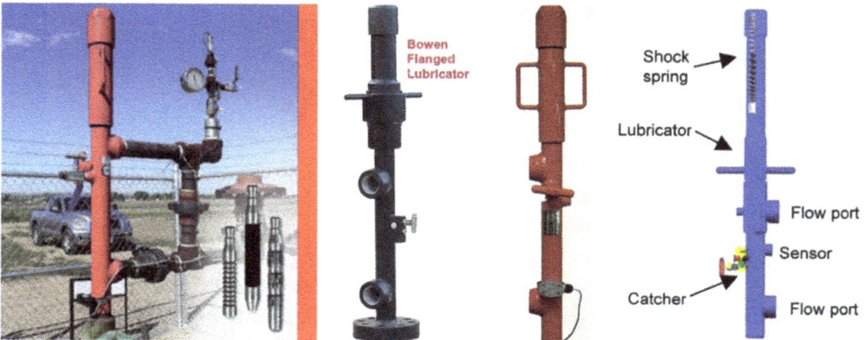

Fig. 6.5 Surface equipment: lubricator

6.1.2.2 Surface Equipment: Controller Valve and Controller

The controller valve (or sales valve), as shown in the schematic of a plunger lift system in Fig. 6.2, is used to control the plunger lift cycle. When the valve opens, the upstroke stage begins and the plunger travels upward. When the controller valve closes, the downstroke begins and the plunger falls. The sales valve is operated by the plunger lift controller. Controllers can employ pneumatic, electric, or even wireless methods to control the plunger lift cycle, but their function can be broken down into the following modes of operation:

- Time control: Timers (either electronic or mechanical) are used to set the open and close duration of the sales valve. As line pressure or well conditions change, the timers need to be manually adjusted to ensure proper plunger lift operation. Typically, the settings are set conservatively to ensure there is more than enough energy to cycle the plunger regardless of changing line pressure and well conditions. This means the well is never producing at its maximum potential.
- Pressure control: This controller monitors tubing or casing pressure, and sometimes line pressure, and operates the valve based on preset pressure triggers. Typically, the well opens when casing pressure has reached a certain high value, and closes when the pressure is reduced to a pre-set low. A more reliable method is to monitor the differential between casing and line pressure to determine when to open the valve. This allows the system to adjust to changing conditions. More aggressive settings can be employed to improve production without increasing the risk of stalling the plunger.
- Full automation: Sophisticated plunger controllers are quite versatile, and can provide all of the above modes of operation. They can also incorporate a variety of other features such as multiple open and close triggers (such as flow rate) and customizable logic that can deal with a variety of abnormal conditions. They are typically part of a SCADA system that allows for remote monitoring, control, and data trending. These systems greatly improve the ability to quickly troubleshoot problems and optimize production. Such systems can be expensive, so they are typically part of a field-wide operation strategy that can make the greatest use of their capabilities.

6.1.2.3 Subsurface Equipment: Plungers

Plungers are used in a plunger lift system as a solid interface to prevent the breakthrough of gas through liquid slugs. This minimizes the fall back of liquid to the bottom of the well during gas production and hence reduces backpressure on the formation and can prolong the life of the well. Plungers are designed in such a way to maintain a good seal between their surface and the tubing to mitigate the liquid falling during the upstroke. However, mechanical friction should be minimal, and enough clearance must be maintained to prevent the plunger from getting stuck in tight spots in the tubing. To achieve these requirements, manufacturers have come up with many different designs presented as follows:

Bar stock plungers: The simplest plunger design is a bar stock plunger (Fig. 6.6). It is a piece of metal (solid or hollow) whose surface is machined with grooves, spirals, or other shapes. The reason of creating different shapes around the surface of the plunger is to create liquid turbulence between its surface and the inner wall of the production tubing. This design is supposed to provide good seal and hence to improve the lift efficiency compared to a plain cylinder shape. This plunger type has low efficiency compared to other plunger designs. However, they are simple, sturdy, and tolerant of high speeds, rough tubing, paraffin, and even a small amount of sand production. This plunger falls at speeds between 700 and 900 ft/min.

Wobble washer plungers: The design of a wobble washer plunger (Fig. 6.7) is similar to that of a bar stock plunger. The only difference is that instead of the fixed groove or spiral shapes of a bar stock plunger, the wobble washer plunger has many loose-fitting washers. The side-to-side movement of its loose washers are intended to improve the seal, but in practice the sealing characteristics are comparable to those of a bar stock plunger.

The disadvantages of the wobble washer plunger are: (1) It is less durable than a bar stock or brush plunger; (2) It is much harder to retrieve if it breaks apart.

Pad plungers: A pad plunger (Fig. 6.8) incorporates spring-loaded metal pads, fitted on a mandrel, that expand to maintain contact with the tubing wall. The pads

Fig. 6.6 Bar stock plunger

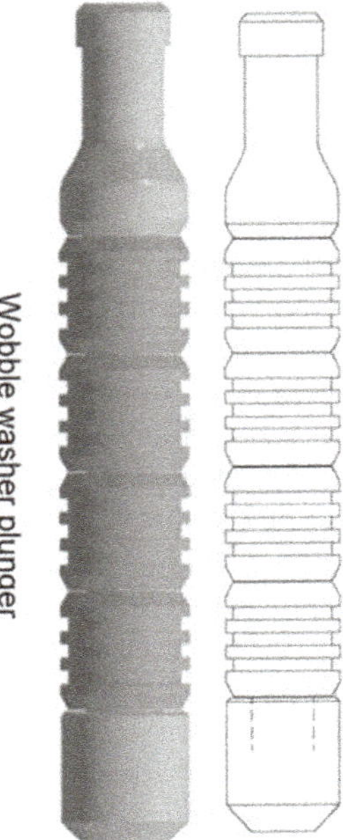

Fig. 6.7 Wobble washer plunger

Fig. 6.8 Pad plungers

are designed to have direct contact with the inner wall of the production tubing and hence to improve the sealing ability. In other words, the spring-loaded metal pads minimize the bypass area for gas slippage or liquid fall back. This type of design provides higher lift efficiency than other plunger designs, making it the primary choice for gas wells with tight producing formations or low reservoir pressure.

The disadvantages of this type of plunger are: (1) There is mechanical friction between the plunger and the tubing wall. This is typically minimal, but can be high if the springs are too stiff. This can reduce their efficiency and can cause the plunger to get stuck in the tubing or master valve. However, this friction decreases over time as the pads wear down. Manufacturers typically have several models or different spring tensions to choose from so that an optimal plunger can be chosen for a given well. (2) They do not tolerate sand production. Sand or trash easily cause the pads to stick, causing the plunger to get stuck. However, paraffin is usually not an issue, and the pads tend to keep the tubing scraped clean. (3) Fall speed is low, typically 250–400 ft/min in gas. The good sealing ability that makes them efficient at lifting fluids also restricts the bypass of gas around them as they fall. Extra fall time must be incorporated into the controller settings to make sure the plunger has made it all the way to bottom before the upstroke. (4) They are less durable than bar stock plungers, but more reliable than wobble washers. Plunger speed on the upstroke should be kept under 1000 ft/min to reduce the risk of a broken plunger.

Sealed pad plungers: A sealed pad plunger (Fig. 6.9) is a variation of the pad plunger. These plungers attempt to seal off the gap between the pad and the mandrel to further reduce gas slippage and improve lift efficiency. The seals may be made up of metal, rubber, polymer, or a tortuous path that creates turbulence behind the pads. The sealing materials are selected to be compatible with formation fluids.

Brush plungers: A brush plunger (Fig. 6.10) incorporates a bristle section made of polymer or steel bristles. As with any polymer, it is important to verify compatibility with wellbore fluids and chemicals before use. These plungers have the most efficient seal because the bristles contact the tubing, and fill in the gap between the mandrel and the tubing wall. The tortuous path through the bristles greatly

Fig. 6.9 Sealed pad plungers

Fig. 6.10 Brushed plungers

reduces gas and liquid slippage. This also means they have very slow fall velocities. However, the biggest advantage of these plungers is that they tolerate sand very well. Their main disadvantage is that the bristles wear very quickly. Within a month or less, the bristles wear down to the diameter of the mandrel, and they lose their seal and behave like a bar stock, with similar low efficiency and fast fall velocities. They are typically used for continuous sand production, or immediately after a well workover to clean up the production tubing for a few weeks before switching over to a pad plunger.

6.1.2.4 Subsurface Equipment: Bottomhole Assembly

The bottomhole assembly works as a shock absorber for the plunger as it reaches the bottom of the well at the end of the downstroke stage. The assembly can be set in a seating nipple, or it can be set anywhere in the tubing using a collar stop, tubing stop, or tubing packer. The latter is less reliable than the other methods. A few of these types are shown in Fig. 6.11. The assembly also includes a standing valve cage, and a bumper spring. "Floating" bumper springs that simply rest upon the

Fig. 6.11 Bottomhole assembly

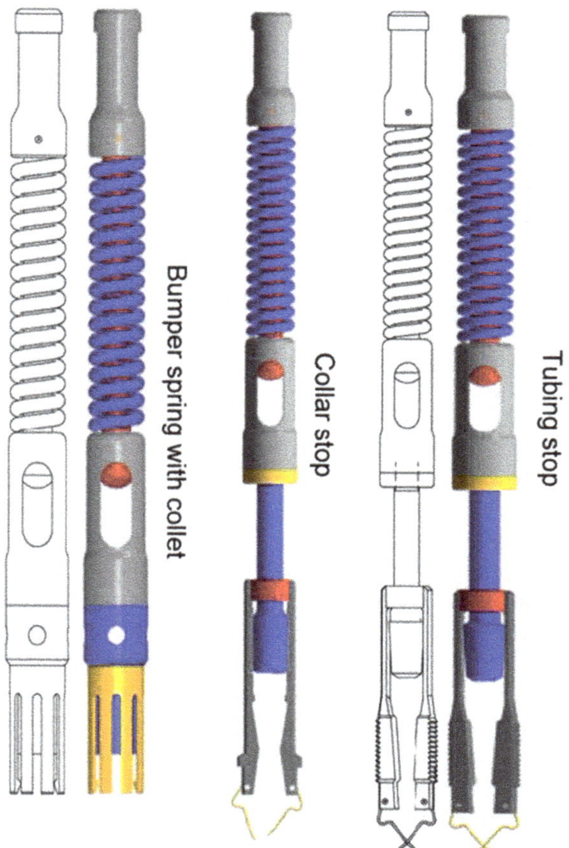

seating nipple are available, and have the advantage that they can be set and retrieved without slickline, but in some circumstances they can travel with the plunger.

- Collar stop: This device lands in the internal recess created by the joints of tubing at the collar. It is set and retrieved by slickline.
- Tubing stop: This slip type stop is utilized when it is necessary to land between collar recesses or if integral joint tubing is encountered.
- Standing valve cage: This is a standard pump standing valve cage. When a ball installed, it acts as a check valve to prevent liquid from dropping out of the tubing during the downstroke and buildup stages.
- Bumper spring: This sits above the standing valve, and it has a fishing neck for retrieval with slickline.

The bottomhole assembly is prone to scale build-up due to the pressure and temperature drop as the fluid and gas make their way through the assembly. A chemical treatment program is sometimes required to mitigate this. Slickline intervention is typically required to inspect and replace the bottomhole assembly.

6.1.3 Plunger Lift Applications

Plunger lift is used mainly to remove liquids from gas wells, to produce either oil or gas in high Gas Liquid Ratio (GLR) wells, and to remove paraffin and hydrate.

6.1.3.1 Remove Liquid from Gas Wells

Almost all gas wells will experience liquid loading toward the end of their life. Liquid loading occurs in a gas well when gas velocity inside the tubing is not high enough to carry liquids out. If no artificial lift is applied to this well, the problem will get worse until the well loads up and dies. Liquid loading is easily diagnosed as a sharp reduction in production for both gas and liquids, and a high differential between casing and tubing pressures. Plunger lift is used to prevent these liquids from accumulating to the point that the well would die or require a lengthy shut-in period to recover.

The well is shut in when the pressure and formation fluids inside the well are low. The well is opened up when enough formation fluids have entered the well and casing pressure has built up enough to lift the accumulated liquids in the tubing along with the plunger. This pressure and velocity must be great enough to overcome the sales line or separate pressure encountered on the trip to the surface. As the plunger arrive the surface, liquids in the tubing are removed and hence the back pressure applied on the formation at the bottom of the hole is minimal. This will enhance the flow from the reservoir to the wellbore and prolong the life of the well.

6.1.3.2 Improve Production in High GLR Wells

Plunger lift can be used to improve production in high GLR oil wells in exactly the same manner as in gas wells. There are many oil wells that are periodically "milked" by shutting them in for days, or even weeks, and then after pressure builds up they are flowed for a short period of time. In many cases, the GLR is high enough to use plunger lift to automate this process and improve the overall oil production of the well.

For either oil or gas well applications, the reservoirs typically fall into one of the following types: Low reservoir pressure but high productivity index and high reservoir pressure with low productivity index.

High GLR wells with low reservoir pressure but high productivity normally do not have enough energy to produce formation fluids by themselves. These wells tend to die out very quickly due to the liquid loading. Here, the plunger lift system is used to move the liquid to surface as quickly and efficiently as possible. Typically, the high productivity of the reservoir causes significant amounts of fluid to enter the wellbore during all stages of the plunger lift cycle, even the buildup stage. Balancing the available pressure energy with the energy required to lift for these types of wells can be a challenge. If the well is shut for too long, or if it is allowed to blowdown for too long, too much fluid will enter the well for the plunger to successfully lift the next slug, and production will drop off. So it is necessary to cycle the plunger frequently enough to remove small amounts of liquids at a time in order to maintain production. Plunger fall speed and control type may need to be considered for successful operation.

Another type of high GLR reservoirs, which are good candidate for plunger lift, are those with high reservoir pressure but low productivity. These reservoirs are normally tight and do not have enough continuous flow to sustain liquid production. Over time, the well loads up with fluid and production drops off with steady bubble flow through the liquid column. Here, plunger lift can be effectively used to lift the fluids, enhance their production, and increase the ultimate recovery factor of these reservoirs. The key is to have a long enough buildup stage to store the energy required to lift the plunger and slug. Buildups of one to several hours are not uncommon for these types of wells, so good sealing plungers with slow fall speeds are acceptable. For optimized plunger lift operation, blowdown is typically short, or non-existent. Additional blowdown only depletes the casing-tubing annulus, depleting the available energy for the next cycle. This makes it necessary to lengthen the buildup stage, which increases the cycle time and reduces overall production.

6.1.3.3 Remove Paraffin and Hydrate

Paraffin is deposited as a microscopic film on surfaces below the cloud point of the oil, typically below 100 °F. The expansion of gas as it travels up the production tubing often causes it to cool below the cloud point partway up the tubing, and this is where the deposition occurs. Over time, the accumulation of paraffin begins to restrict flow, and it may even bridge off the tubing. For wells that have sufficient gas liquid ratios, installing a plunger lift system may be a simple solution to clean the deposition. Operating the plunger system several times a day may completely solve the problem associated with the paraffin deposition.

Hydrate is formed under low temperature and high pressure. The higher the pressure, the higher the temperature at which the hydrates will form. High pressure gas wells are particularly prone to this problem. The problem is compounded if there is a fresh water zone down hole that creates a temperature anomaly. This cooling effect could cause hydrate formations that can block off all flow up the tubing. This problem has been solved in many areas by the installation of a plunger in conjunction with a pneumatic chemical pump connected to the tubing at the surface. A typical cycle would synchronize injection of methanol or alcohol down the tubing when the low-line is shut-in and the plunger is falling. The methanol softens the hydrate plug so that the next cycle of the plunger removes any deposits.

6.2 Review of Inflow Performance of Gas Wells

Similar to other lift methods, petroleum engineers have to understand the inflow performance of the reservoir (IPR) and the outflow performance in the tubing (OPR). As discussed in Sect. 6.1, plunger lift systems are applicable mainly for gas wells hence reviewing the IPR for gas is important before moving to the modeling section.

IPR of a vertical gas well is a relationship describing how gas production rate depends on pressure drawdown and reservoir fluid properties. A simple form of the IPR can be written as followed

$$q = PI \times \Delta P \tag{6.2.1}$$

where PI is defined as the productivity index and ΔP is the pressure drawdown. The productivity of a gas well is determined with deliverability testing. Deliverability tests provide information that is used to develop reservoir rate-pressure behavior for the well and generate an inflow performance curve or gas-backpressure curve.

Recall Darcy's flow IPR of an oil well (for Darcy's flow: rate versus pressure gradient is a linear relationship):

$$q = \frac{kh\Delta P}{141.2B\mu\left(\ln\frac{r_e}{r_w} + S\right)} = PI\left(P_e - P_{wf}\right) \tag{6.2.2}$$

where $\Delta P = P_e - P_{wf}$ is the pressure drawdown defined as the difference between the reservoir and the flowing bottom hole pressures in psi; q is the flow rate in STB/D; μ is fluid viscosity in cp; k is the formation permeability in md; h is the pay zone thickness in ft; B is the formation volume factor to convert STB into reservoir bbl; S is the skin factor.

For natural gas wells, the average gas formation volume is calculated as follows:

$$\bar{B}_g = \frac{0.0283\bar{Z}T}{\frac{\left(P_e + P_{wf}\right)}{2}} \tag{6.2.3}$$

Substituting Eq. (6.2.3) into Eq. (6.2.2) and converting the rate from STB/D to MSCF/D gives an equation for calculating the gas well deliverability under steady state conditions:

$$q = \frac{kh\left(P_e^2 - P_{wf}^2\right)}{141.2\bar{\mu}\bar{Z}T\left(\ln\frac{r_e}{r_w} + S\right)} \tag{6.2.4}$$

Equation (6.2.4) tells us that a gas well deliverability is approximately proportional to the pressure squared difference. The viscosity and compressibility factor, Z, are average properties between P_e and P_{wf}. A similar approximation can be developed for pseudo-steady state.

$$q = \frac{kh\left(\bar{P}_e^2 - P_{wf}^2\right)}{141.2\mu ZT\left(\ln\frac{0.472r_e}{r_w} + S\right)} \tag{6.2.5}$$

Rawlins and Schellhardt [14] proposed a practical correlation to describe the IPR of a gas well if production data are available:

$$q = C\left(\bar{P}_e^2 - P_{wf}^2\right)^n \tag{6.2.6}$$

where C and n are the two constants and $0.5 < n < 1$. One may easily recognize that a log-log plot of q versus $\left(\bar{P}_e^2 - P_{wf}^2\right)$ would yield a straight line with the slope of n and the intercept of C as shown in Fig. 6.12.

Aronofsky and Jenkins [1] developed an analytical equation for predicting the IPR for gas flow through porous media which takes into account the non-Darcy effect (turbulent effect).

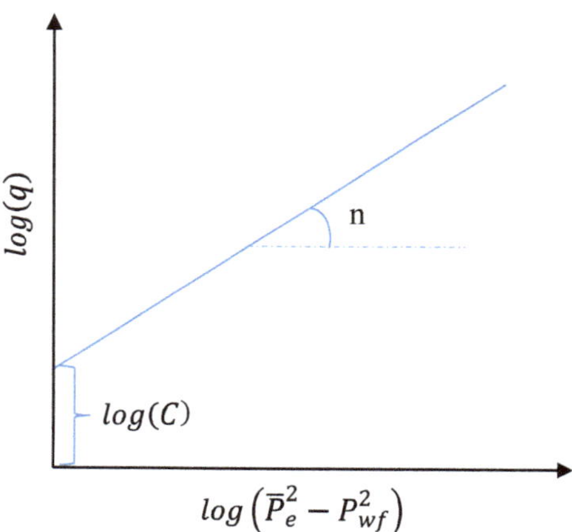

Fig. 6.12 Meaning of constants C and n

$$q\left(\frac{MSCF}{D}\right) = \frac{kh\left(\bar{P}_e^2 - P_{wf}^2\right)}{141.2\bar{\mu}\bar{Z}T\left(\ln\frac{r_d}{r_w} + S + D_q\right)} \tag{6.2.7}$$

where D_q is the non-Darcy coefficient and r_d is the Aronofsky and Jenkins "effective" drainage radius. If $r_d < 0.472r_e$, Eq. (6.2.7) is time-dependent and r_d can be determined as follows:

$$\frac{r_d}{r_w} = 1.5\sqrt{t_D} \tag{6.2.8}$$

$$t_D = \frac{0.000264kt}{\phi\mu c_t r_w^2} \tag{6.2.9}$$

where t_D is the dimensionless time and c_t is the total compressibility of formation rock and formation fluid.

If $r_d \geq 0.472r_e$ then Eq. (6.2.7) is time-independent and the value of r_d is a constant and equal to $0.472r_e$.

If production data are available, Aronofsky and Jenkins equation is normally rearranged as below to determine the non-Darcy coefficient:

$$\left(\bar{P}_e^2 - P_{wf}^2\right) = \frac{141.2\bar{\mu}\bar{Z}T}{kh}\left(\ln\frac{0.472r_e}{r_w} + S\right)q + \frac{141.2\bar{\mu}\bar{Z}TD_q}{kh}q^2 \tag{6.2.10}$$

The first term in the right-hand side of Eq. (6.2.10) represents for Darcy effects and the second term represents for non-Darcy effects. Rearranging Eq. (6.2.10) to form a straight line gives:

$$\frac{\left(\bar{P}_e^2 - P_{wf}^2\right)}{q} = a + bq \tag{6.2.11}$$

where

$$a = \frac{141.2\bar{\mu}\bar{Z}T}{kh}\left(\ln\frac{0.472r_e}{r_w} + S\right) \tag{6.2.12}$$

$$b = \frac{141.2\bar{\mu}\bar{Z}TD_q}{kh} \tag{6.2.13}$$

Using production data plot $\frac{\left(\bar{P}_e^2 - P_{wf}^2\right)}{q}$ versus q to obtain constants a and b. The non-Darcy coefficient D_q can be achieved if b is known.

If the production data are not available, the non-Darcy coefficient can be calculated using an empirical correlation developed by Economides et al. [6]:

$$D_q = \frac{6 \times 10^{-5} \gamma k_s^{-0.1} h}{\mu r_w h_{perf}^2} \tag{6.2.14}$$

where γ is the gas gravity, k_s is the near-wellbore permeability in mD, h and h_{perf} are the net and perforated thicknesses in ft, and μ is the gas viscosity in cp.

6.3 Analytical Modeling of Plunger Lift System

To have a better understanding the physics behind a plunger lift system, analytical modeling the system is a great way to approach before going into practical methods. Mass, momentum, and energy balance will be used to model a complete plunger cycle including four stages: buildup stage, upstroke stage, blowdown stage, and downstroke stage. Readers should be aware of that the system is always under changing or transient conditions regardless to the stage [16]. In other words, an accurate model must take into account the change of physical parameters with time.

6.3.1 *Modeling the Buildup Stage*

When the surface control valve is closed, gas from the reservoir will flow into the casing annulus between the casing and the production tubing leading to an increase in surface casing pressure. How fast the surface casing pressure increases depends on the performance of gas flow in the reservoir (IPR), the total volume of the annulus, and the initial casing pressure (casing pressure at the beginning of the buildup stage) [9]. Theoretically, if the well is shut down long enough, the bottomhole pressure is approaching to the reservoir pressure, P_e, and hence the surface casing pressure will reach a maximum value and remains constant. Applying the mass balance, the maximum casing pressure can be estimated as follows (readers can review Chap. 1 to see the derivation of this equation):

$$P_c^s\Big|_{max} = \frac{P_e}{\exp\left(\frac{0.01877\gamma_g TVD}{Z\bar{T}}\right)} \tag{6.3.1}$$

where the subscript, c, and superscript, s, are for casing and surface; TVD is the true vertical depth. However, the surface control valve is open before the surface casing pressure reaches its maximum value. Production engineers operate the surface control valve based on either constant time intervals or the rate of surface casing pressure increase/decrease.

The most important factor that impacts the valve operations is the IPR as described in Eq. (6.2.7). If the reservoir pressure, permeability, and pay thickness are high, the waiting time for the buildup stage is less or the valve is open sooner. One of the key operating parameters is the surface casing pressure. This casing pressure must be high enough to make sure the plunger can reach to the surface. If the surface casing pressure, P_c^s, is pre-determined to open the control valve, the waiting time for the buildup stage can be estimate as follows:

$$t_{buildup} = \frac{141.2\mu ZT\left(\ln\frac{0.472r_e}{r_w} + S\right)}{kh\left(\bar{P}_e^2 - P_b^2\right)} V_{annulus} \tag{6.3.2}$$

where P_b is the bottomhole pressure corresponding to the predetermined casing pressure.

$$P_b = P_c^s \exp\left(\frac{0.01877\gamma_g TVD}{Z\bar{T}}\right) \tag{6.3.3}$$

6.3.2 *Modeling the Upstroke Stage*

This is the most important stage of a plunger cycle as well as the most difficult stage to model due to the transient nature of the system. Let's first look at the very general energy equation to describe the system. Readers need to be aware of that pressure is defined as energy per unit volume. Therefore, energy and pressure can be expressed interchangeably. At the bottom of the well, the pressures in the casing-tubing annulus and in the tubing at the same depth are the same regardless to the position of the plunger. This fact can be expressed by the following energy equation (or pressure equation):

$$P_c^s + P_c^{h-g} - P_c^{f-g} = P_{t-bp}^{h-g} + P_{t-bp}^{f-g} + P_t^{f-p} + P_p + P_{sl} + P_{t-ap}^{f-sl} + P_{t-ap}^{f-g} + P_{t-ap}^{h-g} + P_{wh} \tag{6.3.4}$$

where P_c^s is the pressure in the casing-tubing annulus at the surface; P_c^{h-g} is the hydrostatic pressure of gas in the casing-tubing annulus; P_c^{f-g} is the frictional pressure loss of gas due to the expansion in the casing-tubing annulus; P_{t-bp}^{h-g} is the hydrostatic pressure of gas in the tubing below the plunger; P_{t-bp}^{f-g} is the frictional pressure loss of gas in the tubing below the plunger; P_t^{f-p} is the mechanical friction between the plunger and in the tubing; P_p, P_{sl} are the pressures to lift the weights of the plunger and the liquid slug; P_{t-ap}^{f-sl} is the frictional pressure loss of the liquid slug in the tubing above the plunger; P_{t-ap}^{f-g} is the frictional pressure loss of gas in the tubing above the plunger; P_{t-ap}^{h-g} is the hydrostatic pressure of gas in the tubing above the plunger; and P_{wh} is the wellhead pressure.

During the upstroke stage, all the physical parameters are in a state of change in such a way that the energy equation described by Eq. (6.3.4) always satisfies. The key physical parameter change includes casing pressure, tubing pressure, plunger velocity, and gas rate. The casing pressure decreases rapidly from a maximum value then reduces fairly constant to a minimum value. The tubing pressure decreases drastically from the maximum value (similar to the maximum casing pressure) to a minimum value controlled mainly by the separator pressure and the frictional pressure drop along the flowline. The plunger velocity accelerates from zero velocity to a maximum value then decelerates approaching the surface. The gas flow rate increases drastically from zero rate to a maximum value. The gas rate will then remain more or less constant until the plunger arrives the surface.

It is now obvious that modeling plunger lift systems should be carried out under transient conditions. Practically, we can assume the plunger and the liquid slug travel at the same velocity. Let's consider a plunger and a liquid slug as a Free Body Diagram (FBD) as shown in Fig. 6.13. Forces acting on this FBD as the surface control valve opens include:

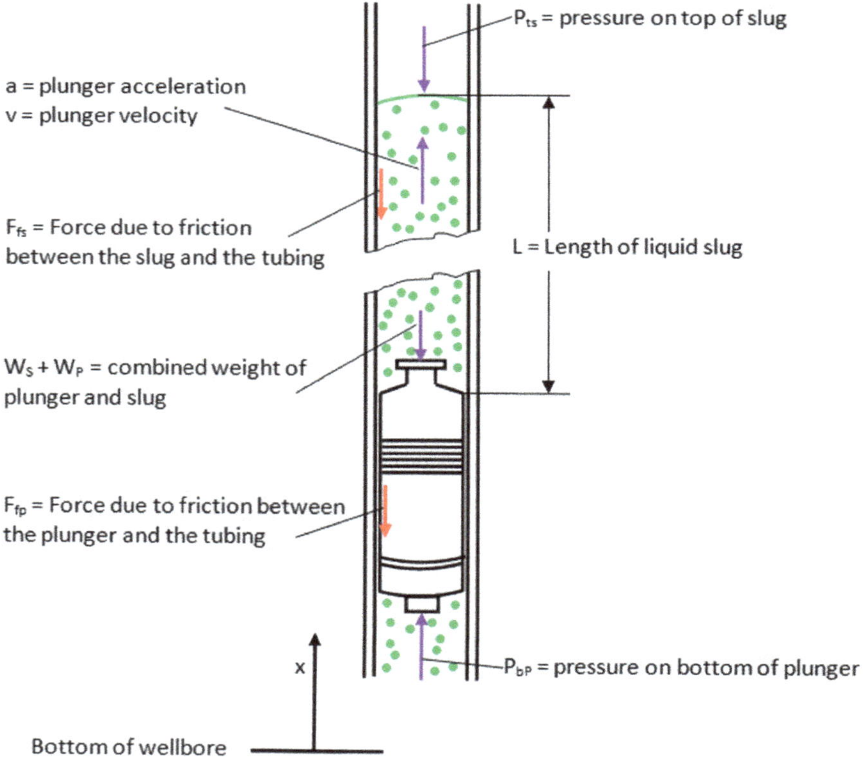

Fig. 6.13 Plunger and liquid slug free body diagram

- Bottom plunger force, F_{bp}: due to the gas pressure acting on the bottom of the plunger. This force is totally dependent on the gas expansion from the annulus to the tubing.
- Plunger friction force, F_{fp}: due to the friction between the plunger and the production tubing.
- Weight of the plunger and the slug, W_p and W_s.
- Slug friction force, F_{fs}: due to the friction between the liquid slug and the production tubing.
- Top slug force, F_{ts}: due to gas pressure acting on the top of the slug.

Applying the Newton's second law into this FBD gives:

$$\left(P_{bp}-P_{ts}\right)A_p-\left(W_p+W_s\right)-\left(F_{fp}+F_{fs}\right)=ma=m\frac{\Delta v_p}{\Delta t} \tag{6.3.5}$$

where A_p is the cross-sectional area of the bottom plunger, m is the total mass of the plunger and the liquid slug, v_p is the plunger travel velocity.

Equation (6.3.5) tells us that, the driven force created by the pressure difference between the bottom of the plunger and the top of the slug $\left(\left(P_{bp}-P_{ts}\right)A_p\right)$ is the summation of the total weight of the plunger and the liquid slug $\left(W_p+W_s\right)$, the total friction forces $\left(F_{fp}+F_{fs}\right)$, and the acceleration.

The gas pressure acting on the bottom of the plunger, P_{bp}, is a function of time and dependent on the gas expansion from the annulus to the tubing. In other words, this pressure depends on the position of the plunger. Assuming gas does not leak through the plunger and the liquid slug, at a given position of the plunger (or a given time) in the tubing and under isothermal conditions, the pressure below the plunger can be calculated as:

$$P_{bp}=\frac{P_{ci}^{b}V_c}{V_c+V_t^{p-b}}-P_g^{p-b} \tag{6.3.6}$$

where P_{ci}^{b} is the initial casing pressure at the bottom of the well calculated using Eq. (6.3.3); P_t^b is the pressure of gas in the tubing at the bottom of the well when the plunger is traveling upward; V_c is the volume of the casing annulus, V_t^{p-b} is the volume of the tubing from the plunger to the bottom of the well; and P_g^{p-b} is the hydrostatic pressure of gas from the plunger to the bottom of the well. Readers should review Example 6.3 in this chapter to have better understanding on how to get P_{bp}.

The pressure acting on the top of the liquid slug, P_{ts}, when the valve is just open can be assumed as the wellhead pressure or $P_{ts}=P_{wh}$.

The total weight of the plunger and the liquid slug (Note: $W=mg$):

$$W_p + W_s = \frac{\pi}{4} D_p^2 (\rho_p h_p + \rho_s h_s) g \quad (6.3.6)$$

where D_p is the plunger diameter, ρ_p and ρ_s are the densities of the plunger and the slug, h_p and h_s are the heights of the plunger and the slug.

The mechanical friction between the plunger and the tubing is expressed as:

$$F_{fp} = f_p N \quad (6.3.7)$$

where f_p is the mechanical friction factor and in the range of (0.1–0.6), f_p is the mechanical friction factor depends solely on types of plunger, clearance between the plunger and the tubing wall, liquid and gas surrounded the plunger. Theoretically, for a vertical well, the normal force is zero and hence the mechanical friction is zero. However, plungers are design to contact partially with the production tubing to mitigate fluids and gas leakage. Therefore, normal force is exist when plungers move upward.

The friction force when fluid flows in a tubular is calculated using the shear stress, τ, as follows:

$$F_f = \tau \times 2\pi RL = \frac{f \rho v^2}{2} 2\pi RL = \frac{2 f \rho v^2}{D} L A_p \quad (6.3.8)$$

For the friction between the liquid slug and the tubing, Eq. (6.3.8) becomes:

$$F_{fs} = \frac{2 f_s \rho_L v_s^2}{D_s} L_s A_s \quad (6.3.9)$$

Practically, we can assume that: $v_s = v_p$, $D_s = D_p$, and $A_s = A_p$. The Fanning friction factor of the liquid slug, f_s, can calculated depending on the flow regime:

For laminar flow:

$$f_s = \frac{16}{Re_s} \quad (6.3.10)$$

For turbulent flow; assuming smooth tubing and if Re < 100,000, the Fanning friction factor can be calculated using Blasius's correlation:

$$f_s = \frac{0.0791}{Re_s^{0.25}} \quad (6.3.11)$$

where Reynold number of the liquid slug is expressed as:

$$Re_s = \frac{\rho_L v_s D_s}{\mu_L} \quad (6.3.12)$$

At any time during the plunger upward travel, the plunger velocity, v_p, and the distance, L, can now be estimated by using the following equations:

$$v_p^t = \int_0^t a dt + v_p^i \tag{6.3.13}$$

$$L_t = \int_0^t v_p^t dt + L_i \tag{6.3.14}$$

where the initial velocity, v_p^i, and the initial distance traveled, L_i, are zero at the bottom of the tubing. Solving the coupled Eqs. (6.3.5), (6.3.13), and (6.3.14) numerically gives the plunger velocity profile during the upstroke stage as shown in Fig. 6.14.

For a quick estimation, assuming $F_{fp} = F_{fs}$ and the flow is under steady state, the steady state plunger velocity can be expressed as:

$$v_p = \sqrt{\frac{D_p(P_{bp} - P_{ts})A_p - (W_s + W_p)D_p}{4f_s\rho_s L_s A_p}} \tag{6.3.15}$$

6.3.3 Model the Blowdown Stage

As the plunger arrives to the surface, production gas from the casing annulus and from the reservoir flow freely in the tubing and in the flowline. How long this blowdown stage lasts before the control valve is closed depends on the amount of gas energy in the casing annulus and the performance of gas flow in the reservoir. Most of plunger lift systems, the gas flow in the reservoir is very small compared to

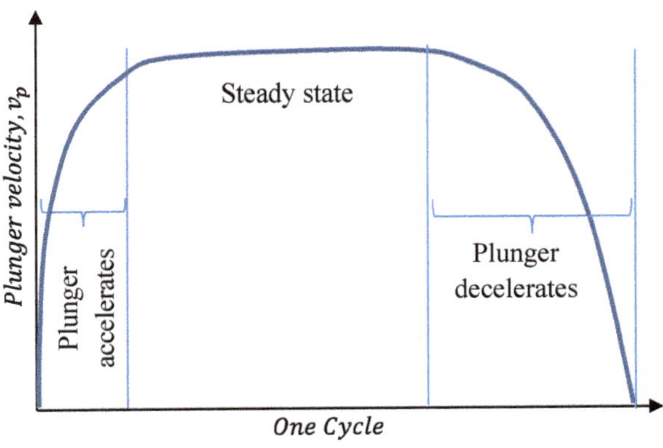

Fig. 6.14 Plunger velocity profile

the gas expansion from the annulus to the tubing and hence neglecting the gas flow from the reservoir is reasonable.

Applying the momentum equation for vertical x-direction to the gas flow from the casing annulus to the production tubing and neglecting the friction term gives:

$$\rho_g\left(\frac{\partial v_g}{\partial t}+v_g\frac{\partial v_g}{\partial x}\right)=-\left.\frac{\partial P}{\partial x}\right|_T+\rho g_x \tag{6.3.16}$$

Under steady state conditions, Eq. (6.3.16) becomes:

$$\rho_g v_g\frac{\partial v_g}{\partial x}=-\left.\frac{\partial P}{\partial x}\right|_T+\rho g_x \tag{6.3.17}$$

Equation (6.3.17) tells us that under steady state conditions and if the friction is neglected, the total pressure drop, $\left.\frac{\partial P}{\partial x}\right|_T$, equals to the summation of the pressure drops due to the gravity and the acceleration. Integrating Eq. (6.3.17) at the bottom and at the surface of the well gives the flowing bottomhole pressure during the blowdown stage:

$$P_{wf}=P_{wh}\exp\left(\frac{MgTVD}{ZR\bar{T}}\right)+P_{wh}+\exp\left[\frac{M}{2ZR\bar{T}}\left(v_{wh}^2-v_{wf}^2\right)\right] \tag{6.3.18}$$

where P_{wf} and P_{wh} are the flowing bottomhole pressure and the wellhead pressure, M is the molecular weight of gas, R is the gas constant, v_{wf} and v_{wh} are gas velocities at the bottom fo the well and at the wellhead. Note that Eq. (6.3.18) is in SI unit. In reality, during the blowdown stage, the wellhead pressure and the gas velocity (gas rate) can be monitored and hence the flowing bottomhole pressure can be estimated using Eq. (6.3.18).

6.3.4 Modeling the Downtroke Stage

When the surface control valve is closed, the pressures on the top and on the bottom of the plunger are the same. The plunger falls back into the well and it breakthrough the gas and the liquid column due to its weight. The plunger will accelerate until its weight and friction forces equal to zero then it will fall at a constant velocity. The transient period is normally very short compared to the total falling time and hence the transient time practically can be neglected when modeling the falling plunger time. The forces acting on the plunger when it is falling down under steady state conditions are:

$$W_p - F_{fm} - F_{ff} = 0 \tag{6.3.19}$$

where the subscripts fm and ff are for mechanical friction and fluid friction. The mechanical friction is due to the mechanical contacts between the plunger and the tubing. The fluid friction is due to the contact between the plunger and either gas, liquid, or mixture. Readers can review equations from (6.3.7) to (6.3.11) for calculating these two friction forces.

As mentioned in Section "Subsurface Equipment: Plungers", each type of plunger is designed to contact the tubing differently. Therefore, the mechanical friction of each type plunger is different. In addition, the mechanism of fluid passing each type of plunger is also different. Hence, each type of plunger has different fluid friction. Analytical modeling these two friction forces for all types of plungers are almost impossible. The industry normally assumes that the velocity of the falling plunger is 1,000 feet per minute (fpm) while in the gas phase and 172 fpm in the liquid phase.

6.4 Approximation Modeling of Plunger Lift System

Due to the complexities by nature of plunger lift systems as well as the complications of solving Eq. (6.3.4) or Eq. (6.3.5), the industry has attempted to introduce approximation solutions for the plunger lift system. The most common assumption to simplify the solution of Eq. (6.3.5) and to make it more practical is that the system is assumed to be under steady state condition by using the average plunger rising velocity. Other common assumptions are neglecting frictions of gas in the casing-tubing annulus and in the tubing; effects of reservoir flow during the upstroke; and leakage of liquid through the plunger (plunger inefficiency).

6.4.1 Foss and Gaul Approximation Model

6.4.1.1 Determination of Average Casing Pressure

Foss and Gaul [7] used field data of more than 100 plunger lift wells at the Ventura Avenue Field to develop semi-empirical equations for predicting the performance of these wells. When modeling the upstroke stage, they focused on the most important location of the plunger, which is near the surface or when the fluid is surfacing and passing through the well head. The casing pressure will be at the lowest value under these conditions. If this minimum surface casing pressure is used to design the plunger lift system, one can make sure that the plunger stall somewhere in the tubing can be avoided. When the plunger is near the surface, the authors assumed that: (1) the hydrostatic pressures of gas in the casing annulus and in the tubing are the same and cancelled out; (2) the casing pressure is at its lowest value; (3) the pressure effect from liquid production in the tubing under the plunger is maximum; and (4) the plunger is under steady state when it surfacing. They also neglected the mechanical plunger friction, frictional pressure loss in the casing annulus, and pressure differences caused by fluid entry below the plunger. They further assumed the plunger rise velocity was a constant and equal to 1,000 ft/min; the plunger fall velocity through gas was 2,000 ft/min; and the plunger fall velocity through liquid was 172 ft/min [12]. With these assumptions, the authors simplified Eq. (6.3.4) and introduced the equation for calculating the minimum casing pressure (plunger is near at the surface) as follows:

$$P_c^{\min} = P_f^t + P_p + P_{sl} + P_f^{sl} + P_{wh} \tag{6.4.1}$$

where $P_f^t = P_{t-bp}^{f-g} + P_{t-ap}^{f-g}$ is the total frictional pressure drop of gas in the tubing (review Eq. (6.3.4)); P_p and P_{sl} are the pressures to lift the weight of the plunger and 1 bbl liquid slug; $P_f^{sl} = P_{t-ap}^{f-sl}$ in Eq. (6.3.4) is the friction caused by 1 bbl liquid slug in the tubing above the plunger; and P_{wh} is the wellhead pressure.

To estimate the average casing pressure, the author further assumed that the summation of the pressure to lift the weight of the slug and the slug friction, $\left(P_{sl} + P_f^{sl}\right)$, is a constant for a give tubing size and liquid type; The oil API is 30°; fluid temperature is 150 °F. Using a simplified equation the gas friction in the tubing, P_f^t, the average casing pressure was given in psi as follows:

$$P_c^{avg} = 1.05\left(1 + \frac{TVD}{K}\right)\left[5 + P_{wh}^{\min} + \left(P_{sl} + P_f^{sl}\right)V_{sl}\right] \tag{6.4.2}$$

where TVD is the true vertical depth of the well in ft; K is a constant depending on the tubing size unitless; V_{sl} is the volume of the slug or the load size in bbl. The values of K and $\left(P_{sl} + P_f^{sl}\right)$ are given in Table 6.1.

6.4.1.2 Determination of Maximum Cycling Frequency

According to the authors, the minimum time required to complete one cycle (maximum cycling frequency) is the summation of the following: (1) time required

Table 6.1 Values of $\left(P_{sl} + P_f^{sl}\right)$ and K depending on the tubing size

Constant values	Tubing size (in.)		
	1.999 (in.)	2.441 (in.)	2.992 (in.)
$\left(P_{sl} + P_f^{sl}\right)$	165	102	63
K	33,500	45,000	57,600

for the plunger to rise to the surface, T_r; (2) time required for the plunger to fall through the gas filled tubing, T_{f_g}; and (3) time required for the plunger to fall through the liquid column in the tubing, T_{f_l}.

$$T_{max} = T_r + T_{f\_g} + T_{f\_l} \tag{6.4.3}$$

With the data measured at the Ventura field, the authors assumed the rising plunger velocity of 1,000 ft/min, falling plunger velocity through gas of 2,000 ft/min. The falling velocity of the plunger in liquid was obtained in the lab with a value of 172 ft/min. With these inputs, the maximum cycling frequency, cycles per day, can be estimated as follows:

$$T_{max}\left(\frac{cycles}{D}\right) = \frac{1,440}{1.5\left(\frac{TVD}{1,000}\right) + \frac{L_{1\,bbl\,load} \times V_{sl}}{172}} \tag{6.4.4}$$

where $L_{1\,bbl\,load}$ is the length of one barrel load and V_{sl} is the load size in bbl. $L_{1\,bbl\,load} \times V_{sl} = h_l$ is the length of the liquid column in the tubing.

6.4.2 *Lea Approximation Model*

Lea [11] presented estimate corrections to Foss and Gaul's calculation by introducing a dynamic model of plunger lift operations.

6.4.2.1 Determination of Maximum Cycling Frequency

The maximum cycling frequency is determined according to Lea:

$$T_{max} = \frac{1,440}{\frac{TVD - h_l}{v_{f\_g}} + \frac{TVD}{v_r} + \frac{h_l}{172}} \tag{6.4.5}$$

Comparing Foss and Gaul's and Lea's equations for estimating the maximum cycling frequency, one can notice that Lea presented the total plunger travel time for one cycle in three components: rising in the tubing, falling through gas and falling through liquid in the annulus. Whereas Foss and Gaul combined the plunger falling velocity through gas and plunger rising velocity into one component.

6.4.2.2 Determination of Average Casing Pressure

Lea rewrote Foss and Gaul's equation for calculating the average casing pressure as follows:

$$P_c^{avg} = \left(1 + \frac{TVD}{K}\right)\left[P_p + P_{wh}^{\min} + \left(P_{sl} + P_f^{sl}\right)V_{sl}\right]\left(1 + \frac{A_t}{2A_a}\right) \tag{6.4.6}$$

where P_p is the pressure to lift the plunger weight; $P_{sl} + P_f^{sl}$ are the pressure to lift 1 bbl of liquid slug and friction in the tubing caused by 1 bbl of liquid slug; V_{sl} is the load size in bbl; A_t and A_a are the cross-sectional areas of the tubing and the annulus.

The minimum casing pressure occurs when the plunger and the slug are approaching the surface and determined as:

$$P_c^{\min} = \frac{P_c^{avg}}{\left(1 + \frac{A_t}{2A_a}\right)} \tag{6.4.7}$$

The maximum casing pressure occurs right before the plunger and the slug begin to move upward and determined as:

$$P_c^{\max} = \frac{P_c^{avg}}{\frac{2(A_t + A_a)}{A_t + 2A_a}} \tag{6.4.8}$$

Combing Eqs. (6.4.6) and (6.4.7), a dimensionless was introduced by Lea:

$$\frac{P_c^{\min}}{P_p + P_{wh}^{\min} + \left(P_{sl} + P_f^{sl}\right)V_{sl}} = \left(1 + \frac{TVD}{K}\right) \tag{6.4.9}$$

where 1/K is expressed based on the average rise velocity as follows:

$$\frac{1}{K} = \frac{\rho_g \bar{f}_g v_r^2}{2gD_t(144)\rho_l} \tag{6.4.10}$$

6.5 Examples

Example 6.1 Calculate the maximum surface casing pressure during the build-up stage for a vertical gas well with the following inputs: gas specific gravity of 0.7; average gas temperature in the annulus of 150 °F; compressibility factor of 0.95; TVD of 8,000 ft; and the bottomhole pressure right before the plunger begins to travel upward is 2000 psi.

Solution

Under static conditions, the maximum surface casing pressure when the plunger is ready to travel upward is less than the bottomhole pressure due to its weight and calculated using Eq. (6.3.1)

$$P_c^{\max} = \frac{P_e}{\exp\left(\frac{0.01877\gamma_g TVD}{Z\bar{T}}\right)} = \frac{2{,}000}{\exp\left(\frac{0.01877 \times 0.7 \times 8{,}000}{0.95(150 + 460)}\right)} = 1{,}668\,\text{psi}$$

The weight of the gas column under these conditions is 2,000 − 1,668 = 332 psi.

Example 6.2 Toward the end of the build-up stage, the surface casing pressure and the tubing pressure were recorded at 800 psi and 600 psi, respectively. This is a gas well with a TVD of 8,000 ft and the gas specific gravity of 0.7. The gas compressibility factor is 0.95. Calculate the height of the liquid column in the tubing. Assuming the liquid slug is water and the mass of the plunger of 10 lbm. The plunger diameter is assumed to be the same as the ID of the tubing, which is 1.995 in.

Solution

Note that for most of places on earth:

1 kg mass = 2.2 lbm or equivalent to 9.81 N = 2.2 lbf (pound force)

Assuming the plunger pumper is installed at 8,000 ft. The bottomhole pressure of the gas column in the annulus is:

$$P_{wf} = P_c^s \exp\left(\frac{0.01877\gamma_g TVD}{Z\bar{T}}\right) = 800 \exp\left(\frac{0.01877 \times 0.7 \times 8{,}000}{0.95(150 + 460)}\right) = 959\,\text{psi}$$
$$= 6613\,\text{kPa}$$

The bottomhole pressure caused by the fluids and the plunger in the tubing:

$$P_{wf} = P_g^{t-p} + \frac{(W_s + W_p)}{A_p}$$

where

The pressure at the top of the slug due to the gas column in the tubing:

$$P_g^{t-p} = P_t^s \exp\left[\frac{0.01877\gamma_g (TVD - L_s)}{Z\bar{T}}\right]$$

The pressure due to the weight of the liquid slug (Density of water: 1,000 kg/m^3 = 62.5 lb/ft^3 = 8.3 lb/gal or ppg):

$$P_s(\text{psi}) = 0.052\rho_s\left(\frac{\text{lb}}{\text{gal}}\right)L_s(\text{ft})$$

The weight of the liquid slug

$$W_s = P_s(\text{psi}) \times A_p\left(\text{in.}^2\right)$$

The weight of the plunger:

$$W_p = 10\,\text{lbm} = 10\,\text{lbf}$$

$$P_{wf} = P_t^s \exp\left[\frac{0.01877\gamma_g(TVD - L_s)}{Z\bar{T}}\right] + \frac{\left(0.052\rho_s L_s A_p + 10\right)}{A_p} = 959$$

Solving this equation with $P_t^s = 600\,\text{psi}$, $\gamma_g = 0.7$, TVD = 8000 ft, Z = 0.95, $\bar{T} = 150 + 460 = 610\,\text{K}$, $\rho_s = 8.3\,\text{ppg}$, $A_p = 3.12\,\text{in.}^2$ gives:

$$L_s = 570\,\text{ft}$$

If the weight of the plunger is neglected, the height of the liquid slug will be 580 ft.

Example 6.3 The 6,000 ft plunger lift vertical well is being shut down and the surface casing pressure is recorded at 800 psi. The tubing ID and OD are 1.995 in. and 2.495, respectively. The casing ID is 4.0 in. Assuming gas in the annulus does not pass through the plunger and the liquid slug. Calculate the pressure at the bottom of the plunger during the upstroke stage and when the plunger is at 4000 ft from the surface. Given: $\gamma_g = 0.7$; $Z = 0.95$; $\bar{T} = 150\,\text{F}$.

Solution

The initial bottomhole casing pressure

$$P_{ci}^b = P_{ci}^s \exp\left[\frac{0.01877\gamma_g TVD}{Z\bar{T}}\right] = 800\exp\left[\frac{0.01877 \times 0.7 \times 6000}{0.95(150 + 460)}\right] = 916\,\text{psi}$$

Total volume of the casing annulus

$$V_c = \frac{\pi\left(ID_c^2 - OD_t^2\right)}{4} TVD = 552{,}763\,\text{in.}^3 = 320\,\text{ft}^3$$

Volume of the tubing from the bottom to the bottom of the plunger when the plunger is at 4000 ft

$$V_t^{p-b} = \frac{\pi ID_t^2}{4}(TVD - 4{,}000) = 75{,}022\,\text{in}^3 = 43.4\,\text{ft}^3$$

Applying Eq. (6.3.6) gives the pressure at the bottom of the plunger when it is at 4,000 ft:

$$P_{bp} = \frac{P_{ci}^{b} V_c}{V_c + V_t^{p-b}} - P_g^{p-b} = 807 - P_g^{p-b}$$

where the hydrostatic pressure of the gas column from the plunger to bottom of the well is calculated as:

$$\begin{aligned} P_g^{p-b} &= P_c^b - P_{bp} = P_{bp} \exp\left[\frac{0.01877 \times 0.7 \times 2,000}{0.95 \times (150+460)}\right] - P_{bp} = 1.046 P_{bp} - P_{bp} \\ &= 0.046\, P_{bp} \end{aligned}$$

Combining these two equations gives the gas pressure below the plunger:

$$P_{bp} = 807 - 0.046\, P_{bp}$$

$$P_{bp} = \frac{807}{1+0.046} = 772\,\text{psi}$$

Example 6.4 Calculate the friction force when the liquid slug traveling upward in a 2.441-ID production tubing during the upstroke stage with the plunger rise velocity of 1,000 ft/min. Assuming the liquid slug density is 62.5 lb/ft^3; the slug length is 650 ft.

Solution

$$\begin{aligned} D_s &= 2.441\,\text{in} = 0.062\,\text{m} \\ v_r &= 1{,}000 \frac{\text{ft}}{\text{min}} = 5.08\,\text{m/s} \\ \rho_s &= 62.5 \frac{\text{lb}}{\text{ft}^3} = 1{,}000 \frac{\text{kg}}{\text{m}^3} \\ L_s &= 650\,\text{ft} = 198\,\text{m} \\ \mu_s &= 1\,\text{cp} = 0.001\,\text{PaS} \end{aligned}$$

Reynolds number of the liquid slug in the tubing:

$$Re_s = \frac{\rho_s v_s D_s}{\mu_L} = 314{,}967$$

The slug is under turbulent flow

Assuming smooth tubing, the Fanning friction factor:

$$f_s = \frac{0.0791}{Re_s^{0.25}} = 0.00334$$

Cross-sectional area of the plunger

$$A_p = \frac{\pi D_p^2}{4} = 4.679\,\text{in.}^2 = 0.003\,\text{m}^2$$

The friction force due to the liquid slug surfacing:

$$F_{fs} = \frac{2f_s \rho_s v_s^2}{D_s} L_s A_s = \frac{2 \times 0.00334 \times 1{,}000 \times 5.08^2}{0.062} 198 \times 0.003$$

$$F_{fs} = 1{,}663\,\text{N} = 374\,\text{lbf}$$

Example 6.5 Calculate the total frictional pressure losses due to the flow of natural gas in the annulus and in the production tubing when the plunger is approaching the surface with a rise velocity of 1,000 ft/min. The well and fluid information is given: TVD = 6,500 ft; tubing ID and OD are 1.995 in. and 2.495, respectively; casing ID is 4.0 in. Gas specific gravity when the plunger approaching surface is 0.7; gas viscosity is 0.012 cp. The relative roughness is 0.001.

Solution
We will solve this problem using SI unit.

$$\begin{aligned} TVD &= L = 6{,}500\,\text{ft} = 1{,}981\,\text{m} \\ ID_t &= 1.995 - \text{in.} = 0.051\,\text{m} \\ OD_t &= 2.495 - \text{in.} = 0.063\,\text{m} \\ ID_c &= 4.000 - \text{in.} = 0.102\,\text{m} \\ A_t &= 3.000 - \text{in.}^2 = 0.002\,\text{m}^2 \\ A_a &= 7.677 - \text{in.}^2 = 0.005\,\text{m}^2 \end{aligned}$$

Using hydraulic pipe diameter definition gives the equivalent pipe diameter for the annulus:

$$D_e = ID_c - OD_t = 0.038\,\text{m}$$

The flow of gas in the tubing and in the annulus is due to the upward travel of the plunger. Therefore, it is reasonable to assume that the gas velocities in the tubing and in the annulus are the same and equal to the plunger rise velocity.

$$v_t = v_a = v_r = 1{,}000 \frac{\text{ft}}{\text{min}} = 5.080 \frac{\text{m}}{\text{s}}$$

Average gas density and viscosity when the plunger is approaching the surface:

$$\rho_g = 0.7 \times 1.225 = 0.8575 \frac{\text{kg}}{\text{m}^3}$$

$$\mu_g = 0.012\,\text{cp} = 1.2 \times 10^{-5}\,\text{Pas}$$

Reynolds numbers of gas flow in the tubing and in the annulus:

$$Re_t = \frac{\rho_g v_t ID_t}{\mu_g} = \frac{0.8575 \times 5.08 \times 0.051}{1.2 \times 10^{-5}} = 18{,}395$$

$$Re_{ann} = \frac{\rho_g v_a D_e}{\mu_g} = \frac{0.8575 \times 5.08 \times 0.038}{1.2 \times 10^{-5}} = 13{,}877$$

Using the Moody friction factor chart presented in Fig. 6.15 gives the Moody friction factors for the flow of gas in the tubing and in the annulus of $f_M^t = 0.028$ and $f_M^a = 0.031$, respectively. Converting the Moody friction factors to the Fanning friction factors gives: $f_F^t = 0.007$ and $f_F^a = 0.0078$.

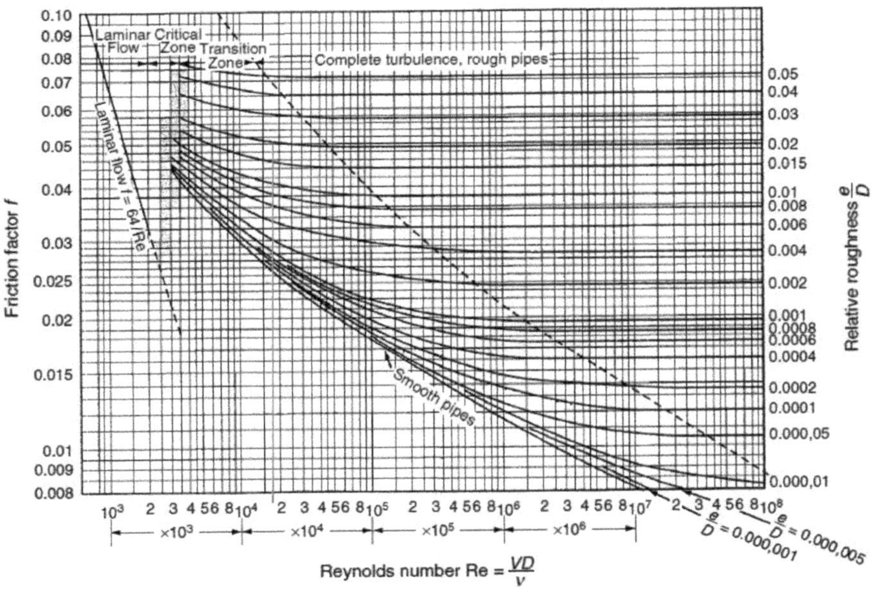

Fig. 6.15 Moody friction factor chart

Using Eq. (6.3.8), the total frictional pressure losses due to gas flow in the tubing and in the annulus:

$$P_t = \frac{2f_F^t \rho_g v_t^2}{ID_t} L = \frac{2 \times 0.007 \times 0.8575 \times 5.08^2}{0.051} 1{,}981 = 12{,}112\,\text{Pa} = 2\,\text{psi}$$

$$P_a = \frac{2f_F^a \rho_g v_a^2}{D_e} L = \frac{2 \times 0.0078 \times 0.8575 \times 5.08^2}{0.038} 1{,}981 = 17{,}776\,\text{Pa} = 3\,\text{psi}$$

From this example, one can draw a quick conclusion that the frictional pressure drops due to gas flow in the tubing and in the annulus of a plunger lift system are small and negligible.

Example 6.6 Given: TVD = 9,300 ft. The tubing ID is 2.441 in. and the flow-line pressure is 100 psig. Using Foss and Gaul approximation model, calculate the average casing pressure, maximum cycling frequency, and the liquid production rate. The load size is expected to be 0.5 bbl/cycle.

Solution

According to Foss and Gaul, the average casing pressure can be calculated using Eq. (6.4.2):

$$P_c^{avg} = 1.05\left(1 + \frac{TVD}{K}\right)\left[5 + P_{wh}^{\min} + \left(P_{sl} + P_f^{sl}\right)V_{sl}\right]$$

where the summation of the pressure to lift the weight of the slug and the slug friction, $\left(P_{sl} + P_f^{sl}\right)$, and the constant K, which is dependent on the tubing size are given in Table 6.1 as: 102 psi and 45,000, respectively.

$$P_c^{avg} = 1.05\left(1 + \frac{9{,}300}{45{,}000}\right)[5 + 100 + 102 \times 0.5] = 197.6\,\text{psi}$$

The maximum cycling frequency (maximum time required to complete one cycle):

$$T_{\max} = \frac{1{,}440}{1.5\left(\frac{TVD}{1{,}000}\right) + \frac{L_{1\text{bbl}load} \times V_{sl}}{172}}$$

where the length of one barrel (5.61 ft^3) load in the 2.441 in. ID tubing ($A_t = 0.03$ ft^2) is:

$$L_{1\text{bbl}\,load} = \frac{5.61}{0.03} = 172.8\frac{\text{ft}}{\text{bbl}}$$

$$T_{\max} = \frac{1,440}{1.5\left(\frac{TVD}{1,000}\right) + \frac{L_{1\text{bbl}\,load} \times V_{sl}}{172}} = \frac{1,440}{1.5\left(\frac{9,300}{1,000}\right) + \frac{172.8 \times 0.5}{172}} = 100\frac{cycles}{D}$$

The liquid production rate:

$$Q_L = 100\frac{cycles}{D} \times 0.5\frac{\text{bbl}}{cycle} = 50\frac{\text{bbl}}{D}$$

Example 6.7 Given: TVD = 9,300 ft; tubing ID = 2.441 in.; tubing OD = 3.0 in.; casing ID = 4.56 in.; flow-line pressure Pwh = 100 psig; densities of gas and liquid slug 0.0625 lb/ft^3 and 52 lb/ft^3, respectively; Moody friction factor for gas in the tubing 0.015; plunger rise velocity 500 ft/min; plunger fall velocity through gas 2,000 ft/min; pressure to lift the weight of the plunger 5 psi. Using Lea approximation model, calculate the average, minimum, and casing pressure, maximum cycling frequency, and the liquid production rate. The load size is expected to be 1.0 bbl/cycle.

Solution

$$TVD = L = 9{,}300\,\text{ft} = 2{,}835\,\text{m}$$

$$ID_t = 2.441\,\text{in.} = 0.062\,\text{m}$$

$$OD_t = 3.000\,\text{in.} = 0.914\,\text{m}$$

$$ID_c = 4.560\,\text{in.} = 0.129\,\text{m}$$

$$A_t = 4.68\,\text{in.}^2 = 0.003\,\text{m}^2$$

$$A_a = 9.260\,\text{in.}^2 = 0.006\,\text{m}^2$$

$$\left(1 + \frac{A_t}{2A_a}\right) = 1.25$$

$$\frac{2(A_t + A_a)}{A_t + 2A_a} = 1.20$$

According to Lea's model:

$$\frac{1}{K} = \frac{\bar{f}_M^g \rho_g v_r^2}{2gD_t(300)\rho_{sl}} = 0.0001\ or\ K = 10{,}463$$

$$\left(1 + \frac{TVD}{K}\right) = 1.89$$

Pressure to lift 1 lbm of liquid slug:

$$P_{sl} = 0.433 \times SG_{sl} \times L_{1\mathrm{bbl}\,load} = 62.44\,\mathrm{psi}$$

Friction in the tubing due to the moving of the liquid slug (v_r = 500 ft/min = 8.33 ft/s; ρ_{sl} = 52 lb/ft^3 = 6.94 lb/gal):

$$P_f^{sl} = \frac{f_F^t \rho_L v_r^2}{25.8 \times ID_t} L_{1\mathrm{bbl}\,load} = \frac{\frac{0.015}{4} 6.94 \times 8.33^2}{25.8 \times 2.441} 173 = 4.96\,\mathrm{psi}$$

$$\left(P_{sl} + P_f^{sl}\right) = 62.44 + 4.96 = 67.4$$

The average, minimum, and maximum casing pressure:

$$P_c^{avg} = \left(1 + \frac{TVD}{K}\right)\left[P_p + P_{wh}^{\min} + \left(P_{sl} + P_f^{sl}\right)V_{sl}\right]\left(1 + \frac{A_t}{2A_a}\right)$$

$$P_c^{avg} = 1.89[5 + 100 + 67.4 \times 1]1.25 = 412.63\,\mathrm{psi}$$

$$P_c^{\min} = \frac{P_c^{avg}}{\left(1 + \frac{A_t}{2A_a}\right)} = \frac{494}{1.25} = 329.41\,\mathrm{psi}$$

$$P_c^{\max} = P_c^{avg} \frac{2(A_t + A_a)}{A_t + 2A_a} = 494 \times 1.20 = 495.84\,\mathrm{psi}$$

The maximum cycling frequency is determined according to Lea:

$$T_{max} = \frac{1,440}{\frac{TVD - h_l}{v_{f\_g}} + \frac{TVD}{v_r} + \frac{h_l}{172}} \tag{6.39}$$

where $h_l = L_{1\,\mathrm{bbl}\,load} \times V_{sl}$ is the length of the liquid column in the tubing. Similar to Example 6.6, the $L_{1\,\mathrm{bbl}\,load} = 173\,\mathrm{ft/bbl}$.

$$T_{max} = \frac{1,440}{\frac{9,300 - 173}{2,000} + \frac{9,300}{500} + \frac{173}{172}} = 59.58\,\frac{cycle}{D}$$

The liquid production rate:

$$Q_L = 59.58\frac{cycles}{D} \times 1.0\frac{\mathrm{bbl}}{cycle} = 59.58\,\frac{\mathrm{bbl}}{D}$$

References

1. Aronofsky J, Jenkins R (1954) A simplified analysis of unsteady radial gas flow 6. SPE-271-G
2. Avery D, Evans R (1988) Design optimization of plunger lift systems. SPE 17585
3. Baruzzi J, Alhanati F (1996) Optimum plunger lift operation. SPE 29455
4. Beauregard E, Ferguson P (1982) Introduction to plunger lift: applications, advantages and limitations. SPE 10882
5. Chava G, Falcone G, Teodoriu C (2009) Plunger lift modeling towards efficient liquid unloading in gas wells. SPE 124515
6. Economides M, Daniel H, Economides C (1994) Petroleum production systems. PTR Prentice Hall
7. Foss D, Gaul R (1965) Plunger lift performance criteria with operating experience—Ventura avenue field. In: 65-124 API conference paper
8. Gasbarri S, Wiggins M (1997) A dynamic plunger lift model for gas wells. SPE 37422
9. Hashmi G, Hasan A (2016) Design of plunger lift for gas wells. SPE-181220-MS
10. Hassouna M, Lufkin Industries Inc. (2013) Plunger lift applications: challenges and economics. SPE 164599
11. Lea J (1982) Dynamic analysis of plunger lift operations. SPE-10253-PA
12. Mower L, Lea J, Beauregard E, Ferguson P (1985) Defining the characteristics and performance of gas lift plungers. SPE 14344
13. Petroleum Recovery Research Center a Division of New Mexico Tech (1997) Petroleum technology transfer council—southwest region. Farmington, NM
14. Rawlins L, Schellhardt A (1935) Backpressure data on natural gas wells and their application to production practices, vol 7. Monograph series, USBM
15. Tang Y, Liang Z (2008) A new method of plunger lift dynamic analysis and optimal design for gas well deliquification. SPE 116764

17\. Weatherford (2007) Deliquification of gas well with plunger lift

16\. Zhu J, Zhu H, Zhao Q, Fu W, Shi Y, Zhang H (2019) A transient plunger lift model for liquid unloading from gas wells. IPTC-19211-MS

Chapter 7
Artificial Lift Selection Methodology for Vertical and Horizontal Wells in Conventional and Unconventional Reservoirs

Nomenclature

AL	Artificial lift
API	American Petroleum Institute
bll	Barrel
BLPD	Barrels of liquid per day
CAPEX	Capital expenditure
Cp	Centipoise
DLS	Dog leg severity
ESP	Electrical submersible pump
ID	Inner diameter
In	Inch
KOP	Kick off point
IPR	Inflow performance relationship (flow in the reservoir)
GL	Gas lift
GLR	Gas liquid ratio
GOR	Gas oil ratio
MD	Measured depth
NPV	Net present value
OD	Outer diameter
OPEX	Operating expenditure
OPR	Outflow performance relationship
P_b	Bubble point pressure
PCP	Progressive cavity pump
PL	Plunger lift
Q	Flowrate
SBHP	Static bottomhole pressure
SCF (MSCF)	Standard cubic feet (1000 SCF)
SRP	Sucker rod pump
TVD	True vertical depth
WC	Water cut

T. Nguyen, *Artificial Lift Methods*, Petroleum Engineering,
https://doi.org/10.1007/978-3-030-40720-9_7

An appropriate Artificial Lift (AL) method is a guarantee of efficient production during the life of wells. There are many factors that need to be considered when selecting an AL method for a particular well and a particular reservoir. Most oil and gas companies have their own philosophies of how to select artificial lift method(s) for a specific application in a form of guideline or recommendation. These selection philosophies are mainly applicable for vertical wells in conventional reservoir and dependent on expert experiences and rule of thumbs. In addition, with the boom in horizontal drilling and hydraulic fracturing for unconventional reservoirs, the industry did not have enough time to understand and study the challenges and uniqueness of applying AL for horizontal wells. In fact, companies did not have many available options rather than the traditional AL systems developed for conventional wells and reservoirs. It is obvious that applying AL technology for vertical wells to horizontal wells without major changes in both technology and AL selection is not a best practice [8]. In many cases, this practice may lead to low and/or very low lift efficiency, struggling of the lift system to perform well in new unconventional applications, and medium to huge economic loss. With lessons learnt in the past 10 years, the industry and academia experts have come up with guidelines and recommendations for using AL for horizontal wells in unconventional reservoirs [15]. This chapter consists of three main sections: the first section will review the characteristics of common AL methods; the second section will cover the AL selection methodology for vertical wells in conventional reservoirs and the third chapter will cover the AL selection methodology for horizontal wells in unconventional reservoirs.

7.1 Characteristics of Common Artificial Lift Methods

Before looking into AL selection methodology for vertical wells in conventional reservoirs and horizontal wells in unconventional reservoirs, this section will review the strengths and the weaknesses of five common AL methods, namely Gas Lift (GL), Electrical Submersible Pump (ESP), Sucker Rod Pump (SRP), Progressive Cavity Pump (PCP), and Plunger Lift (PL). These advantages and disadvantages of each lift method are presented in each chapter from Chaps. 2 to 6.

7.1.1 Gas Lift

The advantages of continuous GL can be summarized as follows:

- GL design and installation are considered to be one of the most forgiving forms of AL methods. As long as gas is injected into the production tubing, the well will normally produce some liquid. The only concern is the efficiency of the gas energy used to lift a unit volume of liquid.

- GL is an excellent method for wells with high sand production and high formation gas-liquid ratio. Sand production may cause severe damages for downhole pumps. In addition, high formation gas-liquid ratio may significantly reduce the efficiency of downhole pumps due to gas bubbles, gas pockets, and gas lock.
- GL system is relatively simple with few moving parts. The only downhole moving part is the valve stem, which can be made and controlled very reliably from the surface.
- Using wireline, coil tubing, or other special tools, GL system can be installed in wells with very high Dog Leg Severity (DLS) up to 75°/100 ft. In addition, valves can be replaced without the need of killing the well or pulling the tubing.
- Operations in highly deviated wells are less complicated than that of other lift methods.
- Subsurface GL components are relatively inexpensive and the surface gas injection control equipment is simple and required very minimum space for installation. Therefore, GL is a good candidate for offshore wells, which have a very limited surface footprint.
- GL is a very flexible method. It can be used as a transition method or as a permanent installation during the life of the well. A well can begin using continuous GL method, then transition to intermittent GL and finally combines GL with PL as a third stage in the production life cycle.

The Disadvantages of Continuous GL are as follows:

- GL system strongly depends on the gas source which is from the formation gas and the outside supplied source. If formation gas is limited then continuous GL method relies mainly on the outside supplied source, which can be expensive and not reliable.
- The infrastructure to build the compression system on the surface may require a high capital expenditure.
- Space for installing gas compressors can be a problem for offshore platforms.
- If the spacing between wells is wide and the number of wells is low, GL application can be limited. In other words, GL system should be applied for small spaced wells as well as high number of wells.
- GL is not recommended for heavy oil wells.
- GL is basically a low energy efficiency method when compared to other lift methods.

7.1.2 Electrical Submersible Pump

The advantages of using ESP in comparison with other lift methods are as follows:

- The most important factor that needs to be considered when selecting ESPs is high liquid rate. ESP can be economically designed for both oil and water wells, at production rates ranging from 200 to 60,000 B/D and at depths of up to 15,000 feet. This strength of ESPs, in turn, makes it become one of the top candidates for unconventional wells which produce at high and very high liquid production rates in the early life of the wells.

- ESP can be applied in crooked or deviated wells which have DLS of less than 9°/100 ft.
- Surface equipment of an ESP system is quite simple compared to that of other lift methods. It has a relatively small surface footprint and hence is appropriate for use in offshore, urban or other confined locations.
- Generally speaking, ESPs provide low lifting costs for high fluid volumes. In other words, ESPs are best suit for high liquid rate.

The disadvantages of using ESP in comparison with other lift methods are as follows:

- For wells stimulated using hydraulic fracturing, the proppant flowback and produced gas have become a major problem for ESP operations. Under harsh environments where gas and solid content are high, the average ESP run life is about 6–9 months. In recent years, the industry has improved the efficiency of the downhole gas separator and hence high GLR is not a big concern. However, ESPs still don't work well with the presence of solid due to the solid erosion at the surface of impellers and diffuser.
- For most cases, an ESP system is applicable mainly to single-zone completions. If the production casing is big enough, then the y-tool can be used to have dual-zone completions.
- As always, ESPs requires a stable high-voltage electric power source.
- Running or pulling the tubing string may cause damage to the power cable installed along the tubing. In addition, cables may deteriorate in high temperature and abrasive conditions. The temperature limit for most of the power cables available on the market is 400 °F (about 200 °C).

7.1.3 Sucker Rod Pump

Advantages of SRP can be listed as follows:

- Because SRP works similar to positive displacement pump, it can reduce the bottomhole pressure to a very low level and hence deplete the reservoir better. This advantage makes SRP stand out to be the best lift candidate toward the end of the life of the well.
- SRP systems are relatively simple in design, operation, and maintenance. Therefore, they are quite understood by the industry in terms of design, operations, diagnose and troubleshooting.

- Because SRP systems are the most common lift methods, replacement parts are normally widely available, compatible and interchangeable.
- The operating cost is quite competitive in comparing to other lift methods.

Disadvantages of SRP are as follows:

- Like any positive displacement pumps, SRP systems do not operate well with the presence of gas and solids. The pump efficiency will be significantly low if the GLR is high. The reason is that part of the stroke length will be used to compressed gas to open the traveling valve causing a shorter effective stroke length or less liquid is lifted. Therefore, it is obvious SRP should not be selected as the first lift method in the lift strategy chain when the well is still under high natural flow potential.
- Wells with medium or high DLS (normally greater than 5°/100 ft) are not suitable for SRP systems. The mechanical friction between rod string and production tubing in high DLS wells may cause rod/coupling failures [2]. In addition, the system requires more surface power to operate due to this friction.
- SRP systems are only applicable for wells with shallow and medium depths. The pump depth is normally limited by the strength of the rod string and coupling materials.
- Due to its high surface space requirement, SRP system is mainly applicable for onshore wells.

7.1.4 Plunger Lift

Advantages of using PL are as follows:

- PL method is the most cost effective method for high liquid-producing gas wells, which continuous production is not available and intermittent flow is the must. As long as the reservoir pressure is high enough to lift the plunger, the system will work.
- Another major advantages over other AL methods, such as intermittent GL or subsurface pumps, are the relatively small investment and reasonable operating and maintenance costs.
- The system does not require major power supply. The main power to drive the plunger is from the reservoir. The only minor surface power needed for operating this system is the power supply for the data acquisition system and for the control valve. Small solar power panels and a battery are enough to provide power for this system.
- PL system is quite reliable. The only part that usually wears is the plunger. If it is inspected on a monthly basis, and wear is evidenced, it can be exchanged or repaired at a minimal cost.

Limitations of using PL are as follows:

- PL is a complex lift method which involves compressible gas expansion, gas and fluid accelerations, and different multiphase flow patterns existing in the tubing below and above the plunger. Modeling, optimizing and troubleshooting can be a challenge.
- It cannot be applied for continuous production wells. The application of PL systems is quite unique and it is mainly used for high liquid-producing vertical gas wells. Highly inclined wells are not recommended for using this lift method.
- If the reservoir pressure is not high enough to lift the plunger and the liquid slugs upward, PL is not applicable.
- This method is not suitable for wells which have high sand production problems.

7.1.5 Progressive Cavity Pump (PCP)

PCP has several advantages over other pumps:

- One of the best AL methods for pumping high to very high viscous fluids (heavy oils) under medium to high temperatures. PCPs are available for a wide range of operating challenges and applications including coal bed methane, medium to light oil, cold production of heavy oil, thermal production of extra heavy oil and bitumen.
- Providing a uniform flow rate without any pulsation;
- Be theoretically able to deliver solids;
- Less moving parts and hence higher reliability;

The limitations of PCP can be listed as follows:

- PCPs are not suitable for high liquid production rates.
- Under high temperature, the rotor elastomer will deform causing a high slippage (low pump efficiency).
- Similar to ESPs, PCPs are limited with high DLS wells.

7.2 Artificial Lift Selection for Vertical Wells in Conventional Reservoirs

AL technology for vertical wells in conventional reservoirs is relatively mature with many proven solutions. Conventional reservoirs are commonly classified as reservoirs in which hydrocarbons were migrated and trapped below low permeability overlying layers or cap-rocks. Hydrocarbons in conventional reservoirs are typically ready to flow into the wellbore. Fracturing is normally not needed when completing these wells.

AL selection should be part of the well planning process to maintain the stability of the well. Future production strategy has a strong impact on selecting a lift system and hence impacting the well design [1]. It is now important for us to review factors that impact the selection of lift methods.

7.2.1 Important Factors Impacting AL Selection

There are many factors that impact the selection of lift methods but we can group them into four categories: reservoir characteristic, wellbore geometry characteristics, surface characteristics, and field operating characteristics [5].

- Reservoir characteristics:
  - *Reservoir flow performance or IPR*: IPR defines the potential production of the well. IPR can give us an idea of how large the production rate envelops may be.
  - *Liquid production rate and water cut*: this is an important factor in selecting a lift method. If high liquid rates are required then submersible pumps are more suitable.
  - *Gas-liquid ratio*: This factor is important to screen out if gas lift or pumping is better. Also the gas-liquid ratio will be used to select/size downhole separators.
  - *Formation fluid viscosity*: This factor is particularly important for selecting among PCP, SRP, and ESP. PCP lift method is better for very high viscous fluids with the presence of sands.
  - *Formation volume factor*: This factor is used to determine how much total fluid must be lifted under downhole conditions to get the desired production rate at the surface.
  - *Paraffin, scale, salts*: These depositions may reduce run life of pumps.
  - *Reservoir drive mechanism*: If reservoirs are water drive, water cut will be high toward the end of the life of the wells. Submersible pumps may be more appropriate. If reservoirs are gas cap drive, the wells will exhibit more gas toward the end of the life of the wells. This is important factor if gas lift is considered. If a submersible pump is considered, a gas separator must be integrated with the pump.
- Wellbore geometry characteristics:
  - *Well depth*: This factor can be used to screen out some lift methods such as SRP. Well depth is also used to calculate how much energy is needed to lift fluids to the surface.
  - *Casing and tubing size*: Small diameter casing limits production tubing size and many other options. Tubing size will greatly impact the liquid loading

efficiency by maintaining high fluid velocity and proper flow patterns in the tubing.
  - *Wellbore deviation and horizontal wellbore*: highly deviated wells may limit applications of SRP or PCP because of excessive drag and torque of the rods which may cause rod failures and tubing wears.

- Surface characteristics:
  - *Power sources*: power source availability may govern the selection of AL method. Electric and natural gas are the two important sources to provide power for lift methods. Diesel and propane may also be considered depending on applications.
  - *Field location*: For offshore wells, lift methods that have small footprint are important. For onshore fields, noise limits, water treatment ability, environmental concerns, well spacing, etc. should be considered.

- Field Operating Characteristics:
  - *Enhanced oil recovery operations*: these operations may change the reservoir pressure, produced fluids, fluid properties and hence may change the lift system.
  - *Field automation*: This factor may impact the selection of surface equipment for artificial lift systems.
  - *Local support services*: SRP requires relatively less maintenance compared to other lift methods. Therefore, service requirements should be considered when selecting AL methods.

7.2.2 Selection of Artificial Lift Method for Vertical Wells in Conventional Reservoirs

There are three main approaches for selecting an AL method: using experience and known information from offset wells; using guidelines from reliable companies; and using commercial software. This section will present a selection methodology which combines these three approaches and presented in Fig. 7.1.

- ***Step 1***: The first step is to screen out the obvious inappropriate lift methods. This stage involves a wide range of artificial lift techniques considering a wide range of factors such as DLS, well depth, liquid rate, gas-liquid ratio, etc. This stage is normally done using a reliable guideline or recommendation. The Weatherford published a very rough guideline in 2005 for the selection of AL. The guideline is shown in Table 7.1.
- ***Step 2***: Once a shortlist of feasible lift methods is achieved, a more in-depth comparison of these options is carried out. Details of this stage can be summarized as follows:

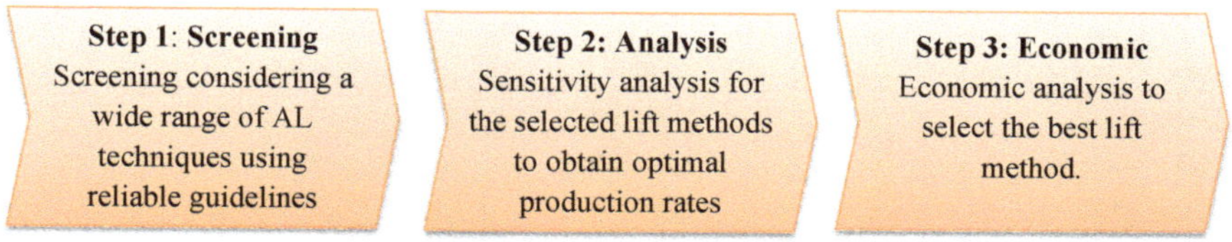

Step 1: Screening
Screening considering a wide range of AL techniques using reliable guidelines

Step 2: Analysis
Sensitivity analysis for the selected lift methods to obtain optimal production rates

Step 3: Economic
Economic analysis to select the best lift method.

Fig. 7.1 Artificial lift selection methodology

Table 7.1 Artificial lift selection guideline [16]

Operating parameters	Positive displacement pumps			Dynamic disp. pumps		Gas lift	Plunger lift
	RSP	PCP	Hydraulic piston	ESP	Hydraulic jet		
Typical TVD, ft	100–11,000	2,000–6,500	7,500–10,000		5,000–10,000	5,000–10,000	Up to 8,000
Typical rate, BLPD	5–1,500	5–2,200	50–500	100–30,000	300–4,000	100–10,000	1–5
Typical temp., °F	100–350	75–150	100–250		100–250	100–250	120
Typical deviation deg/100 ft	0–8	0–8	0–8	0–8	0–8	N/A	N/A
Maximum deviation, deg/100 ft	<15	<15	<15	<15	<24	<70	<80
Gas handling	Fair to good	Good	Fair	Fair	Good	Excellent	Excellent
Solids handling	Fair to good	Excellent	poor	Fair	Good	Good	Poor to fair
Fluid gravity, ° API	>8	<35	>8	>10	>8	>15	N/A
Offshore applications	Limited	Good	Good	Excellent	Excellent	Excellent	N/A
System efficiency, %	45–60	40–70	45–55	35–60	10–30	10–30	N/A

- Obtaining the reservoir characteristics to develop the IPR.
- Obtaining the wellbore geometry characteristics to develop the OPR.
- Using commercial software to plot the IPR and OPR.
- Using commercial software to carry out the sensitivity analysis to achieve the optimal desired rate with available constrains for each selected lift method.

• ***Step 3***: The next step is to perform an economic analysis to guide the ultimate selection of the preferred lift methods. The economic analysis accounts for the capital and operational expenditure (Capex and Opex) associated with each lift technique. The estimation of the net present value also needs to be conducted in this stage.

Table 7.2 Input data for Example 7.1

Data	Value	Unit
TVD	10,925	ft
MD	10,925	ft
Initial GOR	545	SCF/STB
Oil gravity	36	°API
Water cut	48	%
Tubing size, OD	2.875	in
Tubing size, ID	2.441	in
Reservoir thickness	68	ft
Drainage area	160	acres
Drainage radius	1489.5	ft
Wellbore radius	0.354	ft
Reservoir temperature	180	°F
Permeability	6	mD
Reservoir pressure	5000	psi

Example 7.1 Based on the given data presented in Table 7.2 for a given well, select the best artificial lift method.

Solution

***Step 1*: Screening** [3]

Using Weatherford's guideline [16] given in Table 7.1, one can conduct the screening as follows:

- Based on the typical TVD: with the TVD = MD = 10,925 ft, only RSP, ESP, and GL are suitable for this well.
- Based on typical temperature: with the T = 180 °F, SRP, ESP and GL are working well.
- Based on gas handling: With the initial GOR = 545 SCF/STB, SRP may be suffering a little bit because there is no downhole separator. With a gas separator integrated with the pump, ESP may be operating fine. Gas lift would not have any problem with this GOR.
- Based on solid handling: Assuming sand production of this well is low and hence these three lift methods will be working fine.
- Based on fluid gravity: API = 36 °API is good for these three lift methods.
- Based on system efficiency: RSP and ESP are similar. Gas lift is much lower compared to RSP and ESP.

In summary, using Weatherford guideline, one can choose RSP, ESP, and GL for this well.

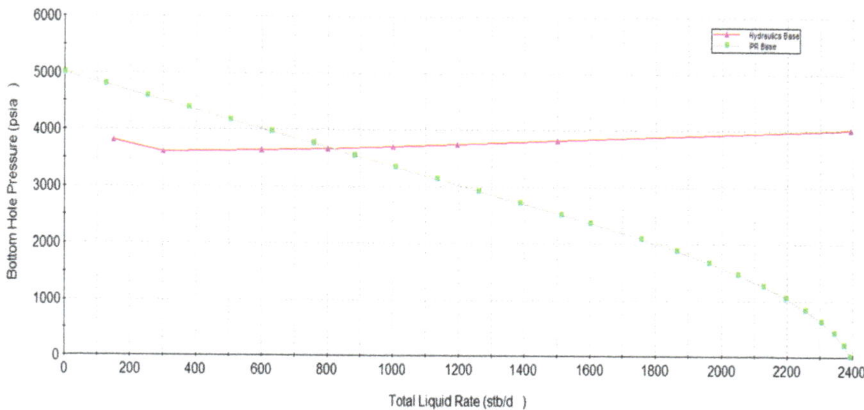

Fig. 7.2 IPR and OPR using SNAP

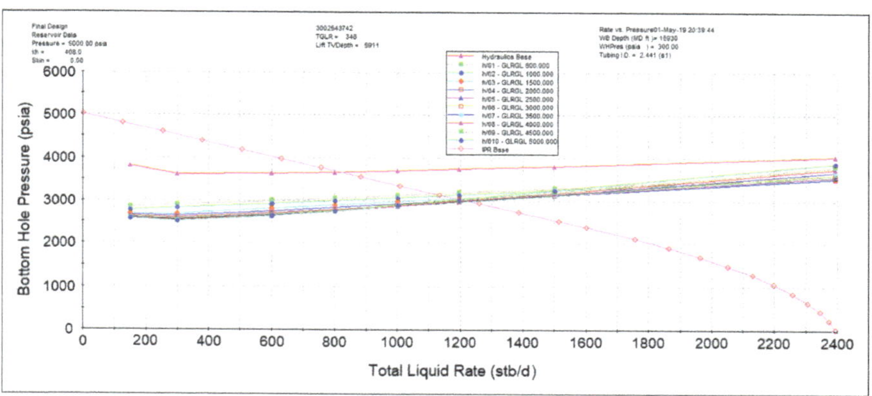

Fig. 7.3 Effects of GLR on the liquid production

Step 2: Analysis

In this step, we used two commercial software to do the design analysis: SNAP is for GL and PIPESIM is for SRP, and ESP.

Figure 7.2 shows the IPR and OPR of this well using the input data. The well can be produced naturally with a liquid rate of 820 STB/D and at the flowing bottomhole pressure of about 3,560 psi. Using SNAP software to carry out the sensitivity analysis for different GLR values, the results reveal that the optimal GLR is about 2,000–2,500 SCF/STB as shown in Figs. 7.3 and 7.4.

The design analysis using SNAP also showed that 8 GL valves are needed for this well. The deepest valve (operating valve) is at the depth of 7,501 ft; The other 7 valves are unloading valves. The valve spacing design is shown in Fig. 7.5.

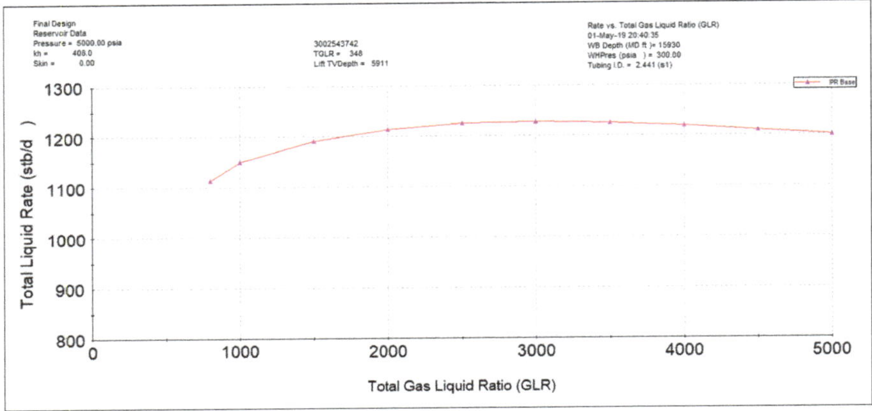

Fig. 7.4 Optimal gas-liquid ratio

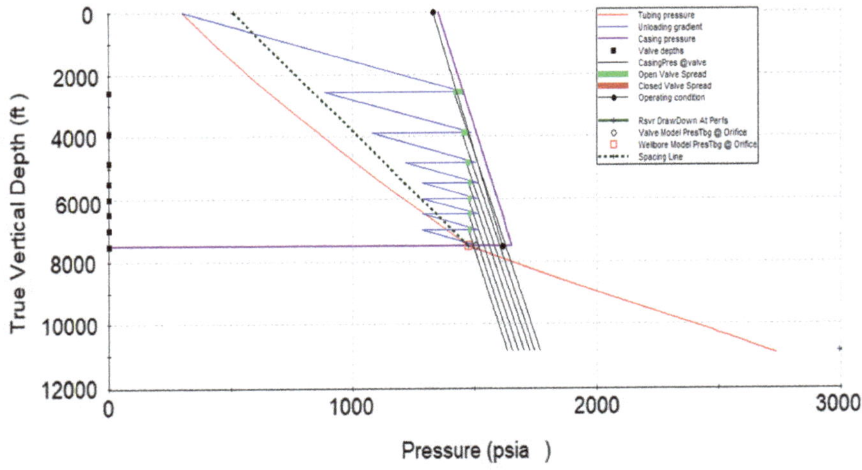

Fig. 7.5 Gas lift valve spacing design using SNAP

The gas lift report shows that if 2,508 MSCF/D (GLR = 2000 scf/stb) is injected into the well, the well will produce a total of 1,338 Barrels of Liquid Per Day (BLPD), which includes 696 STB/D of oil, 642 STB/D of water. The well will also produce 486 MSCF/D of formation gas.

For the ESP design analysis, the commercial software PIPESIM was used. Again, using the input data to run the PIPESIM, one can obtain the IPR and OPR as shown in Fig. 7.6.

It is obvious that ESPs can provide more liquid production than GL system. If the desired liquid production rate is 2,000 BLPD, the well will flow at the

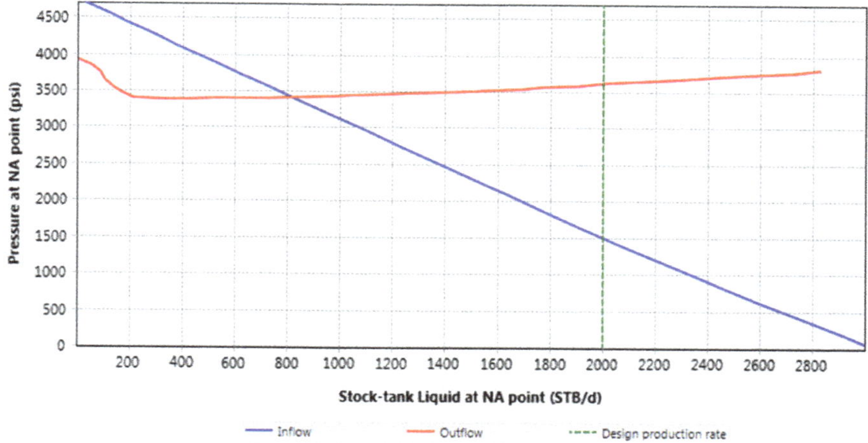

Fig. 7.6 IPR and OPR of this well using PIPESIM

bottomhole pressure of about 1,500 psi and the differential pressure across the pump (pump discharge pressure—pump intake pressure) is about 2,014 psi. From the Schlumberger ESP catalog, REDA DN2150 (387) would be a suitable fit in terms of flowrate and pump outer diameter. The operating range of this pump is from 1,300 STB/D to 2,600 STB/D and the rate at the Best Efficiency Point (BEP) is 2000 STB/D. This pump has an OD of 3.87 in. At the BEP, the head per stage of this pump is 22 ft, the horsepower per stage is 0.553 HP and the pump efficiency is about 59% as shown in Fig. 7.7.

According to the simulation results, 339 pump stages would be required to deliver 2000 BLPD with a total dynamic head of 5749.38 ft. Even though the gas level is not high (3.46%), for safe future operations, a separator is used together with the pump.

For SRP design analysis, the design is not supported in either PIPESIM or SNAP and consequently, the design was performed with an excel spreadsheet using formulas presented in Chap. 5. The final results are shown in Table 7.3.

Until this point, readers can easily recognize that there will be two options for this well: (1) using ESP to maximize liquid production as the first stage of the well then switching to gas lift as the second stage if the surface infrastructure for gas lift system is available; (2) if the first option is not available, ESP can be used as the first stage until the production rate drops around 800 BLPD, the system is then switched to SRP as the second stage to deplete the production. The economic analysis needs to be performed to confirm which one is the best option.

***Step 3*: Economic**

The last step is to perform the economic analysis to guide the ultimate selection of the preferred lift methods. Again, this analysis should consider the CAPEX, OPEX,

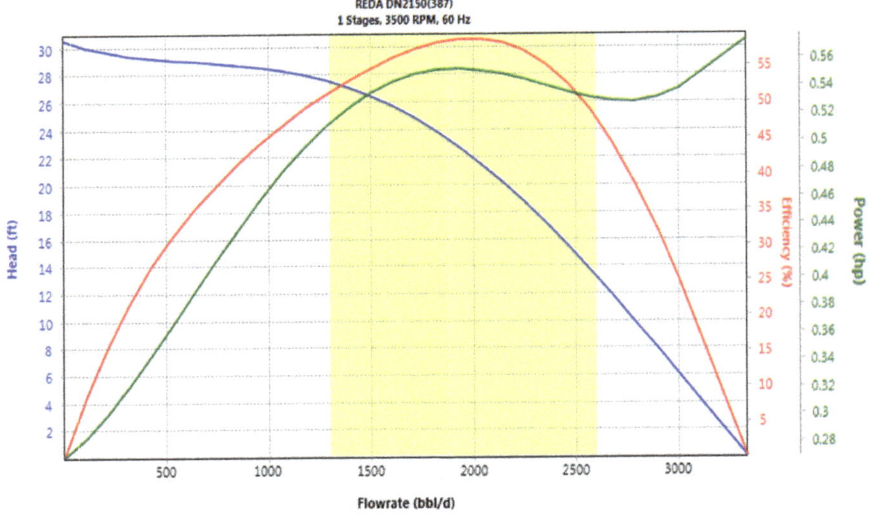

Fig. 7.7 REDA DN2150(387) pump performance for 1 stage at 3500 RPM and 60 Hz

Table 7.3 Results of RSP calculations

Output parameters	Value	Unit
Theoretical pump displacement, PD	803	BPD
Actual pump displacement, Q	773	BPD
Peak polished rod load, PPRL	25,369	lbf
Minimum polished rod load, MPRL	5,646	Lbf
Ideal counterbalance, CB	15,507	Lbf
Peak torque, PT	532,952	In-lbf
Polished rod horsepower, PRHP	324	HP
Nameplate horsepower, HP_{np}	707	HP

Note that the pump is set at the depth of 8200 ft for this calculation

and NPV. This is not the scope of this chapter. Readers should be aware that most commercial software has economic module; users can utilize it to carry out this analysis.

7.3 AL Selection for Wells in Heavy Oil Reservoirs

Exploitation of heavy oil fields is unique for many aspects such as: high to very high viscosity fluids, high sand production due to unconsolidated formations, high depositions, and corrosive environments. Most heavy oil reservoirs have the crude

Table 7.4 Rough guideline for selecting AL methods for heavy oil production [6]

Parameters	SRP	ESP	PCP	Jet pump	ESPCP	Gas lift
Capital cost	Low	High	Low	High	Moderate	High
Operating cost	Low	Moderate	Low	High	Moderate	Moderate
Run life in vertical wells	Average	Average	Average	High	Average	High
Run life in horizontal wells	Low	Average	Low	High	Average	High
Ability to handle sand content	Average	Low	Average	Good	Average	Average
Efficiency	Average	Low	Average	Low	Average	Average
Suitability for thermal production	Applicable	Applicable	Applicable	Applicable	Not Applicable	Applicable
Operational flexibility	Average	Good	Good	Low	Average	Good
Ability to handle gas content	Average	Good	Good	Good	Good	Good
Production handling capacity	Good	Average	Good	Average	Average	Good

oil API ranging from 8–18 °API with a common range of 10–12 °API; well depth from 700–5,000 ft with the common range from 700–3,000 ft; production rates ranges from 30–1,000 BPD with the common range of 30–300 BPD [11]. Due to high viscous fluids, natural flows of these wells are very limited and hence AL is a must to maintain economic productions. The question is that, which AL method(s) is (are) suitable and cost-effective for these wells?

The AL selection methodology for wells in heavy oil reservoirs is similar to that of vertical wells in conventional reservoirs. This methodology also consists of three stages: screening, design analysis, and economic analysis. Mali and Al-Jasmi [6], presented a rough guideline of selecting AL technique for heavy oil production as shown in Table 7.4. This guideline can be combined with the Weatherford's guideline in Table 7.1 to do the screening in the first stage of the selection methodology.

Readers should be aware that GL is not well suitable for heavy oil production because of the challenges of separating gas out of high viscous fluids at surface. As gas is injected into high viscous fluids, it creates an emulsion at which gas remains as micro-bubbles in the fluids [12]. Mechanical separations of these micro-bubbles out of liquids is very limited; the separation efficiency is very low. Thermal separation, which is a high cost operation, must be used at surface. In addition, GL requires large infrastructure to maintain high pressure gas supply and thus it requires high capital investment.

ESP is also not a good lift candidate for heavy oil production. As discussed in Chap. 3, the higher the fluid viscosity, the less the pump efficiency will be. This is because the pump uses the hydraulic energy to overcome the friction due to the fluid viscosity. This inefficiency due to viscous fluids sometimes cause mechanical pump failures and reduce the pump run life. ESP is also well known for poor handling of sand production, which is very common in heavy oil reservoirs. Downhole motor failure is also a big concern when operating ESPs in heavy oil well reservoirs.

PCP system a lot of time is the first choice for handling heavy oil producttion either cold or thermal conditions. Because PCPs work similar to a positive displacement pump, they can provide relatively low flowing bottomhole pressure (high drawdown). Because of this reason, PCP can be operated under a wide range of liquid production rates. When comparing with ESP systems, PCP systems have the motor installed at surface and thus PCP systems can completely avoid failures associated with motor and cables due to high temperatures and mechanical contacts. For high temperature operations, metallic PCPs or/and equal-wall stator PCPs are highly recommended. Readers are recommended to review Chap. 5 of this book to have better understanding of PCP system and how it works.

SRP also a good candidate for heavy oil production because of the following reasons: (1) The operating range of SRP fits quite well with heavy oil production range; (2) Run life and operating cost of a SRP is normally relatively comparative with a PCP system, for given set of operating conditions; (3) Easier to do the maintenance, troubleshoot, and diagnose of a SRP system in comparison to a PCP system.

After the first stage is completed, the design analysis needs to be carried out for the selected AL methods. Finally, the economic analysis is done to confirm the final selection. Example of how to apply this selection methodology is shown in the Example 7.2.

Example 7.2 A well data is given and summarized in Table 7.5. Select the best AL method for this well.

Solution
Based on the TVD, formation fluid temperature, deviation, and fluid gravity of this well, Weatherford's guideline given in Table 7.1 suggests that SRP, PCP, ESP, and hydraulic pumps could be used for this well. Mali and Al-Sasmi's guideline also recommends that ESP and hydraulic pumps should not be used due to its high capital and operating costs and low ability to handle solid. In addition, with the viscosity of 1,400 cp, the ESP's efficiency will be very low compared to other lift methods.

After the screening stage is done, two AL candidates that might fit wells for this application are SRP and PCP. A thorough design analysis should be carried out for these two methods using commercial software similar to what was done in Example 7.1. If commercial software is not available (this is very likely correct for PCP), readers can do the design analysis using excel spreadsheet following examples presented in Chaps. 5 and 6.

Again, the economic analysis must be done at the end to confirm the best AL method selection.

Table 7.5 Results of RSP calculations

Input parameters	Value	Unit
Oil API	12	°API
Static bottomhole pressure, SBHP	1,300	psi
Bubble point pressure, P_b	200	psi
Desired liquid rate, Q	200	BPD
Water cut, WC	50	%
Oil viscosity	1,400	Cp @ 100 °F
Bottomhole temperature	115	°F
Casing size	7	in.
Tubing size	3½	in.
Top of perforation	3050	ft
Pump setting depth	3000	ft

7.4 AL Selection for Horizontal Wells in Unconventional Reservoirs

7.4.1 Basic Concept and Challenges of AL for Horizontal Wells

In this section, we refer to the term "Unconventional Reservoir" to shale plays and tight to very tight reservoirs. They are typically over-pressured with low or very low permeability. To extract hydrocarbons in these reservoirs, horizontal drilling and hydraulic fracturing have become a norm. The initial production of these wells can be easily a few thousand BLPD and drop down quickly even below 100 BLPD within the first year [10]. AL has been the first choice to prolong the life of the well without a major well intervention such as re-fracturing. In addition, the efficiency of drilling and completing a horizontal well has been significantly improved in the past less than 10 years (2010–2018). A typical two miles lateral horizontal well in the Permian Basin, USA is drilled in about two weeks in 2019. In other words, the industry has moved from vertical wells to horizontal wells in a very short period of time. Therefore, oil and gas companies did not have enough time to study and understand the challenges of AL in horizontal wells. They have simply copied and adopted AL technology from vertical wells and applied it for horizontal wells.

The main principle of AL is to reduce the back pressure applied on the face of the reservoir at the bottom of the well. In other words, external energy must be supplied to reduce the fluid hydrostatic pressure (gas lift) or to lift the fluids (submersible pumps) and hence reducing the flowing bottomhole pressure. Therefore, all AL methods should be theoretically applicable to the vertical or inclined well sections instead of the horizontal section. In other words, adopting AL technology for vertical wells to horizontal wells seems to be reasonable. However, there are some unique challenges that must be addressed, when applying AL to

Table 7.6 First year un-choked production decline rates [4]

Shale play	1st year typical un-choked decline rate (%)
Barnett	65
Eagle Ford	60
Haynesville	81
Marcellus	74
Woodford Cana	59
Bakken	69

horizontal wells that have long and very long lateral sections. The challenges can be summarized as follows:

- Because most of the horizontal wells in unconventional reservoirs are stimulated using hydraulic fracturing, the initial liquid production is high then drops down very sharply in a short period of time. Using ONE AL method for the entire life of the well is not suitable. Before the production collapses, these wells can be operated to have natural flow for a few months. If liquid production is high, normally high water cut, an ESP can be used to handle high liquid production rates. If gas production is high, GL can be used to start the production of the well. Sometimes jet pumps are used to recover hydraulic fracturing fluids and solids. SRP or other lift methods may be deployed when the liquid rate is low. According to Pankaj et al. [10], in America, around 40% of wells starting on AL system use GL, 36% use ESP, 13% use SRP, 7% use PL and 4% use jet pumps. The typical first year un-choked production decline rates for different shale plays in the US are presented in Table 7.6.
- Because of the low and very low permeability of unconventional reservoirs, the drawdown is normally kept very high to maintain high production and to meet economic production limits. Gas will start coming out of liquid and create multiphase flow in the horizontal section as shown in Fig. 7.8. The multiphase flow in the lateral section creates liquid slugging, unstable flow rates, and gas channeling which may result in problems for different lift methods in the vertical or inclined section. In addition, shale formations are more compressible than other formations such as carbonate or limestone formations. Under high or very high drawdown, the formation permeability may decrease due to the compaction of the formation. In other words, increasing drawdown by using AL methods may NOT lead to increase in productions. This phenomenon, in fact, has been reported by Oyewole [9].
- There never exists a perfect horizontal section. During drilling, drillers constantly steer the drill bit to make sure it is in the pay zone. The local DLS can be very high causing potential problems when deploying pumps or downhole tools into the horizontal section. In addition, these undulations in the wellbore create pockets of liquid holdup which accelerate slug flow in the lateral section.
- To minimize frictional pressure loss and to reduce the surface pump horsepower, hydraulic fracturing is normally done through large OD casings such as

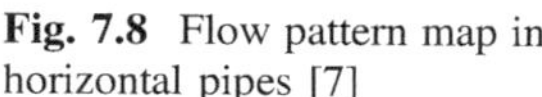

Fig. 7.8 Flow pattern map in horizontal pipes [7]

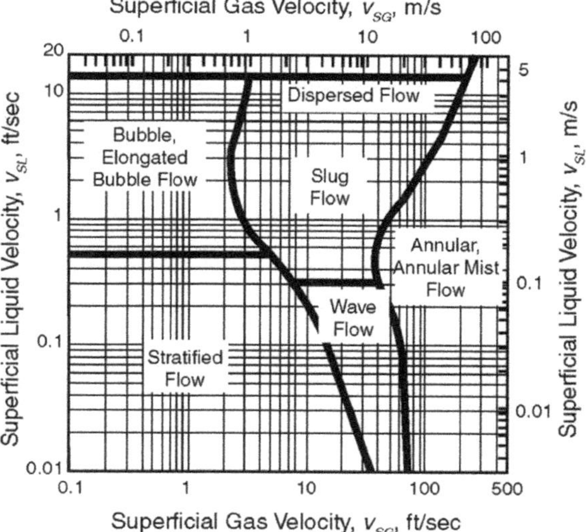

5½ in. In addition, companies have a tendency to complete unconventional wells with large-diameter production tubing to maximize liquid production at the early life of the well. When the production rapidly declines, liquid velocity in the lateral section is far below the critical values resulting in stratified flow, wavy flow, and/or slug flow in the horizontal section as shown in Fig. 7.9. These flow patterns negatively impact the efficiency of the AL installed in the vertical section.

- Theoretically, if the DLS of the build section is small enough, submersible pumps or downhole tools could be deployed into the horizontal section. One of the important questions is that should submersible pumps be run into the lateral section? A few challenges we need to keep in mind before moving into this direction: (1) the productivity index of each stage along the lateral section may be different due to the fact that the efficiency of the stimulation job varies from one stage to another leading to different propped fracture permeability; (2) The deliverability from each stage may be different due to the heterogeneity of the formation; (3) if a submersible pump is run into the lateral section, liquid slugs may cause severe pump inefficiency and even pump damage.
- Another major concern when applying AL for horizontal wells is the liquid loading in the lateral section of gas wells, coaled bed gas wells, etc. Depending on the drilling operations, the lateral section can be classified into four groups: perfect horizontal, toe-up, toe-down, and undulation horizontal section as describe in Fig. 7.10a–d.

  Assuming the lateral section is perfect horizontal as shown in Fig. 7.10a, stratified flow will be the dominated flow pattern. Under steady state flowing conditions, the liquid height in the lateral section will be constant due to the force balance of the system. If the liquid level slightly increases in the annulus,

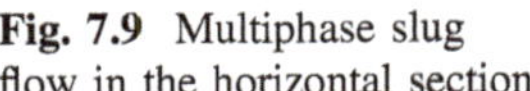

Fig. 7.9 Multiphase slug flow in the horizontal section

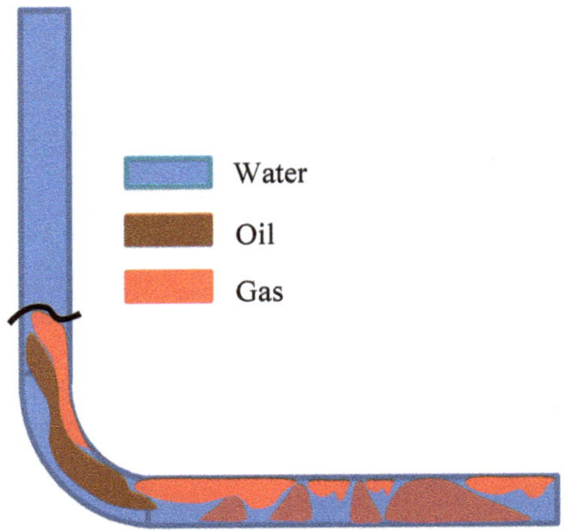

the flowing cross-sectional area of gas reduces leading to an increase in gas velocity. This change causes higher interfacial friction force between liquid and gas leading to a reduction in the liquid level. In other words, the system will adjust by itself to keep the liquid level constant under steady state condition. If the gas velocity is high enough and if there is no liquid accumulation in the build-up and in the vertical section, the gas production will stable.

If the lateral section is toe-up as shown in Fig. 7.10b, liquid will accumulate more at the heel and gas is more at the toe. The well will die if the length of the accumulated liquid is long enough that gas velocity is not enough to break-through.

If the lateral section is toe-down as shown in Fig. 7.10c, liquid will accumulate more at the toe. This well may be quite stable in the early time but the production rate declines faster than that of the toe-up case. This is because liquid will plug-up the section close to the toe and reduce the effective horizontal length.

The most common case is the undulation horizontal section as shown in Fig. 7.10d. As mentioned, during drilling operations, drillers constantly try to steer the drill bit to keep it in the pay zone. Gas pockets accumulated in the lower side of the lateral section accelerate severe slug flow and cause the system to be very unstable.

To unload liquids for these wells, conventional AL methods, which are installed in the vertical, inclined, or build-up section, may not be effective. Liquids in the horizontal section must be controlled properly to maintain the gas production. In other words, submersible pump must be deployed into the lateral section to unload the liquids. There are many concerns when running a pump into a horizontal section including, but not limited to, DLS, hole size, high GLR, high solid concentration, high temperature, location of the pump, and pump run life. This is a one of the challenges that the oil and gas industry is facing. It is

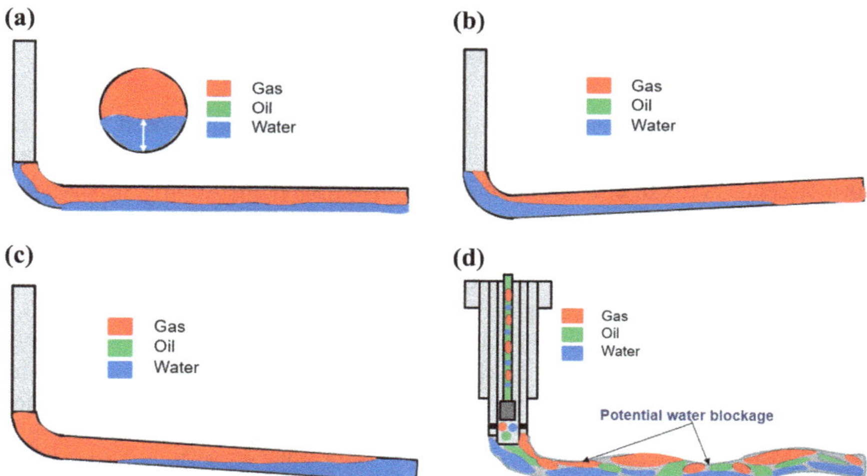

Fig. 7.10 **a** Perfect horizontal section, **b** toe-up horizontal section, **c** toe-down horizontal section, **d** undulation horizontal section

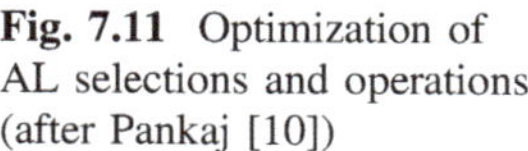

Fig. 7.11 Optimization of AL selections and operations (after Pankaj [10])

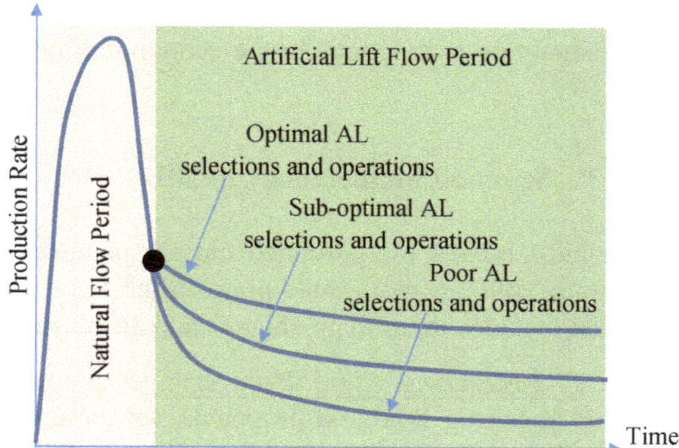

obvious that the industry and academia need to focus on this direction to find a better solution.

In summary, understanding the basic concepts and the challenges when using AL for horizontal wells may lead to better AL technique selections and operations. This, in turn, could prolong the life of the well, AL run life, and ultimately improve the recovery factor. Figure 7.11 shows that after the sharp decline in production of unconventional wells, AL must be deployed to maintain the production. There might be three possibilities leading to three outcomes as shown in Fig. 7.11. If the AL selections and operations are done correctly, production can be maximized or the ultimate recovery factor is improved [14].

7.4.2 AL Selection Methodology for Horizontal Wells in Unconventional Reservoirs

Readers should notice that most of the AL selection techniques used for vertical wells in conventional reservoirs should be considered when selecting AL methods for horizontal wells in unconventional reservoirs. The strengths and weaknesses of each lift method remain the same regardless of vertical/horizontal wells in conventional/unconventional reservoirs. The design of each lift method to obtain the technical parameter (such as GL valve spacing, amount of gas for gas lift system; pump size, pump length, horsepower, etc. for submersible pumps) as described in each chapter in this book remains the same. The two key differences between AL for vertical wells in conventional reservoirs and AL for horizontal wells in unconventional reservoirs are the selection methodology and the operations during the production period. As discussed above, most horizontal wells in unconventional reservoirs are completed with hydraulic fracturing and have high decline production rates in the first year. Therefore, the AL selection methodology for these wells is normally unique and similar to solve the following main problems:

- Recovering fracturing fluids and proppants.
- Handling high liquid production rate in the early production.
- Maintaining productions in the economic limits.

7.4.2.1 Selection Methodology in 2014

Lane and Chokshi [4] presented current practices for selecting AL methods for horizontal wells in most shale plays in the US excluding the Permian Basin. The practice includes three main stages and can be summarized as follows:

Stage 1: Fracturing Fluid Recovery: Jet pumps are the prefer lift method to recover fracturing fluid and proppants. Jet pumps with portable surface units can remove fracturing fluids in less than three days for small volume fractures and in one to three weeks for large volumes. The process begins with lowering the standing valve and jet pump into the well using wireline. The power fluid is pumped through the production tubing and the mixture fluids return in the casing-tubing annulus. During this period choke management can be very useful to mitigate many operational issues caused by slug flow. Choke management can also help to reduce formation damage from excessive pressure drawdown. Choking the well reduces the decline rate which can postpone the need for AL and allows for a broader selection of lift technologies.

Stage 2: Early Production: After the recovery of fracturing fluids is completed, jet pumps can continue producing fluids to surface until production rates decline sufficiently in order to use rod pumps or other lift methods. If extended periods of

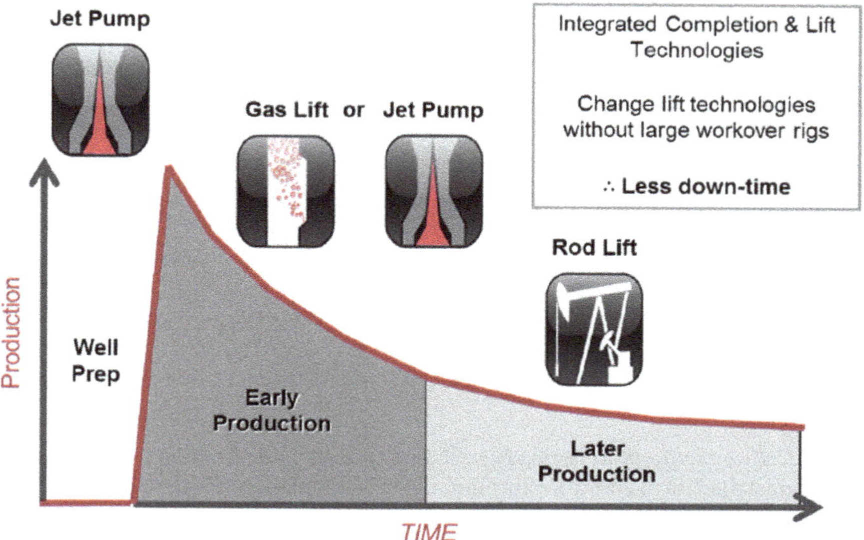

Fig. 7.12 Recommended AL chain for shale plays in the US [4]

high production are expected, the jet pumps are typically replaced with gas lift depending on well spacing, field infrastructure, and availability of dry gas. Another option is to design a hybrid completion which accommodates both jet pumps and gas lift so that the completion does not need to be pulled to initiate gas lift.

Stage 3: Later Production: At some point, most unconventional wells are put on rod lift as production rates and GLR decline to levels where rod lift becomes more effective. In the Eagle Ford, GL is replaced by SRP when GLR declines to between 1,000 and 2,500 SCF/bbl, depending on the operator. Some operators keep the well on GL as long as possible; others move to SRP as soon as liquid rates decline into the operating envelope for rod lift.

The AL selection methodology presented by Lane and Chokshi [4] can be summarized in Fig. 7.12.

7.4.2.2 Selection Methodology in 2016

In 2016, two years after Lane and his team presented their paper, Peter Oyewole presented another AL selection methodology with the focus on the Permian Delaware basin. The Permian Delaware basin is one of the most prolific unconventional oil and gas basins in the US. It has a multi-stacked play that contains several productive zones with the thickness as high as 1,800 ft. When comparing the productive zone thickness with other shale plays in the US such as Bakken of 120 ft, Eagle Ford of 300 ft, the Permian Delaware basin is no doubt a world class oil and gas basin [9]. Another unique characteristic of the Permian Delaware basin is that each production zone has distinct and unique rock and fluid properties. Oyewole [9] said "Non-recognition or appreciation of these changes and complexity has resulted in poor AL selection and performance in some of the unconventional producing zones in the Permian Delaware basin. Incorporating the full understanding of these rocks and fluid phase behaviors into the AL strategy is the bedrock to successfully improve the lift system performance".

Oyewole classified well in the area into five different categories, from well type 1 to well type 2 based on reservoir conditions, formation fluid, liquid production rate, GLR, water cut, and production decline rate. Oyewole then recommended AL strategy for each well type based on the given set of conditions as shown in Table 7.7.

For well type 1, the analysis for selecting AL method(s) can be summarized as follows:

Stage 1: Fracturing fluid recovery and early production: ESPs have been the first choice installed in wells after the initial well completion. With high initial production liquid rates in the first 90–180 days, ESPs stand out to be the best candidate to recover the fracturing fluids and to maintain high liquid production. The only problem that ESPs have faced in this area is the high solid (proppants) and gas content. Under these severe conditions, ESP run life is about 3–6 months. ESP manufacturers have responded by improving the gas separator design, inverted shroud, sand guard, erosive resistant impeller, control animation, etc. ESP Run life reported in 2018 in this area of 8–12 months is quite common.

Stage 2: Later production: After the production rate declines to the economic limits for SRP, ESP is replaced by SRP through a lift conversion from the other form of AL that was initially installed as part of the well completion. The limits of SRP in the area are obvious: low liquid production rates, depth limitations, and low efficiency when handling gas.

GL has also been used in this stage mainly due to the drive for lower operating expenditures. The infrastructure for gas lift system has been expanded rapidly in the region making it more and more favorable.

The author also presented the capital expenditure and operating expenditure for different lift methods as shown in Table 7.8. It is obvious that ESP is the most cost-effective lift method.

7.4.2.3 Selection Methodology in 2018

In 2018, four years after Lane and his team presented their paper and after many lessons learnt, Pankaj and his team presented another AL selection methodology for unconventional reservoirs with a focus for the Eagle Ford basin. Lane and Chokshi [4] summarized the extended problems in Eagle Ford and the focus of their study as follows:

Table 7.7 Recommended AL strategy for well in the Permian Delaware Basin [9]

Well type	Well description	Recommended lift method
1	Undersaturated reservoir	ESP only is highly recommended for the life of well. SRP may be used if well performance does not meet expectation
	Black oil type—PVT	
	High water cut (>80%)	
	Low production decline rate	
	High liquid production (>500 BLPD)	
	Low GLR (<750 SCF/bbl)	
2	Undersaturated reservoir	SRP only is highly recommended for the life of well. There is only managed flow back and managed depletion periods in the well life
	Black oil type—PVT	
	High water cut (>80%)	
	High production decline rate	
	Low liquid production (<500 BLPD)	
	Low GLR (<750 SCF/bbl)	
3	Undersaturated reservoir	Jet pump for early production then SRP for later production when reservoir pressure is low
	Black oil type—PVT	
	High reservoir pressure	
	High TVD	
	Low GLR (<750 SCF/bbl)	
4	Saturated reservoir	SRP only is highly recommended for the life of well. Managed flow back and managed depletion periods in the well life
	Volatile oil type—PVT	
	High water cut (> 80%)	
	High production decline rate	
	Low liquid production (<500 BLPD)	
	Low GLR (<750 SCF/bbl)	
5	Undersaturated reservoir	Plunger lift and gas lift only. Operation will include plunger lift assists gas lift and gas lift assists plunger during the well life
	Volatile oil type—PVT	
	Low water cut (>80%)	
	High production decline rate	
	Low liquid production (<500 BLPD)	
	High GLR (>1.000 SCF/bbl)	
	High gas production	

An unstable flow profile along the multiple clusters produces a complicated problem for artificial lift planning for unconventional oil and gas wells because the multiphase flow may cause gas locks and shorten the life of commonly used ESPs; it may also hinder consistent prediction of the response because the perforation clusters along the wellbore have varying and dynamic inflow PI. Formations such as the Eagle Ford have been challenging for incorporating the appropriate artificial lift strategies because of the deeper wellbores - over 10,000 ft - and the high temperatures of up to 270 °F. Geologically, the lower half of the Eagle Ford has generally higher tendency to produce more gas, which makes it more important to pay attention to well deliverability and multiphase flow stream over time for planning artificial lift. It is seen that as the wells in the Eagle Ford get deeper toward the

Table 7.8 Economic considerations for different lift methods [9]

Lift comparison	Gas lift	ESP	SRP	Jet pump
Artificial lift assembly	$145,387	$49,784	$150,210	$144,500
Work over cost	$19,280	$19,280	$19,280	$19,280
Surface equipment	$57,398	$18,555	$14,268	$24,072
Electrical surface equipment	$8,400	$9,940	$6,875	$9,940
Metering	$62,000	$0	$0	$0
Surface electrical labor	$6,000	$7,900	$6,000	$6,000
Artificial lift labor	$14,642	$4,776	$8,800	$7,620
Total capital cost	$313,107	$110,235	$205,433	$211,412
Incremental capital cost over ESP	$202,872	$0	$95,198	$101,177
Monthly rental per month	$5,136	$6,200	$0	$0
Monthly electric cost per month	$2,850	$3,420	$1,900	$3,135
Failure frequency monthly cost	$6,700	$25,000	$13,500	$5,800
Expected TLOE-exeluding S WD/month	$14,686	$34,620	$15,400	$8,935
Cost savings versus ESP per month	$19,934	$0	$19,220	$25,685
Payout over ESP (months)	10.20	0.00	4.95	3.90

Note TLOE: Lowering normal lease operating expense

south, they tend to produce more gas and have higher initial gas/oil ratios. It is important to understand the influence of reservoir characterization in artificial lift selection methodology to identify the appropriate artificial lift technique and its behavior when deployed in unconventional reservoirs. This study combines reservoir engineering with artificial lift selection and design for four different well trajectories in the Eagle Ford basin to elucidate the various scenarios of well placement affecting well performance.

The common input design parameters of wells in the Eagle Ford basin are given in Table 7.9.

In order to do the AL method screening, the authors used a commercial software to select appropriate AL methods based on the well characteristics such as target liquid rate, maximum expected GLR, DLS, downhole temperature, solid content, etc. This screening stage is similar to the "Stage 1: Screening" presented in the selection methodology for vertical wells in conventional reservoirs. The software evaluates each lift method on a scale ranging from 0 to 100 where 0 is for "zero efficiency" and 100 is for the best efficiency. The results from the commercial software for the screening are presented in Table 7.10.

The final score for SRP is zero meaning that SRP is not applicable to this application. The reason is the limitations of the pump setting depth and the liquid production rates in the Eagle Ford. Setting depth of the equipment is approximately 12,120 ft and initial liquid production rate is about 2,500 BLPD during the recovery of the fracturing fluids. Similar results are for PCP, PL, and rodless PCP (RLPCP).

Table 7.9 Input design parameters of considered wells in the Eagle Ford basin [10]

Input design parameters	Value	Unit
Setting depth	12,120	ft
Initial liquid production rate	2,500	BLPD
DLS @ setting depths	7–9	°/100 ft
KOP	11,600	ft
WC	24	%
GLR	450	SCF/STB
Oil gravity	42	°API
Reservoir fluid temperature	270	°F
Later production rate	300	BLPD

Table 7.10 Screening results using the commercial software (N is for not applicable; Y is for applicable)

Input	SRP	ESP	GL	PCP	RLPCP	PL	Jet P.
Target liquid rate	**N**	Y	Y	**N**	**N**	**N**	Y
Setting depth	**N**	Y	Y	**N**	**N**	Y	Y
Max. expected GLR	Y	Y	Y	Y	Y	Y	Y
DLS impact in operation	Y	Y	Y	Y	Y	Y	Y
Casing OD	Y	Y	Y	Y	Y	Y	Y
API	Y	Y	Y	Y	Y	Y	Y
Anticipated line pressure	Y	Y	Y	Y	Y	Y	Y
Downhole temperature	Y	**N**	Y	Y	Y	Y	Y
DLS Impact in setting depth	Y	**N**	Y	Y	Y	Y	Y
DLS impact pass through	Y	Y	Y	Y	**N**	Y	Y
Scale	Y	Y	Y	Y	Y	Y	Y
Corrosion	Y	Y	Y	Y	Y	Y	Y
Solids	Y	Y	Y	Y	Y	Y	Y
Paraffins and/or asphaltenes	Y	Y	Y	Y	Y	Y	Y
Max. Expected GRL	Y	Y	Y	Y	Y	Y	Y
Offshore	Y	Y	Y	Y	Y	Y	Y
High pressure gas source	Y	Y	Y	Y	Y	Y	Y

(continued)

Table 7.10 (continued)

Input	SRP	ESP	GL	PCP	RLPCP	PL	Jet P.
Power source	Y	Y	Y	Y	Y	Y	Y
High pressure power fluid	Y	Y	Y	Y	Y	Y	Y
Final score	[illegible]	0	31	0	0	0	36

ESPs could be a good candidate based on liquid production rate, setting depth and other factors. However, the final score for ESPs is zero meaning that they are ruled out because of the excessive DLS of 7–9°/100 ft on the selected setting depth of 12,120 ft. For ESPs to work, raising the pump setting depth to above the kickoff point of about 11,600 ft must be considered.

The final scores for GL and jet pumps are 31 and 36, respectively meaning that they are relatively better than other lift methods in terms of operational criteria used in this case. The presence of solids, GLR and water cut really reduce the lift efficiency of these two methods according to the simulation's results. The authors also conducted the sensitivity analysis for the cases of toe-up with trap, toe-down, toe-down with trap. The results are shown in Fig. 7.13a and b for gas production and oil production rates, respectively. Even though the final scores obtained were different, the applicable AL systems remain the same.

Further analysis for the two selected AL methods: GL and jet pumps revealed that when changing the liquid production rates from 2,500 to 500 BLPD, jet pumps would not work appropriately when the rate is at 500 BLPD. The final score from the simulation, in this case, is zero. The simulation's results are shown in Fig. 7.14. Therefore, it is recommended to use GL at the beginning of the wells to recover the hydraulic fracturing fluids as well as to produce in the early life of the wells. Of

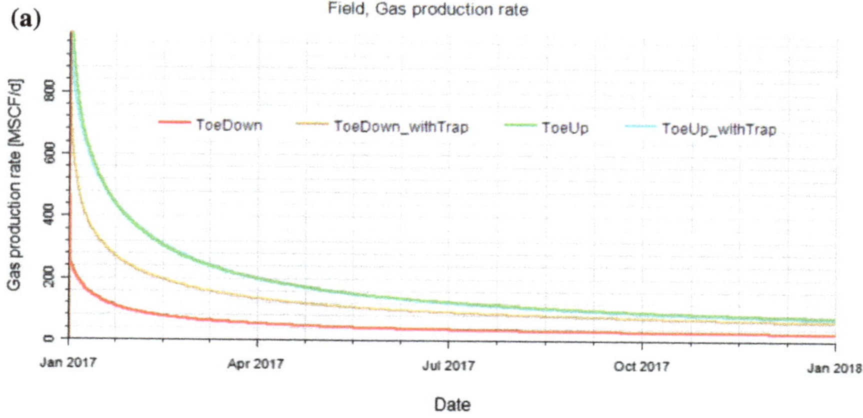

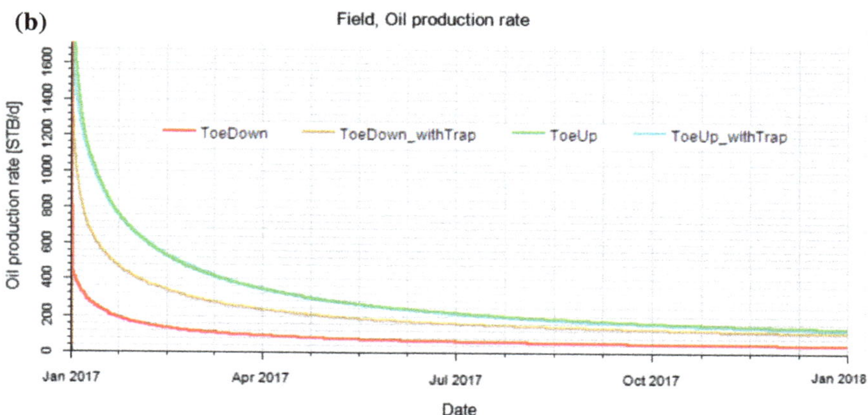

Fig. 7.13 **a** Gas production behavior prediction for four trajectories, **b** Oil production behavior prediction for four trajectories

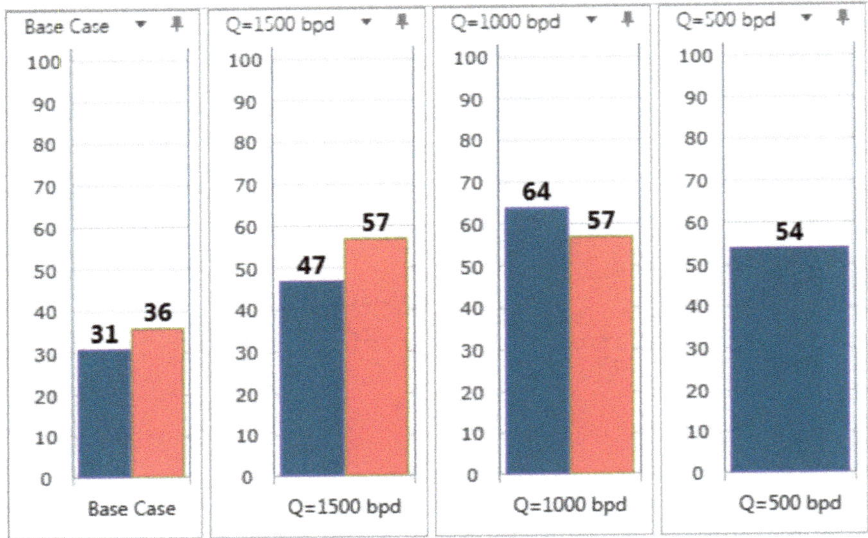

Fig. 7.14 Sensitivity comparison: GL (blue) and jet pumps (red)—base case for Q = 2,500 BLPD

course, if the well depths are not an issue, SRP could be considered for the second stage as the liquid rates fall to preferred rates for SRP.

References

1. Brown K (1982) Overview of artificial lift systems J Pet Technol 0149-2136/82/0010-9979 $00.25
2. Cortines J, Hollabaugh G (1992) Sucker-rod lift in horizontal wells in pearsall field, texas. SPE 24764—Presented at the annual technical conference and exhibition of the SPE in Washington DC
3. Kefford P, Gaurav M (2016) Well performance calculations for artificial lift screening. SPE-181344-MS—Presented at the SPE ATCE in Dubai UAE
4. Lane W, Chokshi R (2014) Considerations for optimizing artificial lift in unconventionals. URTeC 1921823—Presented at the unconventional resources technology conference
5. Lea J, Nickens H (1999) Selection of artificial lift. SPE-52157—Presented at the SPE mid-continent operations symposium—Oklahoma city, Oklahoma
6. Mali P, Al-Jasmi A (2014) Evaluation of artificial lift modes for heavy oil reservoirs. SPE-170040-MS—Presented at the SPE heavy oil conference—Alberta, Canada
7. Mandhane J, Gregory G, Aziz K (1974) A flow pattern map for gas-liquid flow in horizontal pipe. Int J Multiph Flow 1(4):537–553, October 1974
8. Mazzanti D, Dixon D (2016) Artificial lift system for horizontal wells and other wells with problematic lift conditions. SPE-181230-MS—Presented at the SPE North American artificial lift conference and exhibition—Woodlands, TX

9. Oyewole P (2016) Artificial lift selection strategy to maximize unconventional oil and gas assets value. SPE-181233-MS—Presented at the SPE North American artificial lift conference and exhibition—Woodlands, TX
10. Pankaj P, Patron K, Lu H (2018) Artificial lift selection and its applications for deep horizontal wells in unconventional reservoirs. URTeC 2875180—Presented at the unconventional resources technology conference
11. Petrov A, Mikhaylov A, Litvinenko K, Ramazanov R (2010) Artificial lift practice for heavy oil production with sand control. SPE-135973—Presented at the 2010 SPE Russian oil and gas technical conference and exhibition—Moscow, Russia
12. Phan H, Nguyen T, Al-Safran E, Saasen A, Nes O (2017) An experimental investigation into the effects of high viscosity and foamy oil rheology on a centrifugal pump performance. J Pet Sci Technol. https://doi.org/10.22078/JPST.2017.709
13. Shi J, Han Q, Ren X, Zhang X, Zhao R, Li Q (2018) The application of big data analysis in the optimizing and selecting artificial lift methods—SPE-192482-MS—Presented at the middle east artificial lift conference and exhibition—Manama, Bahrain
14. Sickle S, Shelly G, Snyder D (2015) Optimizing compeltions and artificial lift in an unconventional play in the United States—SPE-173974-MS—Presented at the SPE artificial lift conference—Latin America and Caribbean—Bahia, Brazil
15. Valbuena J, Pereyra E, Sarica C (2016) Defining the artificial lift system selection guidelines for horizontal wells—SPE-181229-MS—Presented at the SPE North American artificial lift conference and exhibition—Woodlands, TX
16. Weatherford Artificial Lift Guideline (2005)

GPSR Compliance
The European Union's (EU) General Product Safety Regulation (GPSR) is a set of rules that requires consumer products to be safe and our obligations to ensure this.

If you have any concerns about our products, you can contact us on

ProductSafety@springernature.com

In case Publisher is established outside the EU, the EU authorized representative is:

Springer Nature Customer Service Center GmbH
Europaplatz 3
69115 Heidelberg, Germany

www.ingramcontent.com/pod-product-compliance
Ingram Content Group UK Ltd.
Pitfield, Milton Keynes, MK11 3LW, UK
UKHW021009290726
14059UKWH00001BA/36

* 9 7 8 3 0 3 0 4 0 7 1 9 3 *